Static and Dynamic Properties of the Polymeric Solid State

NATO ADVANCED STUDY INSTITUTES SERIES

Proceedings of the Advanced Study Institute Programme, which aims
at the dissemination of advanced knowledge and
the formation of contacts among scientists from different countries

The series is published by an international board of publishers in conjunction
with NATO Scientific Affairs Division

A	Life Sciences	Plenum Publishing Corporation
B	Physics	London and New York
C	Mathematical and Physical Sciences	D. Reidel Publishing Company Dordrecht, Boston and London
D	Behavioural and Social Sciences	
E	Engineering and Materials Sciences	Martinus Nijhoff Publishers The Hague, London and Boston
F	Computer and Systems Sciences	Springer Verlag Heidelberg
G	Ecological Sciences	

Series C – Mathematical and Physical Sciences

Volume 94 – Static and Dynamic Properties of the Polymeric Solid State

Static and Dynamic Properties of the Polymeric Solid State

Proceedings of the NATO Advanced Study Institute,
held at Glasgow, U.K., September 6-18, 1981

edited by

R. A. PETHRICK

and

R. W. RICHARDS
University of Strathclyde, Glasgow, U.K.

D. Reidel Publishing Company

Dordrecht : Holland / Boston : U.S.A. / London : England

Published in cooperation with NATO Scientific Affairs Division

Library of Congress Cataloging in Publication Data

NATO Advanced Study Institute (1981 : Glasgow, Strathclyde)
    Static and dynamic properties of the polymeric solid state.

    (NATO advanced study institutes series. Series C, Mathematical and
physical sciences ; v. 94)
    Includes index.
    1.   Solid state chemistry—Congresses.   2.   Polymers and polymerization—
Congresses.   I.   Pethrick, R. A. (Richard Arthur), 1942-   .   II.   Richards,
R. W. (Randal William), 1948-   .   III.   Title.   IV.   Series.
QD478.N37   1981   547.7   82-13266
ISBN 90-277-1481-9

Published by D. Reidel Publishing Company
P.O. Box 17, 3300 AA Dordrecht, Holland

Sold and distributed in the U.S.A. and Canada
by Kluwer Boston Inc.,
190 Old Derby Street, Hingham, MA 02043, U.S.A.

In all other countries, sold and distributed
by Kluwer Academic Publishers Group,
P.O. Box 322, 3300 AH Dordrecht, Holland

D. Reidel Publishing Company is a member of the Kluwer Group

Statement by U.S. Army and Navy:
'The views, opinions and/or findings contained in this report
are those of the authors and should not be construed as an
official Department of the U.S. Army, or Navy, position,
policy, or decision'.

CONTENTS

S E M I N A R S

PREFACE

This volume contains the major portion of the
material given at the NATO Advanced Study Institute,
held at the University of Strathclyde, Glasgow, UK.,
September 6th-18th, 1981.   The original idea germin-
ated in a conversation between the organisers on a
cold December night in 1978 in the depths of the
Oxfordshire countryside.   At that time we felt that
the chemical physics of macromolecules in the solid
state was running on two parallel tracks, namely
structure and dynamics.   The contact between the two
appeared to be slight.   We were also concerned that
the degree of special knowledge now required for any
one technique essentially prevented people from
learning the important features of other investigation
methods.   Consequently, we have attempted to bring
together leading authorities on both structural and
dynamic properties of solid polymers in the hope that
the combination of both types of discussion will be
synergistic.

The choice of main subjects is our own and we are
aware that some areas have been omitted.   However, to
be comprehensive would have made an already large
volume enormous.   We therefore chose to concentrate
on what were, in our opinion, the major areas.
Nonetheless, it is apparent that much original
material appears here, especially in those contributions which are more theoretical in concept, the
full experimental implications of which have yet to
be investigated.   The lecture programme of the Study
Institute was organised so that theory and experimental lectures ran in parallel and were accompanied by
short seminars illustrative of the main themes, given
by the participants.   This arrangement has not been
followed in this volume: theoretical papers are

vii

given first, followed by the experimentally oriented
material of some of the seminars, which is included
in a final section.

We are grateful to the NATO Scientific Affairs
Division which provided a grant to finance the Study
Institute.   Additional support was provided by the
European Research Offices of the US Army and the
US Navy whom we also thank.   The Science and
Engineering Research Council of the UK was very
generous in providing additional funds directly to
postgraduate research students enabling them to attend.
Likewise, the National Science Foundation of the
United States provided travel funds to some attendees
from the US.   Mr Peter Nelson and his staff at the
Continuing Education Office relieved us of much
paperwork and coped with all the emergencies, whilst
the Domestic Bursars at the Strathclyde Business
School were always helpful, whatever we requested!

We would also like to record here our personal thanks
to the lecturers on the course who gave their time
willingly both in attending the course and in
providing the texts here.   We enjoyed their lectures
and hope that this book will be of as much interest
to others as it is to us.

R.W. Richards/
R.A. Pethrick

EDITORS NOTE

Regrettably contributions were not received from Professor G. Allen, Professor D.J. Meier and Professor G. Wegner before the manuscript was sent to the publishers. Their contributions are therefore absent.

R.A. Pethrick
R.W. Richards

May 1982

MAIN CONTRIBUTORS

Sir G Allen, FRS — Science & Engineering Research Council, PO Box 18, Swindon, UK.

Dr. M Daoud — SDSRP Orme des Merisiers, BP No. 2, 91190, Gif, France

Sir Sam Edwards, FRS — Cavendish Laboratories, Madingley Road, Cambridge, CB3 OHE, UK

Dr. F Heatley — Department of Chemistry, University of Manchester, Manchester, UK

Dr. J Heijboer — Central Laboratory, TNO, PO Box 71, Delft, The Netherlands

Dr. Julia Higgins — Department of Chemical Engineering, Imperial College, London, UK

Prof. D J Meier — Michigan Molecular Institute, Midland, Michigan 48640, USA

Prof. L Monnerie — ESPCI, Laboratoire de Physieochemie Structurale et Macromoleculaire, 10 Rue Vauquelin, 75231 Paris, France

Prof. A M North — Department of Pure & Applied Chemistry, University of Strathclyde, Glasgow, UK

Prof. P Painter — Department of Materials Science, Pennsylvania State University, PA 16802, USA

Dr. R A Pethrick — Department of Pure & Applied Chemistry, University of Strathclyde, Glasgow, UK

Dr. C Picot — CRM, 6 Rue Boussingault, 67083 Strasbourg, France

Dr. R W Richards — Department of Pure & Applied Chemistry, University of Strathclyde, Glasgow, UK

Dr. D Sadler   H H Wills, Physics Laboratories,
                                Royal Fort, Tyndall Avenue,
                                Bristol, BS8 2LR, UK

Prof. R S Stein   Polymer Research Institute,
                                University of Massachusetts,
                                Amherst, Mass 01003, USA

Dr. J M Vaughan   Royal Signals and Radar Esta-
                                blishment, Malvern, Worcs. UK

Dr. I M Ward   Department of Physics, Uni-
                                versity of Leeds, LS2 9JT, UK

Prof. G Wegner   Albert Ludwigs Universitat,
                                Stefan Meier Strasse 31, D7800
                                Freiburg, West Germany

Prof. G Williams   Edward Davies Chemical Labora-
                                tories, University College of
                                Wales, Aberystwyth, Wales, UK

THE POLYMERIC SOLID STATE

A M North

Department of Pure and Applied Chemistry, University
of Strathclyde, 295 Cathedral Street, Glasgow, U.K.

INTRODUCTION

It is appropriate to start with the simplest question of all;
'Why are we gathered here today to discuss the polymeric solid
state?' It is a simple question and the answer is equally simple.
It is not because a macromolecule shows some chemical or physical
properties which are not shown by small molecules of the same
chemical type, not because study of macromolecules made up of a
large number of units introduces us logically and nicely to
exciting mathematics. The answer is quite simply that the
engineering and technological uses of polymeric materials in
plastics and rubbers is almost exclusively in the solid state.
Therefore, this should be the focus of a meeting which tries to
bring the most exciting, the most fundamental academic principles
to bear on our understanding of this state of matter.

How can we fit this objective into a pattern of study?   One
scheme is suggested in figure 1.

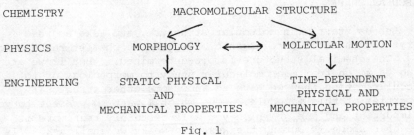

Fig. 1

1

R. A. Pethrick and R. W. Richards (eds.), Static and Dynamic Properties of the Polymeric Solid State, 1–14.
Copyright © 1982 by D. Reidel Publishing Company.

One dimension is the inter-relationship between chemistry, physics and engineering.   I think perhaps the largest number of us have an origin in chemistry.   It is quite appropriate, therefore, to look along a dimension which starts with a consideration of molecules.   If we can determine the molecular structure then we are able to go on to look at molecular behaviour, and are better able to move from the behaviour of molecules in isolation to molecules in a macroscopic structure.   The way that the macroscopic structure is made up leads us naturally to discuss the morphology of polymeric solids.

      This is, if you like, predominantly a static or long time study and many of the things that polymeric solids and polymer molecules do, they do rather rapidly in a fashion that changes with time.   If we are going to be concerned with molecular behaviour, we can simplify our pattern by sub-dividing the behaviour into time-independent and time-dependent properties. Indeed, the whole title of this next fortnight emphasises that division.   There is an intimate connection between the time-independent and time-dependent properties of all polymeric molecules.   I have indicated this in this particular representation by saying that the way in which molecules phase-separate, the way in which the morphology is created, is influenced by whether or not the molecules are moving.   By the same token, the ease or the way in which the molecules move is intimately connected with whether or not they are in hard, soft, crystalline or amorphous phases.   So this little pattern is one way of orienting ourselves into the chemical, physical and engineering properties of a single molecule, molecules in aggregate, and macroscopic solids.

      If I were asked for a single adjective to describe this study, I would say that it is 'use-directed'.   The lectures you are getting will be very esoteric in some instances, very fundamental, but they will never be divorced from the end uses of polymeric materials.   They will be laying the foundation of, and pointing out the connection with, these final engineering properties.

MOLECULAR SHAPE

      Let us start with molecular structure, and take as read everything that the pure organic or synthetic chemist has done for us.   The chemical structure is pre-determined.   But if we are going to go on from macromolecular study to morphology or to molecular motion then we have to know some other structural properties of this molecule.   One which I want to present to you as important and one of which you will certainly hear more is molecular shape.   During the next two weeks we shall ask - 'Is a

macromolecule in the solid wound into the same shape as we have
been able to study for many years in dilute solution?'   Until
recently that question could not be answered, but now with the
help of neutron scattering and because scattering from deuterium
nuclei is quite different in amplitude from the scattering from
protons, we can study per-proto or per-deutero polymers in the
other isotope.   So we can look at the dimensional or shape
properties of polymers in solids, and we can compare then with
what we have known for decades about the same molecules in
solution.   Indeed, it turns out that if we take the pure homopolymer in a purely amorphous state, then that polymer is
generally a random coil with dimensions distinctly similar to
those we can see in the unperturbed state in a neutral solvent,
figure 2.   You will hear more in the next two weeks about this
determination of the shape of molecules in the solid state.

'Unperturbed Dimensions of Polystyrene'

M_N 97,200

$\langle s^2 \rangle^{1/2}$ Neutron Scattering: Solid:
              Perproto Polymer in Perdeutero-   90Å

$\langle s^2 \rangle^{1/2}_0$ Light Scattering: Solution:
              In Hydrocarbon θ Solvent   84Å

after Ballard, Scheltem and Wignall

Fig. 2

This similarity allows us to take some information from
dilute solution and cautiously say that it is likely to exist in
the solid state as well, figure 3.

Polymer	Flexibility
Poly(methyl methacrylate)	Flexible coil
Poly(vinyl chloride)	Flexible coil
Cellulose esters	Stiff coil
Poly(N-vinyl carbazole)(molecular weight $> 10^5$)	Flexible coil
Poly(N-vinyl carbazole)(molecular weight $< 10^4$)	Stiff coil
Poly(n-butyl isocyanate)(molecular weight $< 10^4$)	Stiff rod
Poly(n-butyl isocyanate)(molecular weight $> 10^5$)	Stiff coil
Poly(γ-benzyl L-glutamate) (α-helix)	Stiff rod
DNA (molecular weight $< 10^5$)	Stiff rod
DNA (molecular weight $> 10^6$)	Stiff coil

Fig. 3

Familiar molecules, like polymethylmethacrylate, are random coils capable under many circumstances of quite agitated movement. Cellulose esters are rather less flexible.   Polypeptides are rod-like, both in the solid state and in dilute solution.

CONFORMATIONAL CHANGE

     Shape is all very well, but you will notice that I am starting to introduce words like "stiff" and "flexible."   These are adjectives with connotations of speed and time.   Something flexible is something which can change its shape rather rapidly, and something stiff the opposite.   If we are talking about macromolecular structure, macromolecular shape, we are led very quickly from the time-independent consideration to the time-dependent.   How rapidly does the shape change?   How stiff or flexible, in a time-dependent sense, are these molecules?   We can resolve the change in shape into two idealised components, figure 4.

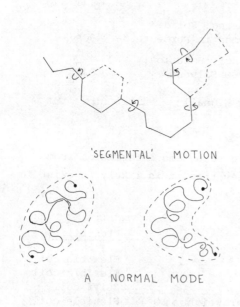

'SEGMENTAL'   MOTION

A   NORMAL   MODE

Fig. 4

We can take a macromolecule and focus our attention on 1, 2 or 3 covalent bonds on that macromolecule to see what is happening on this scale.   This considers a segmental or localised change in shape, and much of the material you will be hearing in the next two weeks will deal with this. You will hear also of shape changes of a rather larger scale, where a whole macro-molecule is changing its shape and we will discuss the so-called normal modes of motion, familiar to spectroscopists.   All the localised partial rotations along the macromolecule add up to an overall gross deformation as if they were all taking place in phase.

     From the gross movement of the macromolecule we build up to the gross deformation of a whole macro structure, which is the engineering end use, figure 5.   If we take any typical linear macromolecule and heat it (change its temperature), then we can

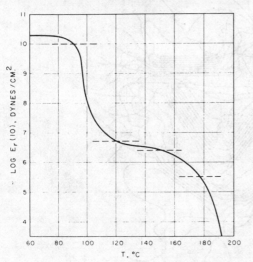

FIVE REGIONS OF VISCOELASTIC BEHAVIOR

Five regions of viscoelastic behavior (for polystyrene sample C.

Fig. 5

go through three states of matter which are familiar to us as the hard glassy, the intermediate rubbery and the very fluid molten states, in which the modulus is high, intermediate and low. The molecules in these states have got virtually no motion, localised segmental motion only (becoming increasingly less localised as the temperature rises) and separation of the centres of gravity of molecules respectively in each state.

Many of your lectures will deal with this picture and will give illustrations of end use.

MORPHOLOGY

As well as existing in a random coil in an homogeneous amorphous state, homopolymer macromolecules may crystallize. When they do there are many differences both in morphology and in macroscopic behaviour between the resulting macromolecular solids and crystals of low molecular weight small molecules.   One reason for these differences is illustrated in figure 6, showing that the single macromolecule may exist in more than one crystallite as well as in the amorphous phases.   Because each segment of a macromolecule is connected to its neighbour, it is not free to wander around to any crystal site, as is the case with small molecules, and so the formation of crystals poses fascinating questions in molecular dynamics and the crystallography of the resulting structure also presents some interesting fundamental questions.

The two-phase morphology exists not only in the crystalliz-ation of a single homopolymer, but can arise also when we have copolymers with different chemical types arranged in blocks in the chain.   The different chemical types in the same chain tend to

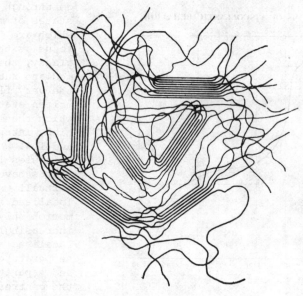

Fig. 6

aggregate in different domains in the solid, Figure 6.   These
domains may be crystalline or simply amorphous regions with
different physical properties from the surrounding matrix.   Such
domains are found in polyurethane, figure 7, and in hydrocarbon
copolymers of inflexible and flexible segmental units.   Thus, block copolymers of styrene and isoprene tend to phase separate into domains which at room temperature are respectively glassy and rubbery.

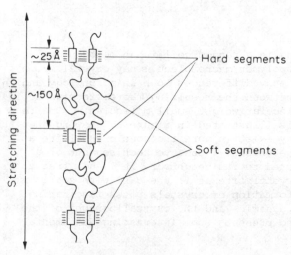

*Flexible and rigid segments in a poly-
  urethane elastomer*

Fig. 7

We can move from the chemical separation associated with
this two-phase morphology to the order (or lack of it) in the
long-range ordering of these domains.   Thus, there is a large
scale analogue of crystallinity which we can find in the styrene -
isoprene block copolymers, figure 8.   For such materials, it is

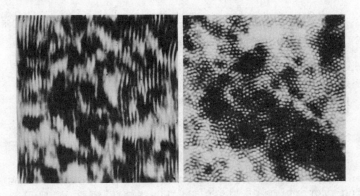

Fig. 8

now possible to characterize the spherical, cylindrical, or
lamellar geometry, figure 9.

STYRENE-ISOPRENE COPOLYMERS

Polymer	Morphology	Dimensions
PS - PI	Cylinders	r 5.4 nm
12,000 - 6,000		s 17.5 nm
(PI - PS)$_4$	Cylinders	r 6.4 nm
Star		s 27.2 nm
PS - PI	Lamellae	d$_s$ 15.6 nm
23,000 - 8,000		d$_I$ 6.2 nm
PS - PI	Lamellae	d$_s$ 15.2 nm
23,000 - 28,000		d$_I$ 21.3 nm

r is the cylinder radius, s is the inter-cylinder
separation, d$_s$ is the thickness of the styrene layer
and d$_I$ is the thickness of the isoprene layer.
Fig. 9

The engineering or physical properties of the macroscopic
material is intimately affected by the geometrical arrangements of
these hard and soft domains as illustrated in figure 10, and as
you will be hearing in greater detail later in this course.

As is illustrated in figure 10, the modulus exhibited by
such a material depends on whether or not we are stressing the
glassy and rubber phases in a parallel or series configuration.
Thus, the observed modulus and, of course, its time-dependent

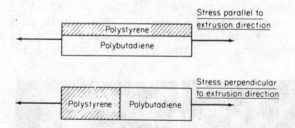

Stress parallel to extrusion direction

Polystyrene
Polybutadiene

Stress perpendicular to extrusion direction

Polystyrene   Polybutadiene

Models for stress relaxation in extruded phase-separated styrene–butadiene–styrene three-block copolymers.

Fig. 10

relaxation may be anisotropic.   Consequently, in discussing the engineering properties of these two-phase systems, we must characterize the morphology before we stand any chance of understanding the gross engineering properties.

Returning to our consideration of modulus against temperature, we can see that the overall transition behaviour between the glassy and rubbery states depends very much on whether the hard and soft segments are very intimately mixed (as in the styrene-butadiene copolymers illustrated) or are phase-separated into blocks and domains (as in the styrene isoprene copolymers illustrated), figure 11.   In this latter case, two separated transitions to an intermediate plateau modulus can be seen. Thus, the temperature-modulus and the time-modulus show quite complex behaviour.   This kind of analysis, again, will be presented to you in more detail in later lectures in the course.

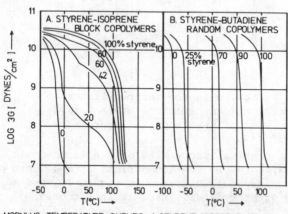

MODULUS TEMPERATURE CURVES A.STYRENE ISOPRENE BLOCK COPOLYMERS   B. STYRENE BUTADIENE RANDOM COPOLYMERS

Fig. 11

MOLECULAR MOVEMENT

The big question about studying the movement of molecules in solids, as opposed to dilute solutions, is how can one molecule wriggle around and move when it is constrained by interactions with neighbouring macromolecules. We must start by defining the interactions and entanglements which constrain this motion and indeed the mathematics of these and of the resultant motion will be presented to you in later lectures this week. A simplistic view of both a non-tangled intermolecular interaction and of a geometrical entanglement is presented in figure 12. Such an entanglement can be given meaningful characteristics such as a lifetime or a strength. Indeed, both the lifetime and the strength can be made effectively infinite by stitching the chains together at an entanglement point by chemical crosslinking. You will be hearing more about how the chains move in between these junction points, which chain sections are characterized by a molecular weight, M_e. Equally, you will be concerned with study of the ways in which the arrays of junction points themselves deform or move.

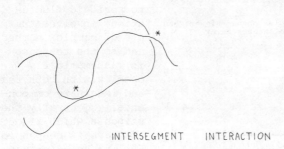

INTERSEGMENT   INTERACTION

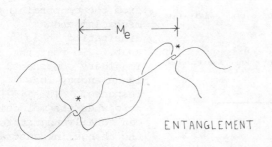

ENTANGLEMENT

Fig. 12

In later lectures in this course, you will move from a theoretical study of the way in which a macromolecule moves in such a constrained situation to experimental methods of observing movement, both of the molecules and of a bulk sample. Many such techniques make observations of what are called relaxation processes in the sample under observation.

MOLECULAR RELAXATION PROCESSES

A relaxation technique is simply one in which we make an experimental observation of the way in which a sample adjusts to a change in the constraints which are acting upon it. Let us suppose that we change the constraints in our sample in a periodic fashion at a series of different frequencies and study the response

of the system to changes in the constraint at varying frequencies.
If we change the constraint very slowly, there is plenty of time
for the macroscopic system to show some sort of response composed,
of course, by an adjustment of the individual macromolecules.
On the other hand, if we change the constraints very rapidly, the
molecules have no time to make the necessary adjustments and the
response observed is much reduced.   Thus, the observed response
drops (goes through a transition) when the frequency of the
change of constraint corresponds to the frequency of the molecular
process responsible for the observation, figure 13.

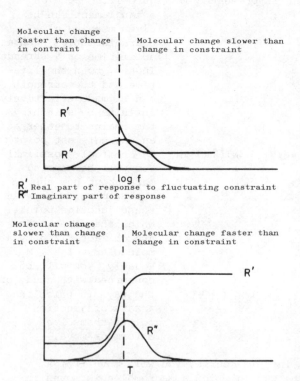

Fig. 13

The mathematics
of systems responding
to periodic fluctu-
ations involves the
use of complex numbers
and so the response
is mathematically
complex with both real
and imaginary compon-
ents.   Generally the
response which is out
of phase is responsible
for absorbing energy
from the constraining
mechanism and
dissipating it as
heat.   So that the
imaginary component
is an energy loss
response.   This is
zero when the molecules
can totally adjust to
the change in constraint;
it is zero when there
is no response and
nothing happens; and
it is a maximum when
there is a phase lag
between the response
of the molecules and
the application of
the constraint.   This phase lag is a maximum when the speeds of
the molecular process and the applied constraint are comparable
and so at this relaxation frequency, the loss, phase lag, or
mathematically imaginary component is a maximum.

    In much of the work you will be covering the relaxation is
studied not by changing the speed of the applied constraint, but
by altering the speed of the molecular process by increasing the
temperature..   Thus we alter the internal molecular process

rather than the external applied process and again get a transition in the observed (mathematically real) response and a maximum in the loss (mathematically imaginary or out of phase) response.

We can apply this to the study of molecular shape change by looking first at the characteristics of localised segmental motion, figure 14. This may be rotation round a single or small number of covalent bonds in which we are interested in the conformational energetics. This can be represented by the enthalpy of the unit varying with rotation angle, giving rise to an enthalpy barrier between stable rotational states, as well as to an enthalpy difference between these states, figure 15. We shall be interested

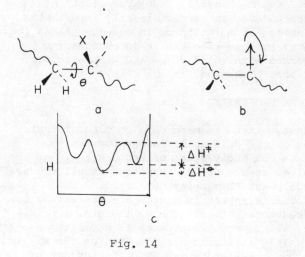

Fig. 14

in consequences of such rotation exemplified by the orientation

SEGMENT ROTATION ENERGIES

(kJ mol^{-1})

	GIBBS-DIMARZIO FLEX ENERGY (FROM T_g)	ACOUSTIC CONFORMATION $\Delta H°$ (FROM SOLUTION)
POLYSTYRENE	6.9	5.4 ± 1.5
S-POLY α-METHYL STYRENE	8.7	8.3 ± 1
POLY METHYL METHACRYLATE		
SYNDIOTACTIC	7.0	6.3 ± 1
ATACTIC	6.8	6.3 ± 1
ISOTACTIC	5.8	3.7 ± 1.5
POLY (2,6 DIMETHYL PHENYLENE OXIDE)	n.a.	0.4 ± 0.1

Fig. 15

of a group dipole in figure 14b.   Thus, the results of the
motions here illustrated might be an increment in the specific
heat of the material and a change in the electrical polarisation
of the bulk sample.   Both of these will be time-dependent and
give us information on how much energy is stored in the molecular
response and how much energy is lost in the phase lag mechanism.

The end uses which follow from such molecular processes are
predominantly electrical and mechanical.   We appreciate, of
course, that biopolymers undergo an extraordinarily important
series of chemical reactions in solution or in aqueous phases (or
none of us would be here) but nevertheless in an engineering
sense we are most concerned with solid state and with electric,
mechanical and viscoelastic phenomena.

ELECTRICAL AND MECHANICAL PROPERTIES

One of the first polymers to be made on a commercial scale
was polyethylene,. One of the principal reasons for its
commercial development in the early years of World War II was the
powers of its potential as an electrical insulator in all the new
electronic and radar equipment then being developed.   The
dielectric properties of an electric insulator are determined by
a number of molecular characteristics., These, in turn, form the
basis of the properties which we observe, such as capacitance or
gross polarisability.   This is itself a reflection of the way in
which the molecules become polarised which forms the subject of
some of your lectures next week.   For example, macromolecules
with dipole vectors can be oriented in an electric field; equally,
mobile charge carriers can be trapped at interfaces between
regions of different density; and at the highest frequencies
there is always a fluctuating polarisation brought about by the
thermal collisions of molecular groupings with their neighbours,
figure 16. Associated with these three quite different aspects of
molecular behaviour are dielectric losses (poor insulation) at
frequencies ranging from 10^{-3} or 10^{-4} Hz for interfacial polarisation up to 10^{12} Hz for the collisional polarisation.   In
applications and end uses these properties show themselves as
failures in base boards or the construction of electronic circuitry
in the one case, and as absorption of submillimetre microwave
communication signals in the other.

While the insulation properties of polymeric materials have
been their principal use up to the present day, there is now a
considerable interest in the conduction properties of macromolecular
solids.   Devices are now constructed which use polymers to do
some of the tasks formerly carried out by silicon, germanium and
other inorganic semi-conducting devices.   So we now have to be
able to discuss these conduction properties in terms of molecular
behaviour and charge transport phenomena.   This is done using the

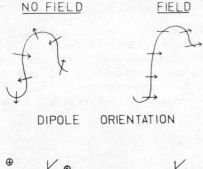

NO FIELD   FIELD

DIPOLE   ORIENTATION

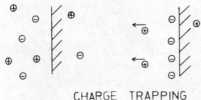

CHARGE   TRAPPING

COLLISIONAL POLARISATION

Fig. 16

HOPPING MODEL

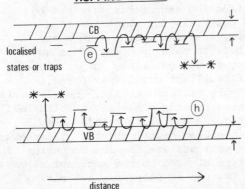

localised
states or traps

distance

✳ deep trap ≡ recombination centre ✳

Fig. 17

language of the solid state physicist in which we construct energy level and band models for the solid. We then ask whether the localised traps and states can be identified as some of the structural (chemical and morphological) characteristics with which we are now familiar. So our time-dependent study of charge carrier movement is interpreted in terms of generating states, hops between shallow traps, and considerable periods of residence in deep traps. Some of the concepts which we have applied to the movement of molecules are then applied to the movement of charge carriers, figure 17.

Finally, I would like to close with a few words on the mechanical properties of polymeric solids. The macroscopic studies of modulus and energy loss relate to the conformational energetics we met earlier. Indeed statistical mechanical series of the glass to rubber transition introduce a 'flex energy' which is the enthalpy difference between stable rotational states and which is comparable to the quantity measured by ultrasonic relaxation in dilute solution, figure 15. I notice that you will be receiving lectures on the glass transition which I am sure will mention this flex energy, and I would like you to

treat these in parallel with the more molecular parameters
presented to you in the lectures on relaxation phenomena.

    In finishing with a mechanical end use, there is a wide
variety of applications to which I could refer.   However, I
thought it appropriate to illustrate just one which will not
appear in your lectures, and this is one particularly relevant to
your visit to Scotland.   It is, in fact, a reference to the golf
ball, figure 18.   As well as illustrating the interest which we

THE GOLF BALL

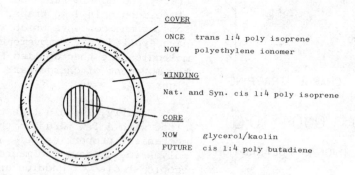

Fig. 18

in Strathclyde take in matters outside the University, it also
shows the importance of constructing engineering devices by
appropriate use of polymeric materials for which the time-
dependent mechanical properties are well known.   Thus our
national symbol of the golf ball is made up today of an outer
cover fabricated from the very robust material, polyethylene
ionomer, a central winding of resilient cis 1:4 isoprene,
surrounding a central core of glycerol and kaolin.   Developments
at the present time are oriented towards achieving a solid ball
which can be moulded in one process and which retains the
durability and resilience of the traditional object.   It seems
likely that such developments will utilise the low mechanical loss
properties of cis poly butadiene blended to the appropriate mass,
resilience and durability by the use of poly(zinc and barium
acrylate and methacrylate).   The best of present day solid golf
balls is now only six to ten yards short (in a standard hit) of
the best wound golf balls.   I trust that you, and people like
you, will be the scientists who will lay the foundation of
completing this development and adding those extra yards to the
pleasure of sporting scientists.

CHAIN STATISTICS OF AMORPHOUS POLYMERS

M. Daoud

LLB, CEN-Saclay, 91191 Gif-sur-Yvette, France

Abstract

The purpose of these lectures is to study the correlations between monomers in an amorphous polymer melt.

Part I deals with the notion of screening of the interactions by concentration. A monodisperse melt is considered as a limit of a concentrated solution. We also study the case of a long chain, made of N steps dissolved in a matrix of shorter chains with P units. Depending on the ratio N/P^2, the long chain is either ideal or swollen.

In Part II the random phase approximation is used to study the scattered intensity by a melt where chains are labelled in different ways, leading to different scattering patterns.

Part III considers the case when some multifunctional units are introduced, thus leading to branched polymers. When the degree of branching is sufficiently high, the statistics is no longer gaussian.

I. LINEAR POLYMER CHAINS IN THE BULK

1. Introduction

In this lecture - and the following ones - we will restrict our attention on linear amorphous chains. Crystalline polymers will be considered elsewhere[1]. So we will consider linear chains made of N statistical segments of length ℓ each. An important property of linear chains is that if we consider a solution which is concentrated enough, the different chains overlap very strongly, at least for usual three dimensional systems. This remains true

15

R. A. Pethrick and R. W. Richards (eds.), Static and Dynamic Properties of the Polymeric Solid State, 15–43.
Copyright © 1982 by D. Reidel Publishing Company.

of course in a melt, which is the main object of these lectures.
The purpose of this lecture is to study the statistics of a chain
in the bulk. We know that two monomers interact with a potential
$V(r)$ shown on figure 1, which has a hard core part. Usually, this
potential is replaced by a point-like potential $v \delta(r)$, where v
is called the excluded volume parameter and is temperature dependent[2])

$$\ell^3 v(T) \; = \; \int \left[1 - e^{-\frac{V(r)}{kT}} \right] d\underset{\sim}{r} \quad . \qquad (I.1)$$

Because of this interaction the statistics of an isolated chain
in a very dilute solution is very far from the Gaussian statistics. A long time ago, Flory[3]) gave a simple argument for chains
in a melt : Because every monomer is surrounded by other monomers,
the effective potential on every monomer is constant, so that the
chain should behave like an ideal chain. Here, we will consider
the problem with two different approaches, either as a concentrated solution where the monomer concentration c goes to ℓ^{-3},
or as the conformation of a very long chain made of N elements
dissolved in a matrix of shorter chains of the same species,
made of P units.

2. Concentrated solutions

Let us first consider a single chain. If there were no interaction between monomers, the chain would be an ideal gaussian chain:
The probability that monomers i and j (i,j = 1,2,...,N) are at a
distance r from each other is

$$P(i,j,\underset{\sim}{r}) \; = \; \left(\frac{3}{2\pi |i-j| \ell^2} \right)^{3/2} \exp \left[-\frac{3r^2}{2 |i-j| \ell^2} \right] \quad . \quad (I.2)$$

Because of the interaction potential we expect some perturbation
for this distribution. Actually, it turns out that the statistics
is completely changed[3][4]). For instance, the mean square end to
end distance is found to be

$$R^2 \; \sim \; N^{2\nu} \ell^2 \qquad\qquad (I.3)$$

where ν is an exponent, first calculated approximately by Flory,
and very close to 3/5 for d=3 (d is the space dimension). This
is the so called excluded volume effect. It has been shown more
recently by de Gennes that the conformation of a chain is related
to another area widely studied, namely critical phenomena[5][4]).

This effect is very strong as long as the molecules are far
apart from each other.

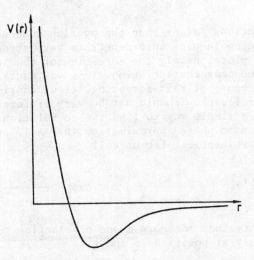

Fig. I.1   The potential V(r) between any two monomers has a hard
           core part preventing them to be on top of each other.

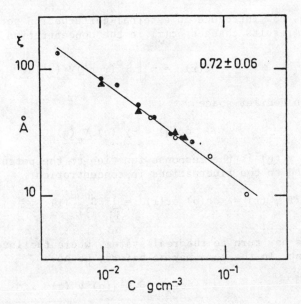

Fig. I.2   The screening length $\xi(c)$ for linear chains in semi-
           dilute solutions it is measured by S.A.N.S. in the
           intermediate range $R^{-1} < q < \xi^{-1}$.

$$c \ll c^* \sim \frac{N}{R^3} \sim N^{-4/5} \ell^{-3} . \qquad (I.4)$$

For concentrations larger than the overlap concentration c^*, the different macromolecules interpenetrate very strongly. A second effect takes place, namely the screening of the interaction due to the high monomer concentration. This was mentioned first by Edwards[6]). Because of this screening, if we consider the chains on a large scale, they should not be very different from gaussian chains. Then a simple way to find the correlation functions is to use the random phase approximation (R.P.A.), starting from the Gaussian statistics. Let us call

$$S_{mn}^{o\alpha\beta}(\underset{\sim}{r},\underset{\sim}{r}') = \left(\frac{3}{2\pi|m-n|\ell^2} \right)^{3/2} \exp\left(- \frac{3(\underset{\sim}{r}-\underset{\sim}{r}')^2}{2|m-n|\ell^2} \right) \delta_{\alpha\beta} \qquad (I.5)$$

the probability that monomers m and n belonging to chains α and β be respectively at positions $\underset{\sim}{r}$ and $\underset{\sim}{r}'$,

and

$$S_{mn}^{o\alpha\beta}(q) = \frac{1}{\ell^3} \int d\underset{\sim}{r} \; e^{-iq\underset{\sim}{r}} \; S_{mn}^{o\alpha\beta}(r) , \qquad (I.6)$$

its Fourier transform.

Let us now introduce an external perturbation potential $V_e(r)$. There results changes $\delta c(\underset{\sim}{r})$ in the concentration

$$\delta c(r) = - \int \chi^o(\underset{\sim}{r}-\underset{\sim}{r}')V_e(\underset{\sim}{r}')d\underset{\sim}{r}' \qquad (1.7')$$

or, in Fourier space

$$\delta c(q) = -\chi^o(q) \; V_e(q) \qquad (I.7)$$

where $\chi^o(q)$ is the response function to the potential and is related to the fluctuations in concentration

$$kT \; \chi^o(r) = <\delta c(o) \; \delta c(r)> = \sum_{\substack{ij \\ \alpha\beta}} S_{ij}^{o\alpha\beta}(r) , \qquad (I.8)$$

let us now turn to the real system, where the interactions are present. In the same way as before, we have

$$\delta c(q) = -\chi(q) \; V_e(q) . \qquad (I.9)$$

In the R.P.A., we suppose that the effect of the interactions is to add to the ideal, non interacting system an internal potential acting uniformly on every monomer

$$\delta c(q) = -\chi^o [V_e(q) + V_{int}(q)] . \qquad (I.10)$$

The internal potential is then calculated by a self consistent equation. It is proportional to the fluctuation in concentration and to the excluded volume parameter.

$$V_{int}(q) = kT \, v' \, \delta c(q) \qquad (I.11)$$

$$v' = v\ell^3 .$$

Combining (10) and (11) we find

$$\delta c(q) = - \frac{\chi^o(q)}{1+v'kT\chi^o(q)} V(q) . \qquad (I.11')$$

Comparing this last relation to Eq.(9) we find

$$\chi(q) = \frac{\chi^o(q)}{1+v'kT\chi^o(q)} . \qquad (I.12)$$

The ideal response function can be calculated from (I.8)

$$kT\chi^o(q) = \sum_{\substack{ij \\ \alpha\beta}} S_{ij}^{\alpha\beta} = N_p \sum_{ij} \exp(-|i-j|q^2\ell^2)$$

$$= 2N_p \frac{N^2}{X^2} (X-1+e^{-X}) \qquad (I.13)$$

where N_p and N are the number of chains per unit volume, the number of elements per chain respectively and $X = q^2 N\ell^2/6$. For $X \gg 1$, we may approximate (I.13) by

$$kT \, \chi^o(q) \approx 2N_p \frac{N^2}{X} = \frac{12c}{q^2\ell^2} \qquad (I.14)$$

where $c = N_p N$ is the monomer concentration and, replacing $\chi^o(q)$ by its expression (I.14)

$$S(q) = \frac{12c}{q^2\ell^2 + 12v'c} . \qquad (I.15)$$

Thus we find a Lorentzian form for the correlation function, and we introduce a screening length[6]

$$\xi = (12v \, c \, \ell^3)^{-1/2}\ell \qquad (I.16)$$

a relation first derived by Edwards.

This notion of screening length is very important because it tells us that for distances larger than ξ, the excluded volume effects are screened, and the chains have an ideal behavior. On the other hand, for distances smaller than ξ these effects are present and the chain has locally a single chain behavior. Note also that ξ depends only on monomer concentration. There is no molecular weight dependence in (I.16), expressing the fact that the screening of the interaction is a pure concentration effect. Different chains with different molecular weights have the same screening length at the same concentration. The chain nature comes in the problem via the cross-over concentration c^* between this regime and the dilute regime, equation (4), and of course through the radius of giration of a chain.

This screening length may be observed experimentally by small angle neutron scattering in the so called intermediate regime, $R^{-1} < q < \xi^{-1}$, where relation (I.15) is valid[7] (see Dr Picot's lectures). These experiments have been performed[8]), and lead to an exponent larger than 1/2 predicted above. One might also notice that when c is decreased and eventually goes to c^* or below, the only length which should stay in the problem is the radius of the chain. Equation (I.16) does not lead to such a result.

One way to overcome this difficulty is to keep the basic two properties of the screening length, namely its existence for concentrations above c^*, and the fact that it depends on concentration [4])[8]) only. Thus we suppose

$$\xi \sim (c\ell^3)^x \ell \qquad (I.17)$$

with x an unknown exponent. The condition which allows us to determine x is that at c^*, the screening should disappear, and thus the screening length sould be of the order of the radius of giration

$$\xi(c^*) \sim R(c^*) \sim N^{3/5} \ell . \qquad (I.18)$$

Combining (17) and (18) we find

$$\xi \sim (c\ell^3)^{-3/4} \ell \qquad (I.19)$$

with an exponent very close to the experimental results. We call a blob this part of the chain where the excluded volume effects are present. Let us call g the number of statistical segments per blob

$$\xi \sim g^{3/5} \sim (c\ell^3)^{-3/4}$$

$$g \sim (c\ell^3)^{-5/4}. \qquad (I.20)$$

As we saw above, for distances larger than ξ, the excluded volume effects are screened. This means that if we now consider the

chain made of blobs, it has an ideal behavior : its radius should
be gaussian :

$$R^2 \sim \frac{N}{g} \xi^2$$

using (19) and (20) we find

$$R^2 \sim N(c\ell^3)^{-1/4} \ell^2 \qquad \qquad (I.21)$$

a relation which has been checked by neutron scattering in the
Guinier range (qR << 1) with solutions where some chains have
been labelled by deuteration[8])[9]), giving a good agreement with
relation (I.21).

If we now look back to the derivation of S(q), we notice that in
fact we should have taken the blobs as units, instead of the
monomers, for doing the calculation of the correlation function.
This will be considered in the next section.

Finally, we can consider the correlation functions, for instance
if we label one chain in the solution

$$S(q) = \int \sum_{ij=1}^{N} c^{-i\underaccent{\sim}{q} \underaccent{\sim}{r}_{ij}} P(r_{ij}) \, d\underaccent{\sim}{r} \qquad \qquad (I.22)$$

where $P(r_{ij})$ is the probability that monomers i and j are at a
distance $\underaccent{\sim}{r}^{ij}$ from each other. If we suppose a more general form
for $P(r)$ than the Gaussian law, Eq.(2)[10])

$$P(r,n) = n^{-3\nu} f\left(\frac{r}{n^{\nu}}\right) \qquad \qquad (I.23)$$

where i and j are at a length n along the chain from each other
and f is a function. Combining (I.22) and (I.23) we find

$$S(q) \sim q^{-1/\nu} \qquad \qquad (I.24)$$

so that the behavior of the pair correlation function is directly
related to the exponent ν. Coming back to what we said above, we
find

$$S(q) \sim q^{-5/3} \qquad (q\xi > 1) \qquad \qquad (I.25')$$

$$S(q) \sim q^{-2} \qquad (q\xi < 1) . \qquad \qquad (I.25'')$$

Note that (25") has the same form as the Lorentzian law, Eq.(15).
These behaviors have been checked by neutron scattering[11]).

As the concentration is increased and one reaches the melt, the screening becomes more and more important. The screening length goes towards the step length ℓ, and the radius of giration goes to its unperturbed value

$$R^2 \sim N\ell^2 .$$ (I.26)

Correspondingly, the intermediate range disappears, leaving just an ideal scattering law[15]), except possibly for small values of the scattering vector, as we will see in next chapter. So the properties of the melt seem to be simpler as those of a solution, as far as the static properties are concerned.

An interesting question worth studying however concerns the case when we dissolve one long chain, made of N units in a matrix made of shorter polymers of the same chemical species, with P segments. When N and P are of the same order, we have a melt of the same type as above. On the other hand, when N is much larger than P, the short chains are very similar to a good solvent, and we should recover the dilute solution limit, Eq. (I.3).

3. One long chain in a solution of shorter chains

We consider a long chain, made of N segments, dissolved in a mixture of a good solvent and shorter chains with P units[3][12]). The monomer concentration c is larger than the overlap concentration c^*. We are interested in the correlation function in the long chain. From what we have seen above, we know that it is more interesting to consider the blobs as elementary units. The excluded volume parameter between blobs is

$$\Omega \sim \xi^3$$ (I.27)

where ξ has been defined in the previous section, and depends only on the monomer concentration.

Let us consider relation (I.11') giving the fluctuations of concentration when a potential V is applied. One interpretation of this relation is to consider that we have an ideal system with an applied screened potential $V'(q)$

$$V'(q) = \frac{V(q)}{1+vkT\chi^o(q)} .$$

If we now consider the excluded volume potential, Ω, then we find that the effect of concentration is to screen out the excluded volume interaction[6])

$$\Omega'(q) = \frac{\Omega}{1+\Omega kT\chi^o(q)} .$$ (I.28)

Note that in the preceding relation, because we consider the blob as the elementary unit, χ^o is the blob-blob correlation function. Introducing the monomer concentration c, the number g of elements per blob and the normalized blob-blob correlation function $P_B(q)$, (I.28) may be written as

$$\Omega'(q) = \frac{\Omega}{1 + \Omega \frac{c}{g} P_B(q)} . \qquad (I.29)$$

Note also that the screening of the interaction is due to the short P chains, and thus that $P_B(q)$ is the correlation function in a short chain.

As we are interested in large distances, of the order of the end to end distance of the long chain, we may take the $q \to 0$ limit of (I.29). Then $P_B(q \to 0)$ is just the number of blobs, Pg^{-1} per (short) chain. We find

$$\Omega'_o = \Omega \frac{g}{P} . \qquad (I.30)$$

Thus we have reduced our initial problem to that of the conformation of a single chain in a dilute solution. We know that the important parameter in this problem[3][13]) is $z = v\ell^{-3} N^{1/2}$, and depending of its value compared to unity, the chain is either swollen or ideal. Here, the excluded volume parameter between blobs is Ω_o, and the number of blobs per chain is Ng^{-1}, leading to

$$z = \frac{\Omega_o}{\xi^3} \left(\frac{N}{g} \right)^{1/2} \sim \left(\frac{Ng}{P^2} \right)^{1/2} . \qquad (I.31)$$

If we consider a melt, g is of order unity and $z \sim \frac{N^{1/2}}{P}$. So that if the chain is short enough, ($N^{1/2} \ll P$), the screening effect is present and the chain is gaussian as we saw in the previous section. If the chain is long enough, $N^{1/2} \gg P$, the short chains act as a simple good solvent and the long chain is swollen :

$$R_N = N^{1/2} f\left(\frac{N^{1/2}}{P} \right) \ell$$

$$\sim N^{3/5} P^{-1/5} \ell . \qquad (I.32)$$

In this last case, $N \gg P^2$, one may also define ideal superblobs : In order to build up a sufficient repulsion, one needs several blobs. For short parts of the chain, the interaction is small and the behavior is ideal. We call superblobs these ideal parts of the long chain. From (I.31), we get the number of elements in a superblob :

$$N_c \sim P^2$$

leading to a size

$$\chi \sim N_c^{1/2} P^2 \sim P \ell \qquad (I.33)$$

All these properties can be seen for instance by neutron scattering experiments. Relation (I.32) might be checked in the Guinier
range, although one needs high molecular weights in order to
check it.

In the regime where the long chain is swollen, one can also look
at the behavior of the correlation function : From what we have
seen above, if we label the long chain, we have in the intermediate range :

$$S(q) \sim q^{-5/3} \quad (R_N^{-1} < q < \chi^{-1}) \qquad (I.34a)$$

$$S(q) \sim q^{-2} \quad (\ell^{-1} > q > \chi^{-1}) \quad . \qquad (I.34b)$$

All these laws have not been conclusively checked as far[14]).

We conclude this chapter by recalling that above the overlap concentration c^* the excluded volume interaction is screened at
large distances. The screening length ξ decreases when the monomer concentration c increases, and is of the order of the step
length in the melt. So for a monodisperse melt, the screening
is total, and the statistics is basically gaussian at least for
d=3, as conjectured by Flory a long time ago. One way to look at
these screening effects in a melt is to vary the length of one
test chain (N monomers) dissolved in a matrix of shorter chains
(P elements) of the same chemical species. If N and P are close
to each other ($N^{1/2} < P$), no excluded volume effect is present
and the test chain is ideal. If N is very large ($N \gg P^2$) the
short chains act like a good solvent. It should then be possible
to check different characteristic behaviors for the correlation
function in the different regimes (relations I.32 to 34). Let us
stress that no sharp transition is ever expected, but smooth
cross-overs, between the different regimes.

II. CORRELATION FUNCTIONS

The basic problem one encounters immediately when one wishes to
measure correlation functions in a melt is a labelling problem.
Let us consider a melt made of identical polymer chains. Then
there are no density fluctuations, and thus no scattered intensity in a scattering experiment. There are different ways of
avoiding this difficulty, by labelling some monomers : In a neutron experiment one replaces some atoms (usually H) by isotopes
(usually D) (physical labelling). In a light scattering experiment, one needs to replace some elements by heavier atoms

(chemical labelling). The merits of each of these methods will be discussed elsewhere[7])[16]). Here, we will be concerned with the calculation of the scattered intensity in different labelling conditions, in order to show that different properties can be observed depending on these conditions. One can either label completely one chain in the melt, or label one (or both) end of every chain or, finally use a mixture of completely labelled and partly labelled chains.

Let us consider a polymer melt, made of chains with N units of length a. We will consider that it is incompressible. As we saw before, the chains obey a Gaussian statistics. Let $P^o_{ij}(r)$ be the probability that monomers i and j, belonging to the same chain, be at a distance r, and $S^o_{ij}(q)$ its Fourier transform.

$$S^o_{ij}(q) = \int P^o_{ij}(r) e^{-iqr} dr \quad . \qquad (II.1)$$

Then

$$S^o_{ij}(q) = \exp(-|i-j|x) \qquad (II.2)$$

with

$$x = q^2 \frac{a^2}{6} \quad .$$

Moreover, two monomers belonging to two different chains are uncorrelated

$$S^{o\alpha\beta}_{ij}(q) = \delta_{\alpha\beta} \exp(-|i-j|x) \qquad (II.3)$$

where α,β are chain indices.

The basic hypothesis we introduce for a polymer melt is that it is an incompressible fluid. This imposes a constraint on the melt, and we wish to calculate the correlation functions with this constraint. In order to do so, we use a method introduced by Edwards in polymer science, and extensively used de Gennes[7]) and Leibler [18])[19]), namely the random phase approximation (R.P.A) : let us apply a perturbation $W^\alpha_m(r)$ on monomer m of chain α, and look at the resulting change in concentration in monomer n of chain β . If we suppose a linear response

$$\delta\phi^\beta_n(r) = - \sum_{\substack{m \\ \alpha}} \int S^{\alpha\beta}_{mn} (r-r') W^\alpha_m(r) dr' \qquad (II.4)$$

or, in Fourier space

$$\delta\phi^\beta_n(q) = - \sum_{\substack{m \\ \alpha}} S^{\alpha\beta}_{mn}(q) W^\alpha_m(q) \qquad (II.5)$$

where the sum is extended to all monomers m and chains α. Note the minus sign, due to the fact that a repulsive (positive) interaction implies a decrease of ϕ_n . $S^{\alpha\beta}_{mn}$ is a response function.

It is related to the density fluctuations, for classical systems

$$kT\ S_{mn}^{\alpha\beta}(q) = <\delta\phi_n^{\alpha}(q)\ \delta\phi_n^{\beta}(q)> \qquad (II.6)$$

In order to calculate $S_{nm}(q)$, we are going to suppose that the response of the system is identical to that of an ideal system subjected to the external potential plus an effective potential $W(q)$ due to the constraint applied to the system (incompressibility).

$$\delta\phi_n^{\beta}(q) = -\sum_{m\alpha} S_{mn}^{o\alpha\beta}(q)\left[W_m^{\alpha}(q)+W(q)\right] \quad . \qquad (II.7)$$

Note that $W(q)$ applies to all the monomers. The incompressibility condition may be written as

$$\sum_{n,\beta} \delta\phi_n^{\beta}(q) = 0 \quad . \qquad (II.8)$$

Combining (7) and (8) we get

$$W(q) = -\frac{\displaystyle\sum_{\substack{mn\\\alpha\beta}} S_{mn}^{o\alpha\beta}(q)\ W_m^{\alpha}(q)}{\displaystyle\sum_{\substack{mn\\\alpha\beta}} S_{mn}^{o\alpha\beta}(q)}$$

$$= -\frac{1}{N_p Nf_D(q)} \sum_{\substack{mn\\\alpha\beta}} S_{mn}^{o\alpha\beta}(q)\ W_m^{\alpha}(q) \quad . \qquad (II.9)$$

Where $f_D(q)$ is the Debye function and N_p the number of chains.

$$f_D(q) = \frac{1}{N}\sum_{ij} S_{ij}^{o}(q) = \frac{2N}{X^2}(X-1+e^{-X}) \qquad (II.10)$$

with $X = Nx$.

Replacing $W(q)$ in (II.7) by its expression in (II.9), and comparing with (II.5), we get

$$S_{nm}^{\alpha\beta}(q) = S_{nm}^{o\alpha\beta}(q) - \frac{1}{N_p Nf_D(q)} S_m^{o\alpha}(q)\ S_n^{o\beta}(q) \qquad (II.11)$$

with $S_i^{o\alpha}(q) = \sum_{j,\beta} S_{ij}^{o\alpha\beta}(q) = \frac{1}{x}\left[2-e^{-ix}-e^{-(N-i)x}\right]$. $\qquad (II.12)$

Relation (II.11) is the basic R.P.A. relation. The first term in the right hand side is the direct correlation between monomers. The second term represents the net repulsion between monomers due to the constraint.

With this relation in hand, we can calculate the scattered intensity in different conditions. The scattered intensity is[7])

$$S(q) = \sum_{ij} A_i^\alpha A_j^\beta S_{ij}^{\alpha\beta} \tag{II.13}$$

where A_i is the scattering amplitude of monomer i.

1) <u>No labelling</u>. If we do not label any monomer, then

$$S^B(q) = A^2 \sum_{\substack{mn \\ \alpha\beta}} S_{mn}^{\alpha\beta}(q) = 0 \tag{II.14}$$

where we have used (10),(11) and (12). This is an expected result because the total density is constant, and thus there are no fluctuations at any scale. In order to get a scattered intensity, we need to label some monomers.

2) <u>Labelling of one chain</u>. Now we label one chain completely, the other chains being unlabelled : $A_i^1 = A$, $A_i^{\alpha\neq1} = B$. Expanding (II.13) we get

$$S(q) = A^2 \sum_{ij} S_{ij}^{11} + B^2 \sum_{\substack{mn \\ \alpha\beta\neq1}} S_{ij}^{\alpha\beta} + 2AB \sum_{\substack{im \\ \beta\neq1}} S_{1m}^{1\beta} \tag{II.15}$$

where ij belong to the labelled chain (1) and m and n to the other chains. The second term in the right hand side of (II.15) is

$$\sum_{\substack{\alpha\beta\neq1 \\ ij}} S^{\alpha\beta} = \sum_{\substack{\alpha\beta \\ ij}} S_{ij}^{\alpha\beta} - 2 \sum_{\substack{\alpha \\ ij}} S_{ij}^{1\alpha} + \sum_{ij} S_{ij}^{11} \quad . \tag{II.16}$$

Using (II.14) and (II.11) and (II.12) it is easy to see that the first two sums in the right hand side of (II.16) are zero :

$$\sum_{(\alpha\beta)\neq1} S_{ij}^{\alpha\beta} = \sum_{ij} S_{ij}^{11} \quad . \tag{II.17}$$

In the same way, the last term in the R.H.S. of (II.15) is

$$\sum_{\substack{1\beta\neq1 \\ i\,m}} S_{im}^{i\beta} = \sum_{\substack{1\beta \\ im}} S_{im}^{1\beta} - \sum_{im} S_{im}^{11}$$

$$= -\sum_{im} S_{im}^{11} \quad .$$

So that (II.15) can be written as :

$$S(q) = (A-B)^2 \sum_{ij} S_{ij}^{11}$$

and, from (II.11),

$$S_{ij}^{11} = S_{ij}^{o} - \frac{1}{N_p Nf_D(q)} \; S_i^{01}(q) \; S_j^{01}(q) \; ,$$

giving

$$\sum_{ij} S_{ij}^{11} = Nf_D(q) - \frac{1}{N_p Nf_D(q)} \; [Nf_D(q)]^2$$

and

$$S(q) = (A-B)^2 (1-\frac{1}{N_p}) \; Nf_D(q) \; . \tag{II.18}$$

So by labelling completely one chain, we get except a constant factor, the Debye function

$$f_D(q) = \frac{2}{X^2} \; (X-1+e^{-X})$$

with $X = \frac{q^2 R^2}{6}$ shown on Fig.(II.1). In the Guinier range, $X \ll 1$,

$$f(X) = 1 - \frac{X}{3}$$

allows for a measurement of the radius R. This shows that it is possible to observe the Gaussian nature of the chains in the bulk by labelling one or some chains completely.

3) Labelling one end[17]). Let us now label one end of every chain. Then $A_1^\alpha = A$ and $A_{i\neq1}^\alpha = B$ and, by expanding (II.13) we get

$$S(q) = A^2 \sum_{\alpha\beta} S_{11}^{\alpha\beta} + B^2 \sum_{\substack{ij=2...N \\ \alpha\beta}} S_{ij}^{\alpha\beta} + 2AB \sum_{\substack{j\neq1 \\ \alpha\beta}} S_{1j}^{\alpha\beta} \; . \tag{II.19}$$

We proceed in the same way as in section (2) : The second term in the R.H.S. of (II.19) is

$$\sum_{\substack{ij=2...N \\ \alpha\beta}} S_{ij}^{\alpha\beta} = \sum_{\substack{ij \\ \alpha\beta}} S_{ij}^{\alpha\beta} - 2 \sum_{\substack{j \\ \alpha\beta}} S_{ij}^{\alpha\beta} + \sum_{\alpha\beta} S_{11}^{\alpha\beta} \; ,$$

and, taking into account (II.14) and (II.11) and (12), we find

$$\sum_{\substack{ij=2,...N \\ \alpha\beta}} S_{ij}^{\alpha\beta} = \sum_{\alpha\beta} S_{11}^{\alpha\beta} \; . \tag{II.20}$$

Similarly $\sum_{\substack{j\neq1 \\ \alpha\beta}} S_{ij}^{\alpha\beta} = \sum_{\substack{j \\ \alpha\beta}} S_{ij}^{\alpha\beta} - \sum_{\alpha\beta} S_{11}^{\alpha\beta} = - \sum_{\alpha\beta} S_{11}^{\alpha\beta} \; . \tag{II.21}$

So combining (II.19),(20) and (21) we find

$$S(q) = (A-B)^2 \sum_{\alpha\beta} S_{11}^{\alpha\beta} \tag{II.22}$$

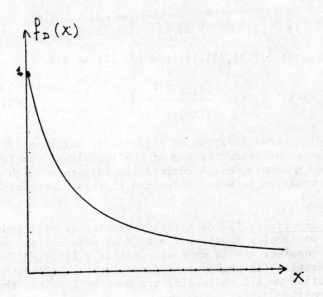

Fig.II.1   The Debye function for ideal non interacting linear
            chains.

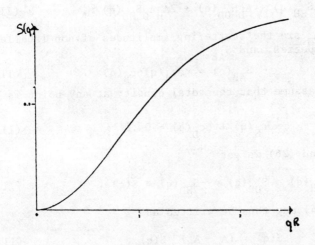

Fig.II.2   The correlation hole is exhibited when one end of
            every chain is labelled.

and, from (II.11)

$$S_{11}^{\alpha\beta} = S_{11}^{o} \delta_{\alpha\beta} - \frac{1}{N_p N f_D(q)} S_1^{o\alpha} S_1^{o\beta}$$

taking into account (II.3),(II.11) and (II.12) we find

$$S(q) = (A-B)^2 N_p \left[1 - \frac{(1-e^{-X})^2}{2(X-1+e^{-X})} \right] . \qquad (II.23)$$

The corresponding curve is shown on Fig.(II.2). Relation (II.23) may be interpreted as being the sum of the individual scattering by the labelled monomers, and a correlation hole term which represents the repulsion between chains due to the incompressibility constraint.

4) <u>Block copolymers</u>[18]). As a final example, we will consider the case when we label a fraction $f = N_D/N$ of some chains. So the system we consider now is made of completely labelled chains, with a concentration (in chains) $(1-\varphi)$, and partly labelled chains, with concentration φ. All the chains are made of N units. The scattered intensity

$$S(q) = \sum_{\substack{ij \\ \alpha\beta}} A_i^\alpha A_j^\beta S_{ij}^{\alpha\beta}(q)$$

is split into three parts as above

$$S(q) = A_H^2 S_{HH}(q) + A_D^2 S_{DD}(q) + 2A_H A_D S_{HD}(q) , \qquad (II.24)$$

where A_H and A_D are the scattering amplitudes of non labelled and labelled species, and S_{AB}

$$S_{AB}(q) = \langle \delta c_A(q) \delta c_B(q) \rangle . \qquad (II.25)$$

As before, we assume that the total density at any point is constant :

$$\delta c_H(q) + \delta c_D(q) = 0 . \qquad (II.26)$$

From (II.25) and (26) we get

$$S_{HH}(q) = S_{DD}(q) = -S_{HD}(q) \equiv \tilde{S}(q) . \qquad (II.27)$$

Relation (II.24) may thus be written as

$$S(q) = (A_H - A_D)^2 \tilde{S}(q) . \qquad (II.28)$$

We calculate $\tilde{S}(q)$ using the same approximation as before : let us introduce two perturbing potentials $U_H(q)$ and $U_D(q)$ acting respectively on unlabelled an labelled units. The resulting changes in densities, in a linear response theory are

$$\delta c_D(q) = - \left[S_{DD}(q)U_D(q) + S_{DH}(q)U_H(q) \right] \qquad (II.29a)$$

$$\delta c_H(q) = - \left[S_{HD}(q)U_D(q) + S_{HH}(q)U_H(q) \right] . \qquad (II.29b)$$

Note that the total constant density assumption brings us back relation (II.27). In the R.P.A., it is supposed that the effect of the constraint is to introduce a potential U(q), acting equally on all the monomers, the chains being supposed to be ideal, non interacting polymers :

$$\delta c_D(q) = - \left[S_{DD}^o(q)[U_D(q)+U(q)] + S_{DH}^o [U_H(q)+U(q)] \right] \qquad (II.30a)$$

$$\delta c_H(q) = - \left[S_{HD}^o(q)[U_D(q)+U(q)] + S_{HH}^o(q)[U_H(q)+U(q)] \right] . \qquad (II.30b)$$

The potential U(q) is calculated by using the constraint (II.26) of a constant total monomer density. We find

$$U(q) = - \frac{S_{DD}^o(q)U_D(q) + S_{HH}^o(q)U_H(q) + S_{HD}^o(q)[U_H(q)+U_D(q)]}{S_{DD}^o(q)+S_{HH}^o(q)+2S_{HD}^o(q)} . \qquad (II.31)$$

Replacing in (II.30a) U(q) by the preceding expression, and comparing with (II.29) we find

$$S_{DD}(q) \equiv \tilde{S}(q) = \frac{S_{DD}^o(q)S_{HH}^o(q)-[S_{HD}^o(q)]^2}{S_{DD}^o(q)+S_{HH}^o(q)+2S_{HD}^o(q)} . \qquad (II.32)$$

Thus we have reduced the problem of calculating the scattered intensity in the real system, Eq.(III.28), to the evaluation of the functions S_{DD}^o, S_{HH}^o and $S_{HD}^o(q)$ in system of ideal non interacting chains.

In order to calculate these functions, we introduce a variable μ_i

$$\mu_i = \begin{cases} 0 & \text{if } i \text{ is an unlabelled monomer} \\ 1 & \text{if } i \text{ is a labelled monomer.} \end{cases}$$

Then

$$S_{DD}^o(q) = \sum_{\substack{ij \\ \alpha\beta}} \mu_i \mu_j \, S_{ij}^{o\alpha\beta}(q) \qquad (II.33a)$$

$$S_{HH}^o(q) = \sum (1-\mu_i)(1-\mu_j) \, S_{ij}^{o\,\alpha\beta} \qquad (II.33b)$$

$$S^o_{HD}(q) = \sum_{} \mu_i (1-\mu_j) \; S^{o \alpha\beta}_{ij} \tag{II.33c}$$

$$S^{o\alpha\beta}_{ij} = \delta_{\alpha\beta} \; e^{-|i-j|x}$$

giving

$$S^o_{DD}(q) = \varphi \, fN \, f_D[fN] + (1-\varphi)N \, f_D(N) \tag{II.34}$$

with

$$fN \; f_D(fN) = \frac{2}{x^2} \, [Nfx - 1 + e^{-Nfx}] \tag{II.35}$$

$$x = q^2 a^2 / 6$$

$$S^o_{HH}(q) = \varphi(1-f)N f_D[(1-f)N] \tag{II.36}$$

$$S^o_{HD}(q) = \frac{1}{2} \, \varphi N [f_D(N) - f f_D(fN) - (1-f) f_D[(1-f)N]] \; . \tag{II.37}$$

Combining (II.35) to (37) we find

$$S_{HH}(q) + S_{DD}(q) + 2S_{HD}(q) = N f_D(N)$$

and

$$\widetilde{S}(q) = \varphi(1-\varphi)N f_D[(1-f)N] + \varphi^2 \widetilde{S}_1(q) \tag{II.38}$$

$$\widetilde{S}_1(q) = \frac{1}{f_D(N)} \left\{ f_D(f_N) f_D[(1-f)N] - \frac{1}{4}[f_D(N) - f f_D(fN) - (1-f) f_D[(1-f)N]]^2 \right\} \tag{II.39}$$

The first term in (II.38) includes the mixing effect and vanishes for a one component system, i.e. when $\varphi = 0$ or 1. $\widetilde{S}_1(q)$ includes both the direct correlations of monomers on the same chain and the correlation hole effect. Note that for $q=0$ the first term is maximum whereas the second vanishes. Thus for $\varphi \neq 0$ (or 1), all the forward scattering is due to the mixing effects. Fig.(3) shows the corresponding curves for $f = \frac{1}{2}$ and for different concentrations φ .

For $\varphi = 1$ (only diblocks), the curve has a maximum, corresponding to the case studied in section 2. (Actually, there is a convolution of the correlation hole by a Debye function). When we

decrease φ_1 there is a contribution for $q = 0$ which is important.
The curve exhibits still a maximum which goes to $q = 0$ when φ is
decreased. For $\varphi = 1/2$, $\tilde{S}(0)$ is a maximum. Decreasing φ further
decreases the intensity. For $\varphi = 0$, there is no scattered intensity
(section 1).

Conclusion

We conclude this part by insisting on the three points.

1) The Gaussian nature of the chains can be checked by
scattering experiments. Because of the incompressibility of the
system, one needs a labelling of the chains.

- labelling one chain completely leads to a direct check
of Debye's scattering law, as well as to a measurement of the
radius of gyration,

- labelling parts of the chains leads to another interesting effect, namely the correlation hole which is a measure of
the repulsion between chains due to the constant density hypothesis.

2) The correlation hole effect is shown as a maximum for
$q \neq 0$ when parts of the chains are labelled ($\varphi = 1$ in section 3).
Let us insist that there is no ordering in the solution, where
different chains are supposed to interpenetrate entirely. Thus
this maximum does not imply any spatial ordering effect.

3) Finally, the random phase approximation proves to be a
very nice theoretical tool to investigate the properties of bulk
polymeric material, where the properties are close to those of
ideal Gaussian chains.

III. INFLUENCE OF BRANCHING

1. Introduction

So far we consider only linear polymers, made of bifunctional
units. One may ask for the influence of the presence of multifunctional units, and thus some degree of branching, on the statistics of the chains. In this part, we will be concerned with
polymers branched in a random way (to be defined below). We will
suppose there are no closed loops (at least no large loops), and
we are going to take into account the interactions between monomers. An important simplification that makes the calculations
much easier is to suppose that the interaction is the same
between any pair of monomers (di or trifunctional). The approach
we use is a Flory theory, which we know is not exact, but gives
results which are very close to the exact results. Let us first
consider a tree like molecule, with a total number of elements N.
We suppose that during the formation of the molecule, the

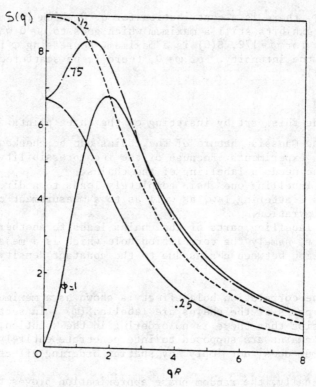

Fig. II.3   Influence of the concentration φ of diblock chains in
             a mixture of labelled and partially labelled chains,
             for $f=1/2$ (from ref. 18).

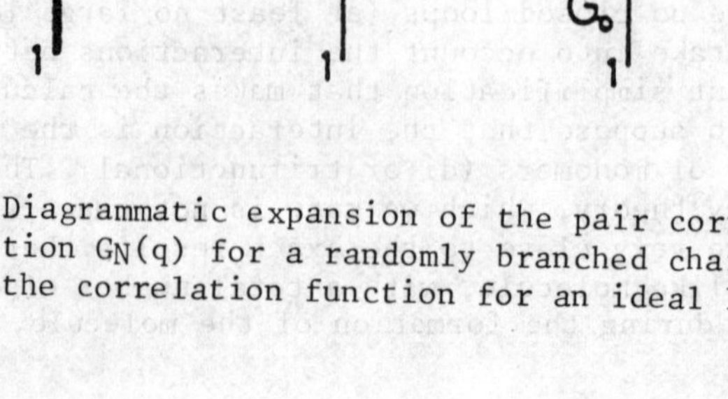

Fig. III.1   Diagrammatic expansion of the pair correlation func-
             tion $G_N(q)$ for a randomly branched chain. $G^0(q)$ is
             the correlation function for an ideal linear chain.

activity Λ^2 of the trifunctional units has been kept constant.
(This can be achieved in a poly condensation process). This is
what we call a randomly branched molecule. Moreover, we suppose
that the monomers do not interact. This is the ideal case, first
considered by Zimm and Stockmayer[20]. It is interesting to calculate the probability $G_N(\underset{\sim}{r})$ that a branched molecule has one end
point at the origin and another at point $\underset{\sim}{r}$. Let $G_N^0(\alpha, q)$ its
Fourier-Laplace transform :

$$\widetilde{G}^0(\alpha, q) = \int_0^\infty dN e^{-\alpha N} \int d\underset{\sim}{r}\, e^{i\underset{\sim}{q}\underset{\sim}{r}}\, G_N^0(\underset{\sim}{r}) \quad . \tag{III.1}$$

$$= \frac{1}{\alpha + q^2 \frac{\ell^2}{6}} \quad . \tag{III.2}$$

It is easy to build a diagrammatic expansion for $\widetilde{G}(\alpha, q)$. This is
shown on Fig.(III.1), where a branch point has a weight Λ^2. Thus
we find

$$\widetilde{G}(\alpha, q) = \widetilde{G}^0(\alpha, q)\,[1 + \Lambda^2 \widetilde{G}(\alpha, q)\, \widetilde{G}(\alpha, 0)] \quad . \tag{III.3}$$

Taking (III.2) into account, we can solve (III.3)

$$\widetilde{G}^{-1}(\alpha, q) = q^2 \frac{\ell^2}{6} + \frac{1}{2}\,[\alpha + (\alpha^2 - 4\Lambda^2)^{1/2}] \quad . \tag{III.4}$$

Taking the inverse Laplace transform we get the probability

$$G_N(q) = \sum_0^\infty (-1)^n\, (q^2 \frac{\ell^2}{3})^n\, \frac{n+1}{(2\Lambda)^{n+1}}\, \frac{I_{n+1}(2\Lambda N)}{2\Lambda N} \tag{III.5}$$

where $I(x)$ are modified Bessel functions, and the partition function

$$Z_N = G_N(q=0) = \frac{I_1(2\Lambda N)}{2\Lambda N} \quad . \tag{III.6}$$

It is possible to get simple expressions for $G_N(q)$ in the two
limits when ΛN is very small or very large : in the high branching
limit ($\Lambda N \gg 1$) it reduces to

$$G_N(q) \approx \frac{Z_N}{(1 + \frac{q^2 \ell^2}{6\Lambda})^2} \tag{III.7}$$

in the low branching limit, we get

$$G_N(q) = e^{-q^2 \frac{N\ell^2}{6}} \tag{III.8}$$

which is identical to the probability for a linear Gaussian chain.

Relations (III.7) and (8) show the difference between the statistics of linear and highly branched chains even in the ideal non-interacting case. A basic difference lies in the probability $G_N(q)$ which for branched chains, is no longer a Gaussian. One can calculate the radius of gyration of a branched chain. This was done first Zimm and Stockmayer. Their result is[21][22] :

$$R_o \sim \left(\frac{N}{\Lambda}\right)^{1/4} \ell \tag{III.9}$$

where ℓ is the step length (supposed to be the same for both di and trifunctional units).

Note that for $\Lambda \sim N^{-1}$, we recover the usual linear Gaussian chain behavior

$$R_\ell \sim N^{1/2} \ell , \tag{III.10}$$

and we may distinguish between linear chains for $\Lambda N \ll 1$ and branched chains for $\Lambda N \gg 1$.

A basic weakness of relation (III.9) is that for large polymers, it leads to a density inside the chain which is too high (for $d=3$).

We consider a single chain (dilute solution) and introduce the excluded volume interaction. As a result, the chain swells and the radius is found to be[23][24]

$$R \sim N^{1/2} \Lambda^{-1/10} \ell \tag{III.11}$$

Relation (III.11) is basically valid in the dilute range, as far as the molecules are far apart from each other. This leads us to the concentration c^* above which the preceding condition is no longer valid

$$c^* \sim \frac{N}{R^3} \sim N^{-1/2} \Lambda^{3/10} \ell^{-3} . \tag{III.12}$$

As for linear chains, we expect concentration effects to become important above c^*, leading to a different behavior of the radius with molecular weight and to the existence of another length related to concentration effects.

2. Concentrated solutions

For concentrations above c^*, the first question is to determine how the radius varies with molecular weight. For this, we are going to use a Flory type approach. The basic hypothesis is to suppose that the molecules do not overlap. Then we may write a

Flory free energy[25])

$$F = \frac{R^2}{R_o^2} + v\frac{N^2}{R^d} + w\frac{N^3}{R^{2d}} + \ldots + t\frac{N^{\alpha+1}}{R^{\alpha d}} + \ldots \quad (III.13)$$

where the coefficient $w, \ldots t$ are supposed to be positive constants. For high concentrations, the higher order terms are the
most important. Minimizing the free energy gives[25])

$$R \sim N^{1/d}\ell \quad . \quad\quad\quad (III.14)$$

Thus we find that the chain is compact at $d=3$

$$R \sim N^{1/3}\ell \quad .$$

The above analysis was carried for the case when we have only
trifunctional units ($\Lambda=1$). If we now consider the more general
case when $\Lambda < 1$, we may give a blob analysis[8)25]). We have seen
above that we have linear parts between trifunctional units.
On the average, these linear parts, which we call blobs, have
$n \sim \Lambda^{-1}$ elements. Their radius is

$$\xi \sim n^{1/2}\ell \sim \Lambda^{1/2}\ell \quad . \quad\quad\quad (III.15)$$

The molecule may be considered as a compact ensemble of blobs

$$R \sim (\frac{N}{n})^{1/d}\xi$$

$$\sim N^{1/3}\Lambda^{-1/6}\ell \quad . \quad\quad\quad (III.16)$$

Again, one may check that when $\Lambda \sim N^{-1}$, relation (III.16) crosses
over smoothly to the ideal linear chain behavior. So when Λ is
small enough ($\lambda N \ll 1$), even when the chain is branched, we expect
it to behave basically as a linear chain, except for corrections
which might be calculated in an expansion in ΛN.

For highly branched chains ($\Lambda N \gg 1$), we find a rather compact
structure. We may evaluate the number of different molecules
inside a volume $\Omega \sim R^3$. The density due to one chain is

$$c_S \sim \frac{N}{R^3} \sim \Lambda^{1/2}$$

and thus there are

$$N_S \sim \Lambda^{-1/2} \quad\quad\quad (III.17)$$

chains inside this volume. The number N_S is the degree of overlap

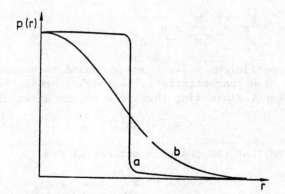

Fig. III.2   The two possible concentration profiles :

(a) a sharp drop in the density profile leads to
$S(q) \sim q^{-4}$

(b) a gradual decrease leads to $S(q) \sim q^{-3}$.

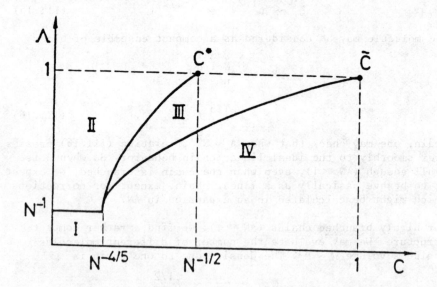

Fig. III.3   Phase diagram of the randomly branched chains. In
region III there is no overlap, whereas there is some
in region IV.

	R	ξ_L	ξ_c
I	$N^{3/5}$		
II	$N^{1/2}\,\Lambda^{-1/6}$	$\Lambda^{-3/5}$	
III	$(\dfrac{N}{c})^{1/3}$	$\Lambda^{-3/5}$	$c^{-1}\,\Lambda^{1/5}$
IV	$N^{1/3}\Lambda^{-1/6}c^{-1/8}$	$c^{-3/4}$	$\Lambda^{-1/2}c^{-1/8}$

TABLE 1 : Characteristics of the different regimes.

(see fig.III.3)

of the molecules. It is interesting to note that it depends only on the degree of branching Λ, and is independent of the molecular weight.

The overlap increases when Λ decreases. For completely branched chains, $\Lambda=1$, there is no overlap as we have already seen above. On the other hand, for $\Lambda \sim N^{-1}$, the chains overlap very strongly, and

$$N_S(\Lambda \sim N^{-1}) \sim N^{1/2}$$

as is known for linear molecules.

Finally, we look at the correlation functions in the different domains.

- In the Guinier range, one measures the radius of gyration

$$S(q) = 1 - q^2\,\frac{R_G^2}{3} \ . \tag{III.18}$$

- In the other limit, if we look at small distances, i.e. large values of q, we are going to probe mainly the linear parts, and we expect a scattering law of the same type as in part I

$$S(q) \sim q^{-2} \quad (\ell^{-1} < q < \xi^{-1}) \ . \tag{III.19}$$

- In the intermediate range, the branched nature of the chain is going to be observed. The scattering pattern however depends strongly on the exact density profile inside the chain : A sharp drop in the density in a limited interface, as shown on

Fig. (III.2a) leads to a characteristic $S(q) \sim q^{-4}$ behavior. On
the other hand, a more gradual decrease, as shown on Fig. (III.2b)
would lead to $S(q) \sim q^{-3} \Lambda^{1/2}$. As we do not know the precise concentration profile, we will not continue this discussion. But
neutron scattering experiments on a melt would make easily this
difference and give a more precise idea about the concentration
profile of a molecule in the melt.

3. The semi-dilute regime

Note that relation (III.16) does not cross-over properly to the
expression (III.11) for the radius of a single chain at c^* : In
fact there is a regime between c^* and a concentration $\tilde{c}$ where
there is no overlap between the chains. In this regime ($c^* < c < \tilde{c}$)
one may define concentration blobs, which are themselves branched
parts of the molecule. As for linear chains, the radius ξ_c of
these blobs does not depend on the molecular weight. It can be
obtained simply if we assume a scaling relation : We have for
the different lengths

$$L = N^{1/2} \Lambda^{-1/10} \ell \, f(\frac{c}{c^*}) \tag{III.20}$$

where $f(x)$ is a universal function which we suppose to have a
power law behavior for concentrations above c^*

$$f(x) \sim x^{\alpha}$$

the exponent α being determined by the condition that ξ_c is
independent of N. We find $\alpha = -1$

$$\xi_c \sim c^{-1} \Lambda^{1/5} \ell^{-2} . \tag{III.21}$$

Inside this concentration blob, we have linear parts which we
already called linear blobs, with $n \sim \Lambda^{-1}$ elements. As excluded
volume effects are present within these linear blobs, we have

$$\xi_L \sim n^{3/5} \ell \sim \Lambda^{-3/5} \ell . \tag{III.22}$$

Finally, the molecule is a compact ensemble of concentration
blobs :

$$R \sim (\frac{N}{g})^{1/3} \xi_c \tag{III.23}$$

where g is the number of elements per concentration blob

$$\xi_c \sim (\frac{g}{n})^{1/2} \xi_L . \tag{III.24}$$

Combining (III.21) to (24) we find

$$R \sim (\frac{N}{c})^{1/3} \quad . \tag{III.25}$$

Note that in the concentration range we are discussing, (c above c^*, but not too high, see below) we have $\xi_c > \xi_L$. In this range, there is no overlap of the molecules.

Concentration $\tilde{c}$ is defined as that for which $\xi_L \sim \xi_c$. Using (III.21) and (22) we find

$$\tilde{c} \sim \Lambda^{4/5} \ell^{-3} \quad . \tag{III.26}$$

For concentrations above $\tilde{c}$, the situation is somewhat different from what we have seen above : The concentration blob itself is linear. From what we have seen in Part I we have then

$$\xi_c \sim (c\ell^3)^{-3/4} \ell \tag{III.27}$$

and, for the linear blob

$$\xi_L \sim \Lambda^{-1/2} (c\ell^3)^{-1/8} \ell \quad . \tag{III.28}$$

Finally, the macromolecule is a compact succession of blobs

$$R \sim (\frac{N}{n})^{1/3} \xi_L$$

$$R \sim N^{1/3} \Lambda^{-1/6} (c\ell^3)^{-1/8} \ell \tag{III.29}$$

which goes to relation (III.16) when $c\ell^3$ goes to unity.

Thus we find different regimes depending on c :

- For $c^* < c < \tilde{c}$, a non overlapping regime where the concentration blob is branched.

- For $c > \tilde{c}$, a regime where concentration blob is linear, and where there is some degree of overlap. This can be calculated in the same way as we did for a melt. We find

$$N_S \sim \Lambda^{-1/2} (c\ell^3)^{5/8} \quad . \tag{III.30}$$

These different results are summarized in Table I. The corresponding phase diagram is shown on Fig.(III.3).

As a summary, we have seen that as far as ΛN is less than unity, the behavior of a randomly branched chain is basically the same as that of a linear polymer. When ΛN is larger than unity, the statistics is completely changed. Even for ideal non interacting chains, we no longer have a Gaussian form for the pair distri-

bution function. Note that when N is very large, this happens
as soon as Λ is non zero. Moreover, the radius of gyration, relation (III.9) leads to an unrealistic density for space dimensions below d=4 (see below). As a result the branched chains do
not have their ideal behavior in the melt but are rather collapsed.
The hypothesis of interpenetration which is valid for linear chains
holds less and less for branched chains as the degree Λ of branching increases. For Λ=1, i.e. no difunctional units, the chains
do not interpenetrate . In the same way, when Λ is large but
smaller than unity, the overlap decreases strongly when we add
some solvent $\underset{\sim}{:}$ There is no more overlap when the concentration
is less than $\underset{\sim}{c}$ (see Fig.III.3).

Finally, one may ask about the usefulness of the ideal chain. In
fact, as one can see clearly on the Flory free energy, the exponents depend very strongly on the dimension d of space. On the
other hand we know from critical phenomena[26]) that for large
enough space dimensions the ideal result is correct : Above a
critical dimension d_c, we can neglect the interactions. The simplest way to evaluate d_c is to estimate the interaction terms,
using the ideal value for the radius R. We make this estimation
for both linear and branched chains. The ideal radii are

$$R_o^L \sim N^{1/2}\ell$$

$$R_o^B \sim N^{1/4}\ell .$$

For a chain in the melt, using the higher order interaction terms
in (III.13), we are led to :

$$d_c^L = 2$$

$$\text{and} \quad d_c^B = 4 .$$

This shows the basic difference between linear and branched
chains in the usual d=3 cases. Whereas for linear chins we are
above the critical dimension, and thus we can neglect the interactions in the melt, this is not true for branched chains. In
the latter case we are below the critical dimension and the
interactions have to be taken into account.

So far a lot of work has been devoted to linear chains, with a
good agreement between experiment and theory. Branched chains
have not been as extensively studied and it would be very interesting to see how the above results compare with experiments.

REFERENCES

[1]) G. Wegner, Lecture
     D. Sadler, Lecture
[2]) S.F. Edwards, Proc. Phys. Soc. 85, 613, 1965
[3]) P.J. Flory, Principles of Polymer Chemistry, Cornell Un.
     Press, 1953
[4]) P.G. de Gennes, Scaling Concepts in Polymer Physics, Cornell
     Un. Press, 1980
[5]) P.G. de Gennes, Phys. Lett. 38A, 339, 1972
[6]) S.F. Edwards, Proc. Phys. Soc. 88, 265, 1966
[7]) C. Picot, Lecture
[8]) M. Daoud, J.P. Cotton, B. Farnoux, G. Jannink, G. Sarma,
     H. Benoît, R. Duplessix, C. Picot and P.G. de Gennes, Macro-
     molecules 8, 804, 1975
[9]) R.W. Richards, A. Maconnachie and G. Allen, Polymer 19, 266,
     1978
[10]) J. des Cloizeaux, J. Physique 37, 431, 1976
[11]) B. Farnoux, Ann. Physique (Paris) 1, 73, 1976
[12]) P.G. de Gennes, J. Pol. Sci. Pol. Symposium 61,   , 1977
[13]) H. Yamakawa, Modern theory of polymer solutions, Harper and
     Row, N.Y. 1972
[14]) R. Kirste, B. Lehnen, Makromol. Chem. 177, 1137, 1976
[15]) J.P. Cotton, D. Becker, B. Farnoux, G. Jannink, R. Ober,
     Phys. Rev. Lett. 32, 1170, 1974
[16]) R.S. Stein, Lecture
     J.S. Higgins, Lecture
     J.M. Vaughan, Lecture
[17]) P.G. de Gennes, J. Physique 31, 235, 1970
[18]) L. Leibler, H. Benoît, Polymer 22, 195, 1981
[19]) L. Leibler, Macromolecules 13, 1602, 1980
[20]) B.H. Zimm, W.H. Stockmayer, J. Chem. Phys. 17, 1301, 1949
[21]) P.G. de Gennes, Biopolymers, 6, 715, 1968
[22]) G.R. Dobson, M. Gordon, J. Chem. Phys. 41, 2384, 1964
[23]) J. Isaacson, T.C. Lubensky, J. Physique Lett. 41, 469, 1980
[24]) G. Parisi, N. Sourlas, Phys. Rev. Lett. 46, 871, 1981
[25]) M. Daoud, J.F. Joanny, J. Physique, to be published
[26]) H.E. Stanley, Introduction to phase transition and critical
     phenomena, Clarendon Press, 1971 (new edition in press).

LECTURES ON POLYMER DYNAMICS

S. F. EDWARDS

Cavendish Laboratory
Cambridge CB3 OHE

Introduction

Polymerized materials share many features with unpolymerized systems,
for example the compressibility of polymer melts is low as is that of
a liquid of the unpolymerized monomers; another example is the
behaviour at sufficiently high frequencies so that internal or highly
local properties of the monomers are involved. But there are other
properties in which polymerized materials are spectacularly different,
in particular visco-elastic properties, and it is these which make
polymers so interesting in technology, basic chemical physics, and
in biology. In these lectures I shall confine myself entirely to
these properties which come from the long chain nature of the polymer,
so that the detailed chemistry of their structure will be compressed
into an absolute minimum of physical quantities. The basic idea is
that the physical laws which describe polymers will contain only a
very few numbers which come from the monomer structure, the laws
themselves being otherwise identical. Thus we will argue that the
equilibrium properties of a polymer solution will depend only on an
effect step length ℓ and an interaction ω whilst in a melt not even
ω is involved. The dynamics of polymers will need a time scale,
say a friction coefficient, but the simpler theory requires no other
specification. To give one immediate example, it is well known that
the elastic modulus of melt is a function of frequency but that there
is a wide band of frequencies over which it is approximately constant.
It can be argued that a polymer chain is hemmed in by its neighbours
in the melt and provided that the period associated with the
frequency in the experiment is short compared to the time it will
take for the polymer to wriggle into a new environment, this modulus
will be the same as if chains were virtually infinitely long. The

45

R. A. Pethrick and R. W. Richards (eds.), Static and Dynamic Properties of the Polymeric Solid State, 45–80.
Copyright © 1982 by D. Reidel Publishing Company.

modulus will then only
depend on how much
polymer there is in
the system, and can
be argued to follow
a power law, roughly
like the density
squared. Here is a
log-log plot[1] which
confirms the state-
ment.

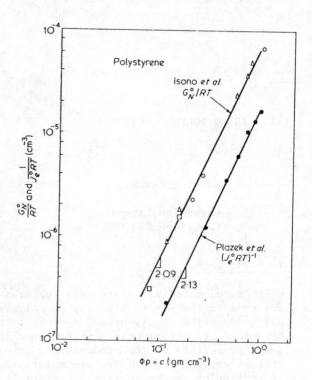

The density in this
diagram is the
number of Kuhn
lengths per unit
volume, where the
Kuhn length is
defined by saying
that a polymer in
a melt will behave
with Gaussian
statistics, thus
have a mean square
distance between
the two points.
If these points
are labelled $s = n\ell$
and $s^1 = n^1\ell$, a
relation

$$(\underset{\sim}{R}(s) - \underset{\sim}{R}(s^1))^2 = |s - s^1|\ell \tag{1.1}$$

must be obeyed and ℓ is the Kuhn length. Thus a certain number of
monomers make up a Kuhn length and instead of referring to the Mth
monomer we can refer to the Mth Kuhn length, and simpler still to
an arc length s along the chain. The interesting quantity is

$$\rho = \frac{\text{Total arc length}}{\text{Volume}} \tag{1.2}$$

which may be very different from the mass density. For example ρ is
relatively low for the heavy complicated geometry of polystyrene and
relatively high for the extremely flexible poly di-methyl siloxane
(pdms).   For the calculation of Kuhn lengths the reader is referred
to Flory's book; there will be no further discussion here. Our
polymers are now chains labelled $\underset{\sim}{R}(s,t)$ which are random walks which
in the absence of interactions or in a melt obey gaussian statistics.

To be consistent we can only discuss long range effects and will always consider chains so long that sections of these chains are still long enough to obey gaussian statistics. Thus it is well known that for large enough separation $s_1 - s_2$ however complex the local definition of a random walk is, the distribution settles down to the Rayleigh form.

$$P(\underset{\sim}{R}_1, \underset{\sim}{R}_2; s_1, s_2) = \left[\frac{2\pi\ell}{3} \left| s_1 - s_2 \right| \right]^{-3/2} e^{-\frac{3}{2\ell} \frac{(\underset{\sim}{R}_1 - \underset{\sim}{R}_2)^2}{(s_1 - s_2)}} \tag{1.3}$$

ℓ being the Kuhn length. This distribution can be extended to a set of points

$$P(R_1, R_2 \cdots; s_1, s_2 \cdots)$$

$$= \prod_i \left[\frac{2\ell\pi}{3} \left| s_i - s_{i+1} \right| \right]^{-3/2} e^{-\frac{3}{2\ell} \sum_i \frac{(\underset{\sim}{R}_i - \underset{\sim}{R}_{2+i})^2}{s_i - s_{i+1}}} \tag{1.4}$$

To handle any theoretical analysis of a physical problem one wants to find a coordinate system which describes motions which are independent. For example in the long wave excitations of a solid one discusses elastic waves, not atoms, because the kind of motions are experiences which can be synthesised easily from the waves, whereas atoms carry far too much information in an unhelpful way. So it is also for polymers. Suppose, as it will turn out without loss of generality, that the intervals $S_i - S_{i+1}$ are all taken equal say to ℓ and introduce the fourier components.

$$R(q) = \sum_m e^{iqs_m} R_m = \sum_m e^{iqm\ell} R_m$$

$$R_m = \frac{\ell}{L} \sum_q e^{iqm\ell} R_q \qquad q = \frac{2\pi\ell}{L} \cdots 2\pi$$

L-total length = NL; N-no. of points. \tag{1.5}

Then if we let q be very small one can see that the probability distribution of the $R_1 \ldots R_N$ becomes a probability

of the R(q) and this becomes

$$R(q) = \int e^{iqs} R(s) \frac{ds}{\ell} \qquad R(s) = \frac{\ell}{2\pi} \int e^{-iqs} R_q d_q$$

$$P([R(q)]) = (\text{Normalis}^n) e^{-\frac{3\ell}{2} \int \frac{q^2}{2\pi} R_{(q)} R_{(-q)} dq} \tag{1.6}$$

a formula strictly valid for small q values, failing for values
of q^{-1} of the order of Kuhn lengths.

This is the basis of polymer theory: the random walk statistics
can be put in a normal mode form and in this form is <u>gaussian</u>.
Such distributions appear everywhere in physics, for example a
classical electromagnetic wave in a cavity will have the vector
potential A_k satisfying a distribution

$$\exp(- \Sigma \underset{\sim}{k}^2 |A_k|^2 /kT) \quad \binom{Rayleigh}{- Jeans} \tag{1.7}$$

equally the distribution of elastic modes of the Debye model of a
solid. But notice that these distributions have kT in them. The
random walk does not; its randomness is intrinsic and not due to
thermal excitation. The mathematics of handling such things
however is much the same; (indeed simpler because the three
dimensionality of the <u>k</u> vector gives troubles in statistical
mechanics at low temperatures which need quantum mechanics to
resolve them). It is important to realise that it is consistent
to work to order q^2 in this distribution, since higher terms refer
to details of the chain which are part of our model, not part of
the true chemistry. These modes were first introduced by Rouse who
thought of the 'S_i' as labels of 'balls', the intermediate polymer
being 'springs' between them. However once the distribution is
obtained no model is required any more than one needs to argue
about how a crystal is constructed if one refers to the elastic
waves it can sustain.

So much for non interacting chains. When an interaction is present,
as must always be the case, the statistical distribution in equilibrium will be given by

$$\exp - \underset{\alpha\beta}{\Sigma} W_{\alpha\beta}(R_\alpha - R_\beta)/kT \tag{1.8}$$

where $\alpha\beta$ label all molecular forces. Such forces will be both
attractive and repulsive of magnitudes depending on the molecules
involved and on what solvent is present, if there is one.

In general one can argue that W has two effects. One is to affect
the static distribution by the expression above. The other is a
dynamical effect related to the infinitely hard nature of W at
close encounters. A chain cannot pass through itself, and this
is true however narrow the core is. This topological constraint
is lost if W is finite however large. However the static
properties of W are unaffected by a softening of the core since if
the core is softened the system is hardly ever in that region.
Thus we find that we can see an approximation emerging which can
greatly simplify our analysis. The real potential can be ideal-
ised quite differently.

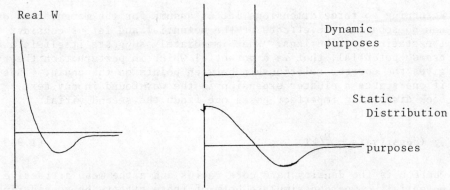

Of course that dynamics of a chain will have to be studied in the
context that the distributions are altered by the potential but the
crucial encounters which are dominated by the integrity of
topological constraints are not frequent and can be studied on the
assumption that the 'soft' potential gives distributions whilst the
hand potential can be shrunk down to effectively zero width but
infinite high potential of zero width contributes nothing to the
static distribution.

It is important to realise the uniqueness of three dimensions. In
one dimension, if one thinks of the polymer as spheres strung out
on a line, only near neighbours interact and the static and dynamic
characters or 'soft and hard' potentials are <u>identical</u>.   In fact
one just has a change from $R^2 \propto s$ to $R^2 \propto s^2$ however weak of
narrow the potential is. In two dimensions it is similar. If
chains cannot cross a random walk will look like

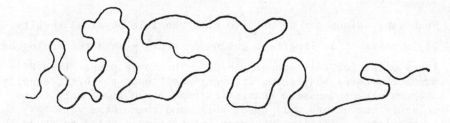

and however narrow the core potential it has the same effect on
the static distribution as a wide softer potential. However a
wide soft potential will not have the same dynamic effects as a
narrow hard potential.   For four dimensions one is in a marginal
situation but above four dimensions the potential has negligible
static effects. I do not understand the dynamical effects though
I have read that one can always unravel threads in four dimensions
e.g. untie knots, so perhaps the dynamics are uninteresting also.

Returning to three dimensions let us assume for the moment that one
can so separate the effects of the potential and let us consider
the static distributions. What immediately suggests itself is a
pseudo potential, that is a potential which in perturbation theory
gives the correct relationship between points on the chain. Thus
if one starts a cluster expansion in the way found in any text
book discussing imperfect gases one finds the second virial
coefficient to be

$$\text{(density)}(a - {}^{b}/kT) \qquad\qquad\qquad (1.9)$$

where b is the density hard core radius and a the mean attractive
potential. One can simulate both of these effects by considering
a pseudo potential $kT\omega\ \delta(r_1-r_2)$ between atoms 1 and 2 so that

$$\int w\delta(r_1-r_2)d^3r_{12} = w$$

$$= (b-akT) \qquad\qquad (1.10)$$

and the Boltzmann factor $e^{-V(r_{12})/kT}$ to be replaced by $e^{-w\delta(r_1-r_2)}$

$$(1.11)$$

If $\omega\delta(r_1-r_2)$ is treated as the soft part of the potential it gives
the right answer. For polymers the analogous term will be

$$\sum_{\alpha\beta} w_{\alpha\beta} \iint ds_\alpha\ ds_\beta\ \delta(\underset{\sim}{R}_\alpha(s_\alpha) - \underset{\sim}{R}_\beta(s_\beta)) \qquad\qquad (1.12)$$

or for one chain just

$$w \iint ds\ ds'\ \delta(\underset{\sim}{R}(s) - \underset{\sim}{R}(s'))\ ds\ ds' . \qquad\qquad (1.13)$$

Such expressions can be used to give sensible physical results. The
Boyle point of an imperfect gas occurs at $a = \dfrac{b}{kT}$ and the analogous
Flory point for polymers at $\omega = 0$. The Boyle point gives PV = NkT.
Since the three body terms are very small near a = b/kT, equally
one can expect Gaussian statistics near $\omega = 0$ since only three
molecule encounters will upset this, and these have very low
probability. (It is however an interesting problem to study the
detail of the Flory point but we will not do it here).

We have now four quantities to describe polymers in a solution
or melt, assuming they all have the same length L.

These are:

  ℓ Kuhn length
  ω interaction
  L length of single chain
  ρ length density of the chains.

To these we must add k the inverse of length over which we are studying.

I will consider initially the static consequences of these quantities. The interaction ω will be taken positive since if it is negative the polymer collapses and the collapsed state lies outside our self enforced discipline of only studying behaviour dominated by the long connectivity of the chain. The effect of the repulsion ω will be to swell the chain so that the end to end distance

$$<(\underset{\sim}{R}(L) - \underset{\sim}{R}(0))^2> > L\ell. \tag{1.14}$$

Suppose it is $L\ell_1$ then $\ell_1 = \ell_1(\omega,L,\ell)$.

However if we look at intermediate points

$$<(R(s_1) - R(s_2))^2> = L\ell_1 \tag{1.15}$$

We must generalize ℓ_1 to be

$$\ell_1 = \ell_1(\omega,L,s_1,s_2,\ell). \tag{1.16}$$

If a chain is very long, in particular infinite, then

$$\ell_1 = \ell_1(\omega,s_1-s_2,\ell) \tag{1.17}$$

which can be alternately expressed by a former transform, then from (1.6)

$$<R(q)\ R(q')> = \delta(q+q')\ \frac{2}{3\ell}\ q^{-2} \tag{1.18}$$

making

$$\ell_1 = \ell_1(\omega,q,\ell). \tag{1.19}$$

These we emphasize are merely definitions sufficiently wide to cover all evantualities. However it has been known for many years that for the end to end distance of a single chain

$$\ell_1 \sim (\frac{\omega^2 L}{\ell})^a \quad \text{as} \quad L \to \infty \tag{1.20}$$

where a is approximately, perhaps exactly, 6/5. The 6/5 is an attractively simple number and follows from the dimensionality, for one can argue that the terms in the exponent will be in balance for long chains thus the distribution has dimension $\dfrac{(\text{Distance})^2}{\ell(\text{Av length})}$ and the interaction $\dfrac{\omega(\text{Arc length})^2}{(\text{Distance})^3}$

i.e.
$$\frac{R^2}{\ell L} \propto \frac{\omega L^2}{R^3} \tag{1.21}$$

$$R^2 \propto L(\omega^2 \ell^{-1} L)^{1/5} \tag{1.22}$$

but modern theories do not quite reach this number in three dimension. The argument is clearly correct in one dimension

$$\frac{R^2}{\ell L} \sim \frac{\omega L^2}{R} \to R \sim L \tag{1.23}$$

In the rest of these lectures I will always assume 6/5 is correct as an asymptotic limit. (I don't believe that experimentally one will ever need an accuracy which will distinguish $L^{.2}$ from $L^{.18}$ but it is nevertheless a fascinating theoretical problem.) Likewise for intermediate points one has

$$\ell_1 \sim |s_{12}|^{1/5} \quad \text{or} \quad \ell_1(q) \sim q^{-1/5} \tag{1.24}$$

All these formulae for ℓ_1 can be found as solutions of simple extrapolation formulae which connect the perturbation expansion in $\omega, \ell,$ L first given by Fixman and now known to high order, with the large L limit, for example

$$\ell_1^2 - \ell_1 \ell = \omega \ell \left(\frac{L}{\ell_1}\right)^{1/2} \qquad (k^2 L^{6/5} < < 1)$$

$$\tag{1.25}$$

$$\ell_1^2 - \ell_1 \ell = \frac{\omega \ell}{\ell_1 k} \qquad (k^2 L^{6/5} > 1)$$

where single if complicated expressions can be given for the right hand sides of these equations. These expressions have a history back to the original paper of Flory.

Another aspect of swelling is found in correlation functions. We have used the symbol ρ for the mean density. Let us generalize to using $\rho(\underset{\sim}{r})$ as a local density, so that

$$\rho = \frac{1}{V} \int d^3 r \rho(\underset{\sim}{r}). \tag{1.26}$$

Then

$$\rho(\underset{\sim}{r}) = \int_0^L \delta(\underset{\sim}{r} - \underset{\sim}{R}(s))\frac{ds}{\ell} \tag{1.27}$$

which has a correlation function

$$<\rho(\underset{\sim}{r})\rho(\underset{\sim}{r}')> = \iint \frac{ds}{\ell} \frac{ds'}{\ell} <\delta(\underset{\sim}{r} - \underset{\sim}{R}(s)) \; \delta(\underset{\sim}{r}' - \underset{\sim}{R}(s'))> \tag{1.28}$$

which will be a function of $\underset{\sim}{r} - \underset{\sim}{r}'$.   Hence if

$$\rho_{\underset{\sim}{k}} = \int e^{i\underset{\sim}{k}\cdot\underset{\sim}{R}(s)} \; ds/\ell \tag{1.29}$$

the former transform of (1.28) is

$$<\rho_{\underset{\sim}{k}} \; \rho_{\underset{\sim}{k}'}> = \delta(\underset{\sim}{k}+\underset{\sim}{k}') \iint <e^{i k(R(s) - R(s'))}>\frac{dsds'}{\ell^2} \tag{1.30}$$

For a free polymer, by a simple integration

$$<e^{i\underset{\sim}{k}(\underset{\sim}{R}(s) - \underset{\sim}{R}(s'))}> \text{ is } e^{- k^2 \frac{\ell}{6} |s-s'|} \tag{1.31}$$

and when an interaction is present we can argue that a useful definition is to put

$$<e^{i\underset{\sim}{k}(\underset{\sim}{R}(s) - \underset{\sim}{R}(s'))}> = e^{-\frac{k^2\ell_1}{6} |s-s'|} \tag{1.32}$$

where

$$\ell_1 = \ell_1(\omega, L, s, s', \underset{\sim}{k}, \ell). \tag{1.33}$$

Clearly if $\underset{\sim}{k} \to 0$ we return to our earlier definition because one can expand in $\underset{\sim}{k}$ but for larger k one finds that

$$\ell_1 \sim |k|^{-1/3} \; \omega^{1/3} \; \ell^{1/3} \tag{1.34}$$

The effect of swelling will then be to alter

$$e^{- \frac{3}{2\ell} \frac{k^2}{L}} \text{ to } e^{- \frac{3R^2}{2(\alpha L^{6/5} \omega^{2/5} \ell^{2/5})}} \tag{1.35}$$

for small R and

$$e^{- \beta R^{5/2}/L^{3/2} \; \omega^{1/2} \; \ell^{1/2}} \tag{1.36}$$

for large R.

(There are also other effects near the origin which will not be discussed here for they are more difficult to reach). The correlation is

$$\langle \rho_{\underset{\sim}{k}} \rho_{\underset{\sim}{k}'} \rangle = \delta(\underset{\sim}{k}+\underset{\sim}{k}') \int_0^L \int_0^L \frac{dsds'}{\ell^2} \; e^{-k^2\ell|s-s'|/6} \tag{1.37}$$

which contains $(1 - k^2L\ell/18)$ for $k^2L\ell$ small (Debye)   (1.38)

and $\dfrac{1}{k^2}$ for $k^2L\ell$ large (de Gennes).   (1.39)

When we have the interaction present we get

$$k^2L\ell \text{ replaced by } k^2L^{6/5} \omega^{2/5} \ell^{3/5} \text{ for } k^2L^{6/5}\omega^{2/5} \text{ small } \tag{1.40}$$

and

$$k^{-4/3} \text{ if } k^2L^{6/5}\omega^{2/5} \text{ is large} \tag{1.41}$$

but both in the limit of $\dfrac{\omega^2 L}{\ell^3}$ large.

If now our chain is surrounded by other chains in solution or melt, one will have the full complexity of it being a function of $L, \omega, \rho, \underset{\sim}{k}$ and s or q, but is in many interesting cases simpler than this complex form suggests. The correlation function now appears in two forms: for a single chain, as before, but with the interactions of others, and the full density correlation

$$\langle \iint \frac{ds_a ds_b}{\ell^2} \; e^{i\underline{k} \cdot (R_a(s_a) - R_b(s_b))} \rangle \tag{1.42}$$

Experimentally these two quantities can be discovered from scattering experiments if one chain alone can be detected, say by a D,H replacement in neutron scattering, and the full scattering.

The interesting thing about (1.42) is that it gives rise to screening. For example in the simplest theory, if one expands the interaction, the expression k^{-2} of eq. (1.39) becomes replaced by a series

$$\frac{1}{k^2} - \frac{\omega\rho}{\ell k^4} + \frac{\omega^2\rho^2}{\ell^2 k^6} + \ldots,$$

$$= \frac{1/k^2}{1 + \frac{6\omega\rho}{k^2}} = \frac{1}{k^2 + \xi^{-2}} . \tag{1.43}$$

This simple argument has much wider validity and will work even if the full complexity of the interaction is studied. The simple argument above is only valid for concentrated solutions where it gives

$$\xi^{-2} = \frac{6\omega\rho}{\ell} \tag{1.44}$$

and for $k^2\xi\ell < 1$ one has

$$\ell_1^2 - \ell\ell_1 = \alpha\omega\xi$$

so that

$$\ell_1 = \ell(1 - 0(\frac{\omega\ell^2}{\rho})^{1/2} \ldots) \tag{1.45}$$

But for small ρ, $\ell_1 \gg \ell$ and a self consistent argument is required. This can be obtained from a scaling argument which is developed in detail in de Gennes' book. A simple version is to say that the formula

$$<(\underset{\sim}{R}(s_1) - \underset{\sim}{R}(s_2))^2> \sim \omega^{2/5} |s_1 - s_2|^{6/5} \ell^{3/5} \tag{1.46}$$

valid for s_{12} large compared to (ℓ/ω) will not be valid for distances where chains overlap where it must latch on to

$$|s_1 - s_2| \xi^{-1/3} . \tag{1.47}$$

Now the density of material inside a sphere of radius

$$<(R^2)>^{1/2} \text{ is } s_{12}(R^2)^{-3/2} = s^{4/5} \tag{1.48}$$

Hence $\ell_1 \sim s^{1/5} = \rho^{-1/4}$ \hfill (1.49)

and $\xi \sim \rho^{-3/4} .$ \hfill (1.50)

To get an extrapolation formula which contains this limit is not easy (it is not good enough to replace ℓ by ℓ_1 in (1.44), but one is now just reaching that point and the extrapolation is

$$\xi^{-2} = (\frac{6\omega\rho\ell/\ell_1}{1 + 27\omega\xi/8\pi\ell_1^2}) \tag{1.51}$$

$$\ell_1^3 - \ell\ell_1^2 = \alpha\omega\xi\ell \tag{1.52}$$

with appropriate extensions to the case where

$$k^2\xi^{1/3} \ell\omega^{1/3} > 1. \tag{1.53}$$

The osmotic pressure can be calculated in terms of ξ,ℓ,ℓ_1 and is given in reference (2) using these extrapolations.

In summary then we have

1) Random walks without interaction are described by gaussian distributions, diagonal in the fourier coefficients of their intrinsic equations.

2) Interactions can be considered to have two effects, a swelling which cominates the static (thermodynamic) functions and a dynamic effect which is dominated by entanglements, and these effects can be taken separately.

3) The static properties can be usefully expressed by an effective Kuhn length which is a function of all the variables of the system but simplifies in many cases. When many chains are present the effective step length has to be supplemented by a screening length which has reasonably simple expressions for it, particularly in special regimes.

Chain dynamics

Different approaches to the problem give different insights. Suppose we start with the simplest idea of a chain on which random forces cause sudden hops. This point R_n hops to a new position as in the diagram.

$$\Delta R_n = R_{n+1} - 2R_n + R_{n-1} \tag{2.1}$$

so that if $p_n(t)$ is the function of time which has a spike at n whenever R_n is moved

$$\frac{dR_n}{dt} = p_n(t) \ (R_{n+1} - 2R_n + R_{n-1}).$$ (2.2)

There will be some average $p_n(t)$, call it $\bar{p}$, and denote the
fluctuation by f. In the fourier transform we have

$$\frac{\partial R(q)}{\partial t} = \bar{p} \ q^2 R(a) + fluct^n.$$ (2.3)

or to tie up with later derivations, we can use

$$\nu \ \frac{\partial R(q)}{\partial t} = 3kTq^2 \ell \ R_{(q)} + f_q(t)$$ (2.4)

where ν is a friction coefficient and f_q will be treated as a random
driving force.   Note that if one writes this in terms of s one is
back to a Fick equation

$$\nu \ \frac{\partial R}{\partial t} - 3kT\ell \ \frac{\partial^2 R}{\partial s^2} = f(s,t)$$ (2.5)

but it must be remembered that as in elastic waves or heat conduction,
the equation is not valid at atomic dimensions. It is however just as
valid to use this equation for a long polymer as it is to use the
wave equation for sound propagation.   Fourier transforming also in
time gives

$$(i\nu\omega + 3kTq^2\ell) \ R(q,\omega) = f_{q\omega}$$ (2.6)

so that $<R_{q\omega} \ R_{q'\omega'}>$

$$= <f_{q\omega}f_{q'\omega'}> \ (i\nu\omega + 3kTq^2\ell)^{-1} \ (+i\nu\omega' + 3kTq^2\ell)^{-1}.$$ (2.7)

In the general spirit of fluctuation – dissipation theory one can
consider f as fluctuating instantaneously, and in addition unformly
along the chain.   It is now convenient and valid to replace the sum
over q by an integral over q to cover all $q = 0$, and a separate term
for $q = 0$.   The $q = 0$ mode does not appear in the distribution
function (1.6) because it corresponds to the centre of mass whereas
the distribution only concerns relative points along the chain.   But
in Brownian motion the centre of mass moves and this term contributes.
The continuum now gives simple δ functions because of invariance in
time and position along the chain, end effects being omittedly this
approximation.   Hence

$$\langle f_{q\omega} f_{q'\omega'} \rangle = A(q) \ \delta(\omega+\omega') \ \delta(q+q') \tag{2.8}$$

and

$$\langle R_{q\omega} R_{q'\omega'} \rangle = \frac{A(q) \ \delta(\omega+\omega') \ \delta(q+q')}{(\nu^2\omega^2 + (3kTq^2\ell)^2)} \tag{2.9}$$

$$\langle R_{0\omega} R_{0\omega'} \rangle = \frac{A(0) \ \delta(\omega+\omega')}{\nu^2\omega^2} \tag{2.10}$$

But if one integrates over ω and ω' we must get the static value

$$\langle R_q R_{q'} \rangle = \frac{\delta(q+q')}{q^2\ell} \tag{2.11}$$

which is equivalent to

$$\langle (\underset{\sim}{R}(s) - \underset{\sim}{R}(s'))^2 \rangle = \frac{2}{\pi} \int \frac{\sin^2 q(s-s')dq}{q^2\ell} \tag{2.12}$$

$$= \ell |s-s'|. \tag{2.13}$$

Doing the integral we find A must be $\kappa T\nu$. This allows us to calculate

$$\langle (R(st) - R(s't'))^2 \rangle \tag{2.14}$$

and in particular $s = s'$ contains terms in

$(t-t')^{1/2}$ and $(t-t')$.   (See de Gennes' book for details)

This has the simple interpretation that a point s moves relative to its neighbours rather more slowly than that in Brownian motion for the correct unity of the chain must be preserved; the _precise_ effect is to change a Brownian $|t-t'|$ to t-t'.   In addition there is a slow drift on the whole chain which is Brownian, but down by a factor (ℓ/L).

Consider how this result would be modified if the polymer lay in a pipe.   Suppose the coordinate describing distance down the pipe is S, and that the pipe has a random walk configuration of steplength a.   Then after N steplengths, S-S' = Na

$$(R(S) - R(S'))^2 = Na^2 = a|S-S'| \tag{2.15}$$

Suppose that we have a single Brownian particle in the pipe and that after a time t it has drifted a distance S.   Then

$$<S^2> = Dt \tag{2.16}$$

so that to an observer who does not know it is in a pipe but measures its coordinate R

$$<(R(t) - R(0))^2> = |S|a \tag{2.17}$$

$$= Da\sqrt{t}. \tag{2.18}$$

Further, consider instead of a Brownian particle, a points on a polymer as described above, in a tube.
Then

$$<(R(S,t) - R(S,0))^2> = |S|a$$

$$= Da(\varepsilon_1 \sqrt{t} + \varepsilon_2 t\ell/L). \tag{2.19}$$

So if one can consider a polymer entangled with its surroundings as a polymer in a rigid pipe (of infinite length), it will diffuse a distance in space like

$$\varepsilon_1 \sqrt{t} + \varepsilon_2 t/L. \tag{2.20}$$

The pipe of course is defined by the polymer itself. As it diffuses down the pipe it creates and annihilates pipe at the ends according to which way its random motion is taking it. In the very long term there will be a further term which is the diffusion of the centre of mass on a length scale larger than that of the pipe i.e. the polymer so finally we expect

$$\varepsilon_3 \sqrt{\varepsilon_1 \sqrt{t} + \frac{\varepsilon_2 t}{L} + \varepsilon_4 t/L^2} \tag{2.21}$$

These different times scales were originally predicted by de Gennes. Here are computer simulations of the problem of this dynamics in a fixed rectangular. They seem reasonably fitted by $\frac{1}{4}, \frac{1}{2}, 1$ laws using a 4000 bead chain (3).

Of course this computer experiment really is governed by the theory alone; whether polymer melts are has to be answered by experiment! But it looks reasonable at least for very long chains. Dense assemblies of short chains are much more difficult to interpret.

An alternative and more
general approach to the
Langevin equation is
to derive the diffusion
equation. Consider an
analogy. If a single
Brownian particle
diffuses in space it
satisfies Fick's
equation.

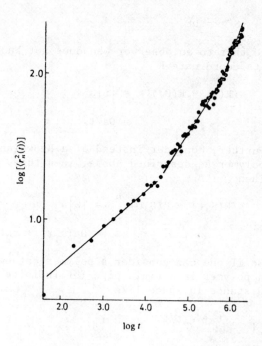

$$(\frac{\partial}{\partial t} - D\nabla^2)P = 0 \tag{2.22}$$

In equilibrium $P = (\text{Volume of box})^{-1}$.

Supposed instead it experienced a potential well of $\frac{1}{2}m\omega^2 r^2$. Then equilibrium would be

$$\exp(-\frac{1}{2}m\omega^2\underset{\sim}{r}^2/kT). \tag{2.23}$$

It is well known that the appropriate modification of (2.22) is

$$(\frac{\partial}{\partial t} - D\frac{\partial}{\partial \underset{\sim}{r}} (\frac{\partial}{\partial \underset{\sim}{r}} + \frac{m\omega^2 \underset{\sim}{r}}{T}))P = 0 \tag{2.24}$$

of which (2.23) is clearly the equilibrium solution.

Every fourier component $R_q(t)$ diffuses independently and equilibrium is

$$e^{-\frac{3\ell}{2} \Sigma q^2 |R_q|^2/2\pi}$$

Hence we can expect (and indeed can easily derive) $P(..r_q..;t)$

$$\left(\frac{\partial}{\partial t} - \sum_q \frac{\partial}{\partial r_q} \; D\left(\frac{\partial}{\partial r_{-q}} + \frac{3\ell q^2}{4\pi} r_q\right)\right.$$

$$\left. + \frac{\partial^2}{\partial r_0^2} \; \frac{D\ell}{L} \right)P = 0 \tag{2.25}$$

Notice that if we study $<R_q>$ we find by multiplying (2.25) by r_q and averaging

$$\int r_q \; P(R_q)\Pi dr = <R_q> \tag{2.26}$$

$$\left. \begin{array}{l} \displaystyle\int r_{q'} \; \frac{\partial P}{\partial r_{q''}} \; \Pi dr dq' = 1 \\[4mm] \displaystyle\int r_{q'} \; \frac{\partial^2 P}{\partial r_{q''}\partial r_{q''}} \; dq''\Pi dr = 0. \end{array} \right\} \quad \text{by parts} \tag{2.27}$$

Hence

$$r \quad <R_q> \; + 3kTq^2\ell <R_q> = 0. \tag{2.28}$$

This is our previous equation, but here we should derive the term $kTq^2<R>$ via a different route to the simple model above. We have the distribution

$$\exp\left(-\frac{3\ell}{2} \int q^2 |R_q|^2 dq/2\pi\right). \tag{2.29}$$

For a configuration R_q we have already integrated out many degrees of freedom (remember Rouse's way of taking R_q as the Fourier components of 'balls'. The 'springs' are integrated out). Thus we can derive the entropy by Boltzmann's formula

$$S = k \frac{3\ell}{2} \int dq |R_q|^2/2\pi \tag{2.30}$$

$$+ \text{ constants}$$

and the free energy $F = -TS$. This acts as a potential energy and hence to a term $\partial F/\partial R_q$ in the equations of motion,

$$kTq^2 R_q.$$

To calculate the state at time t if known at t' we need the Green function

$$(\frac{\partial}{\partial t} - D_0 \frac{\partial^2}{\partial r_0^2} - \int dq \frac{\partial}{\partial r_q} D(\frac{\partial}{\partial r_q} + \frac{3q^2 \ell}{2\pi} r_q))$$

$$G(..r_q..r_q'...; \ t-t')$$

(2.31)

$$= \delta(t-t') \prod_q \delta(r_q - r_q').$$

(This equation is exactly soluble in terms of elementary functions, but we do not need it.)

What will happen to these discussions if the chain has excluded volume? The swelling alone can be described by replacing ℓ by ℓ_1 in this case in the form

$$\ell_1(q) = q^{-1/5} \omega^{2/5} \ell^{4/5}$$

(2.32)

but of course that ensures only the correct equilibrium form. In addition D will be altered in general, but we see that it does not in fact depend on ℓ at all, so that pseudo potential does not affect it. The topological problem has been totally ignored so far, and as we have said above will be present even if only a thread of hard core remains.

How can we ensure that, in the motion of the polymer, it does not pass through itself. It appears that the mathematical problem of specifying all the variants classifying the topology of curves is unsolved and only the simple Gaussian invariant

$$\oint \oint dr_1 \times dr_2 \cdot \nabla \frac{1}{r_{12}}$$

(2.33)

is at all amenable. Even here self entanglements are not covered since for a single curve a definite value to the Gaussian integral can only be obtained if it is considered the limit of a band, and such a band can have twists which have no physical significance. Of course in a melt this effect is negligible just as swelling effects are negligible the contribution to the Gaussian integral from $S_i = s_i$, being negligible in the sum

$$\sum_{ij} \iint dr_i(s_i) \times dr_j(s_j) \ \nabla \frac{1}{|r_i(s_i) - v_j(s_j)|}$$

(2.34)

Nevertheless this approach leaves one with the worry that a treatment is not complete. I believe that there is a much simpler and more consistent approach to the whole problem. The problem of dynamics

has to be solved by approximations as we shall see and as is well
known: its equations are very non linear.   But there is no
physical situation in which polymers can exist in a vacuum, they
must be surrounded by something, either a solvent or a melt.   If one
wished to ask if one configuration of a polymer could be distorted
into another, one way to deal with the problem would be to imagine
the polymer is some elastic matrix and ask if one can distort this
matrix to transform the one configuration into the other.   This
suggests the observation that the problem is unnecessary, it is
subsumed by the problem of solving the general equations of motion
which we have to solve anyway.   To some extent one can think of
the polymers as lines of colour in a liquid.   If one stirs the liquid
the lines move but they never cross.   The same happens in an elastic
medium.   As a method of solution becomes more and more accurate it
preserves more and more of the topological invariants.

We don't need to know what the invariants are for it would be
inconsistent to use them to a greater accuracy than we are solving
the equations of motion.   Thus the prescription being advanced is
that we solve the coupled equations of polymers and (visco-elastic)
liquid with the strict Stokes boundary condition

$$\overset{o}{R}_\alpha(s_\alpha,t) = \underset{\sim}{u}(R_\alpha(s_\alpha,t),t) \tag{2.35}$$

ensuring that when

$$R_\alpha(s_\alpha) = R_\beta(s_\beta)$$

$$\overset{o}{R}_\alpha(s_\alpha) = \overset{o}{R}_\beta(s_\beta) \tag{2.36}$$

We will show in the next lecture that what characterizes the
topological constraint is the vanishing of the diffusion coefficient
when two points of polymer came together.   This ensures they do not
cross and can be used as a basis for the phenomenological pictures
of tubes which can be used as a basis of visco-elastic theory.

The polymer-liquid system.

We saw in the last lecture that a polymer chain whose motion was free
from any surrounding medium had an equation of motion

$$\nu\overset{o}{R}_q + 3kT\ell q^2 R_q = f_q(t)$$

where ν is a frictional drag, $3kTq^2\ell$ is the entropic contribution
from the random walk nature of the chain and $f_q(t)$ a random force,
taken to fluctuate instantaneously but with a mean square value
determined by the equilibrium properties of a random walk.   In fact
a chain will always be surrounded by a medium which can be a solvent,
or the melt of other chains.   One needs therefore to solve the

coupled equations of chains and liquid.   Let us start with the
simplest case of a liquid whose equations can be linearised i.e.
Stokes flow with the Stokes boundary conditions.   The boundary
condition is that, if $u(r,t)$ is the fluid velocity

$$\overset{o}{R}_\alpha(s_\alpha,t) = u(R_\alpha(s_\alpha,t),t).$$ \hfill (3.1)

The formal way to study this is to introduce a Lagrange multiplier
into a Rayleighan function (4).   The resulting equations are quite
plausible so we simply state them here: Let $\sigma_\alpha(s_\alpha,t)$ be the

Lagrange multiplier which plays the role of a 'skin friction' along
the polymer, $u(r,t)$ is the fluid velocity f the random (and/or steady)
force on the polymer (other than that coming from the liquid).   F the
random (and/or steady) force on the liquid (other than that coming
from the polymer).   Then taking for simplicity an incompressible
fluid whose intrinsic viscosity is η and defining

$$\phi_\alpha(r,s_\alpha,t) = \delta(r - R_\alpha(s_\alpha,t))$$

ignoring non linear terms of u . ∇ u type one has

$$\frac{\partial u}{\partial t} - \eta_0 \nabla^2 u + \frac{\nabla p}{\rho_f} + \Sigma \phi_\alpha \sigma_\alpha = F$$ \hfill (3.2)

$$\overset{o}{R}_\alpha = u(R_\alpha,t)$$ \hfill (3.3)

$$d\omega\, u = 0 \qquad \text{(which is used to remove P)}$$ \hfill (3.4)

$$3kTq^2\ell R_q = \sigma_\alpha + f_\alpha.$$ \hfill (3.5)

There is no frictional drag term $\nu\overset{o}{R}$ in these equations since the
frictional drag now comes from the fluid and will emerge as the
calculation proceeds.

If we define $\boldsymbol{6}$ by

$$(\frac{\partial}{\partial t} - \eta_0\nabla^2) \boldsymbol{6} = \delta(t-t')\delta(r-r')\underset{\sim}{0}$$ \hfill (3.6)

where 0 is the Oseen tensor which in fourier terms is

$(\delta^{\mu\nu}-k^\mu k^\nu/k^2)$, μ,ν cartesian indices

$$\boldsymbol{6} = \underset{\sim}{0}e^{-\eta k^2(t-t')} = \underset{\sim}{0}\; e^{-\dfrac{(r-r')^2}{4\eta(t-t')}}(4\pi\eta(t-t'))^{3/2}$$ \hfill (3.7)

then one can formally solve the equations:

$$u = \mathbf{G} F - \int \mathbf{G} \Sigma \phi_\beta \sigma_\beta = u_0 - \int \mathbf{G} \Sigma \phi \sigma \tag{3.8}$$

where u_0 is the velocity field in the absence of polymers,

$$\overset{o}{R}_\alpha = \int \phi_\alpha u = \int \phi_\alpha u_0 - \int \phi_\alpha \mathbf{G} \phi_\beta \sigma \tag{3.9}$$

Write

$$\int \phi_\alpha \quad \phi_\beta = \mathbf{G}(R_\alpha R_\beta tt') $$
$$= \mathbf{J}_{\alpha\beta}(s_\alpha s_\beta tt') \tag{3.10}$$

$$\mathbf{J}^{-1}_{\alpha\beta}(\overset{o}{R}_\beta - u_{0\beta}) = \sigma_\alpha \quad u_{0\beta} = \int u_0 \phi_\beta \tag{3.11}$$

Now define

$$(\delta_{\alpha\beta} \frac{\partial}{\partial t} - \mathbf{J}.3kTq^2\ell)\mathbf{\mathcal{G}}_{\beta\gamma} = \delta_{\alpha\gamma} \quad s_\alpha - s'_\alpha)\delta(t-t') \tag{3.12}$$

so that

$$R_\alpha = \int \mathbf{\mathcal{G}}_{\alpha\beta}(u_{0\beta} - f_\beta). \tag{3.13}$$

Then finally

$$\frac{\partial u}{\partial t} + \eta_0 \nabla^2 u + \Sigma \phi_\alpha \mathbf{J}^{-1}_{\alpha\beta}[\mathbf{\mathcal{G}}_{\beta\gamma}(\phi_\gamma u_0 - f_\gamma) - u_{0\beta}] = F \tag{3.14}$$

Although this analysis may seem rather formal it is a complete
description of the dynamics of non interacting chains in a fluid,
and the interaction can be easily included by the appropriate changes
$\ell \to \ell_1$ etc. for the entanglements are properly treated provided the

hydrodynamics is properly treated.   For the present we can assume
there are no f terms.   Then the equation has the form

$$\frac{\partial u}{\partial t} - \eta_0 \nabla^2 u + \underset{\alpha\beta}{\Sigma} \Lambda_{\alpha\beta}(rr')u_0(r') = F \tag{3.15}$$

We aim to get an effective equation like

$$(\frac{\partial}{\partial t} - (\eta_0 + \delta\eta)\nabla^2)u = F \tag{3.16}$$

in the long wave length limit, or perhaps more generally put

$$(\frac{\partial}{\partial t} + \eta_0 k^2 + \Sigma(k,\omega))u_k = F_k \tag{3.17}$$

where

$$\Sigma(\underset{\sim}{k},\omega) \overset{\sim}{=} \delta\eta k^2 \tag{3.18}$$

The first approximation is to study the case of very dilute solutions which will have Λ diagonal and allows u_0 to be replaced by u. Then J is diagonal and for a first approximation we average it (the pre-averaging approximation) to get

$$\mathbf{J}(q,\omega) = \int e^{iq(s-s')+i(t-t')} < \frac{e^{ijR(st)-R(s't')+i\omega'(t-t')}}{i\omega' + \eta j^2} > d^3 j d\omega' \tag{3.19}$$

For $\omega \sim 0$ one has

$$J(q) = \int \frac{d^3k(k^2\ell/6)}{q^2+(k^2\ell/6)^2} \frac{6}{k^2\ell} \sim \frac{1}{\sqrt{|q|}} \tag{3.20}$$

and

$$\Sigma(k) = \rho \int \frac{k^2\ell/6}{q^2+(k^2\ell/6)^2} \frac{1}{J(q)} dq \tag{3.21}$$

which as $k \to 0$ gives the coefficient of k^2 to contain

$$\int_{2\pi/L}^{2\pi/\ell} dq\ q^{-3/2} \tag{3.22}$$

Hence $\frac{\delta\eta}{\eta_0} \propto {}^{1/2} \rho$, where ρ density, M molecular weight ($\propto$L) a

result due to Zimm   (3.23)

The equation of motion of the single chain is now

$$\overset{o}{R}_q - J(q)q^2 kTR_q = f_q \tag{3.24}$$

so that hydrodynamic interference alters the chain dynamics substantially, for example the $t^{1/2}$ law becomes modified to a $t^{1/3}$ law (with excluded volume a $t^{13/33}$ law!).

To give full weight to the topology of the problem one must avoid the preaveraging approximation. It is clear that this approximation will not alter the $\rho M^{1/2}$ but only affect its coefficient. But the dynamics of the polymer is affected in an interesting way.   The unaveraged equation is

$$\frac{\partial R(s,t)}{\partial t} = \int G(R(st) - R(s't'), \; t-t') \; \mathcal{F}(s't')ds'dt' \qquad (3.25)$$

where $\mathcal{F}$ is the random and entropic force.

Consider the separation of two points on the same polymer

$$R(s_1 t) - R(s_2 t) = R_{12}$$

$$\frac{\partial R_{12}}{\partial t} = \int_{o} \left[G(R_1 - R_3, \; t-t') - G(R_2 - R_3 t - t') \right] \mathcal{F}(s_3 t')ds_3 dt' \qquad (3.26)$$

clearly $R_{12} = 0$ when $R_1 = R_2$ as expected.

If one averages away all degrees of freedom except R_{12}, after some algebra the equation for small R_{12} comes out to be

$$\frac{\partial R_{12}}{\partial t} - \alpha |R_{12}| \mathcal{F}_{12}, \qquad (3.27)$$

α numerical constant $(\alpha \eta / \ell)$ and

$$\mathcal{F}_{12} = (\text{Random force}) + 3R_{12}kT/\ell s_{12}, \qquad (3.28)$$

because the distribution with R_{12} alone must be

$$(\frac{3}{4\pi \ell s_{12}})^{3/2} \exp(-\frac{3}{2\ell} \frac{R_{12}^2}{S_{12}}) \qquad (3.29)$$

which has $\mathcal{F}_{12}$ as its entropic force. It follows that the diffusion equation for R_{12} alone will be

$$(\frac{\partial}{\partial t} + \frac{\partial}{\partial r_{12}} \alpha |r_{12}| \; (\frac{\partial}{\partial r_{12}} + \frac{3r_{12}}{\ell s_{12}})) \; P(r_{12},t) = 0 \qquad (3.30)$$

It is worth digressing for a moment to consider the implications of a vanishing diffusivity. Suppose we study a one dimensional problem

$$(\frac{\partial}{\partial t} - \frac{\partial}{\partial x} \; D(x) \frac{\partial}{\partial x}) \; P(x,t) = 0 \qquad (3.31)$$

where D approaches, but never equals, zero. The equilibrium value of P in cyclic boundary conditions is unity and D has only dynamic effects. Now consider $D(x)$ to be zero over some ranges of x. There is now no way to diffuse from one region to the next, so $P(x)$ will

break up into regions, the total probability in each region being
arbitrary, fixed by initial conditions alone.   Since these regions
are finite P(x) will no longer be a constant within them but be zero
at their boundaries.   The pictures will then be

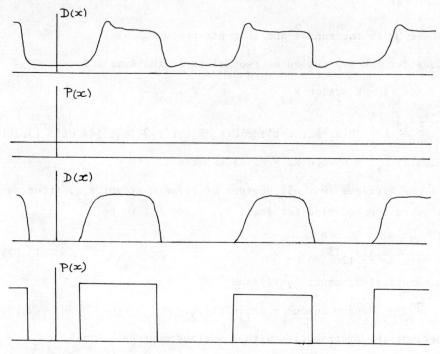

Returning to our polymer in three dimensions, the picture we have
then is that in a configuration like

it cannot reach

by a diffusion process

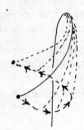

but can do it if $|r_{12}|$ is kept away from zero

The diffusion equation (3.30) for r_{12} alone is of course just a part
of the general diffusion equation for liquid polymer.   Just as
previously one had the alternatives of the Langevin equation (2.5)
or the diffusion equation (2.25), in a liquid polymer derivation above
we used Langevin methods, but there must be a diffusion equation
servable by eliminating the random forces. Since thermal equilibrium
for the liquid will be

$$\exp(-H/kT) = \exp(-\sum_{k} u_k u_{-k}/kT) \qquad (3.32)$$

and for the polymer $\exp(-\frac{3\ell}{2} \sum_{q} u_q u_{-q}\, q^2/2\pi) \qquad (3.33)$

and $\overset{o}{R} = \phi u$, the diffusion equation has to have the form

$$(\frac{\partial}{\partial t} - \sum_{k} kT \frac{\partial}{\partial u_k} (\eta k^2) (\frac{\partial}{\partial u_{-k}} + \frac{u_k}{kT}) + \int \phi\, u\, \frac{\partial}{\partial r(s)}\, d^3 r ds$$

$$+ \int \frac{\partial}{\partial u} \phi kT \frac{\partial^2 r}{\partial s^2}\, d^3 r ds)\, P = 0 \qquad (3.34)$$

$$P_{equil} = (\text{Normalisation}) \exp\left[-\frac{1}{2} \int u^2(r)/kT - \frac{3}{2\ell} \int r^{12}(s)ds\right]. \qquad (3.35)$$

If the force f, the independent random force acting on the polymer
is included, an equation has a further term

$$\int dq \frac{\partial}{\partial r_q} D_q^{ird} (\frac{\partial}{\partial r_{-q}} + \frac{3q^2 \ell r_q}{2\pi})P \tag{3.36}$$

but a term like this arises when u is averaged out of (3.34) which is what in effect the hydrodynamic analysis has done, leaving a D_q which is in fact $J(q)$.

The problem of a superior treatment of the single chain lies in a fuller treatment of these equations, but it is perhaps rather unrewarding since it is very hard work and Zimm's result, being dimensionally sound, will not be upset, only the front coefficient will change in the static limit. The dynamics will change but it is not easy to see an experiment which will distinguish between dynamics permitting the forbidden motion (iii) and the correct dynamics which only permits (iv). It will be seen that in Langevin dynamics the boundary condition that $\overset{o}{R} = u$ is sufficient to ensure a zero flux at any point where polymers meet, and although our analysis is of chains in solution, it is quite general so that any field u(r) will ensure this situation. But there is no need to invent a field as an aid to computation if such a field already exists in the experimental situation.

For many chains let us start with the preaveraging approximation. We then get a mean hydrodynamic equation

$$(i\omega + \eta_0 k^2 + \Sigma(\underset{\sim}{k},\omega))u_{k\omega} = F_{k\omega} \tag{3.37}$$

where in the simplest approximation and with

$$\Sigma(k,\omega) = \rho \int \frac{k^2 \ell/6 \; dq}{q^2 + (k^2 \ell/6)^2} \sqrt{\frac{1}{\frac{j^2 \ell/6}{j^4 \ell^2/36 + q^2} \frac{1}{\eta_0 j^2 + \Sigma(j)}}} . \tag{3.38}$$

Consider the region of j which matters in the denominator integral. The final integral is dominated by $q \sim 0$, leaving an integral which is hwoever not dominated by $j \sim 0$. It turns out that although obviously as $k \to 0$, $\Sigma \sim k^2$, over the range on integration of significance in the denominator Σ is approximately a constant. This will be $\eta \xi_H^{-2}$ where ξ_H is a hydrodynamic screening length. When excluded volume effects are included one has

$$\xi_H^{-1} = \frac{1}{2}\rho \; \ell\ell_1 \tag{3.39}$$

The region where chains are heavily overlapped and $\rho \to 0$, the scaling regime, one finds $\xi_H = 32\xi/9$. This result is at first sight

surprising since these are unrelated phenomena. The reason is the
hydrodynamic interactions do not in themselves introduce a new length
scale, and since in the scaling regime the physics all hangs on the
one length ξ, ξ_H must also come out in terms of ξ. This will not be
true in a more complicated field system for example a higher frequence
experiment or away from the scaling region. The most spectacular
version of this latter point is near the θ regime where $\xi \to \infty$ like
$(T-\theta)^{-\frac{1}{2}}$ but $\xi_H \sim \rho^{-1}\ell^{-2}$. For the region where $\underset{\sim}{k}$ really is near zero
i.e. the $\underset{\sim}{k}$ that appear in the hydrodynamics equation one has

$$\Sigma \propto \rho n_0 k^2 \xi^{-1} \int dq/q^2 \tag{3.40}$$

i.e.

$$\eta = \eta_0 (1 + \beta M \rho^2) \tag{3.41}$$

where β is a numerical constant. Historically the linear M was the
first, discovered by Rouse who assumed a constant bead friction. The
screening length ξ_H provides such a system, but its origin is of
course quite different which is seen by the ρ^2 dependence.

There is experimental evidence of the viscosity changing from $M^{1/2}\rho$
to $M\rho^2$ (or their excluded volume extensions), but at higher M of ρ a
much stronger dependence manifests itself say $M^a \rho^b$ where a > 3 and
ρ > 4. Intuitively the preaveraging approximation is the cause of the
trouble, for it allows chains to pass through one another. This may
not be too bad in weakly systems (and for really dilute systems it
only makes a numerical constant difference as we have seen) but is
clearly wrong at high densities when a tube picture coupled with
reptation motion seems sensible.

To make progress, study the diffusion of a single chain when one
imagines all the chains frozen except that one, R_1; call the others
R_α. There will still be σ_1 and σ_α, but now the equations become

$$0 = \overset{o}{R}_\alpha = \phi_\alpha u = \phi_\alpha G \phi_1 \sigma_1 + \underset{\beta \neq 1}{\Sigma} \phi_\alpha \mathbf{6} \phi_\beta \sigma_\beta \tag{3.42}$$

Define $\quad h_\alpha = \mathbf{6}(R_1 R_\alpha) \quad h_1 = \mathbf{6}(R_1 R_1)$

$$H_{\alpha\beta} = \mathbf{6}(R_\alpha, R_\beta) \quad \alpha, \beta \neq 1. \tag{3.43}$$

Then $\sigma_1 \quad$ Force on chain 1 = F_1

$$\overset{o}{R}_1 = h_1 \sigma_1 + h_\alpha \sigma_\alpha \tag{3.44}$$

$$0 = h_\alpha \sigma_1 + \Sigma\, H_{\alpha\beta} \sigma_\beta \tag{3.45}$$

$$\text{or } \overset{o}{R}_1 = (h_1 - \underset{\alpha\beta}{\Sigma}\, h_\alpha H_{\alpha\beta}^{-1} h_\beta) F_1. \tag{3.46}$$

The h and H are functionals of the fixed chains R_α and the mobile chain R_1.

$$F_1 = f_1 + 3kTq^2 R_1/2\pi \tag{3.47}$$

As usual if we write the equation as

$$R_1 = D_1 F_1$$

then D_1 is just the diffusion coefficient, and the diffusion equation will be

$$(\frac{\partial}{\partial t} - \underset{q}{\Sigma}\, \frac{\partial}{\partial R_{1q}}\, kTD_{1q}(|R|)\, (\frac{\partial}{\partial R_{1q}^*} + \frac{3q^2\ell}{2\pi}\, R_{1q}))P = 0 \tag{3.48}$$

It is easy to prove that $\overset{o}{R}_1 = 0 \; (= \overset{o}{R}_\alpha)$ when $R_1 = R_\alpha$ so D vanishes whenever R_1 touches the other chains. It will be a maximum at various points away from the others. In order to draw a picture and without loss of generality, imagine that the R_α are a uniform rectilinear cage, and look for points which minimize D. These will be the open circles

(The black dots represent the other polymers. In 3 dimensions these are sections of lines, in two dimensions they are dots. The open circles however represent points both in 3 and 2 dimensions. A locus joining these lines we call a primitive path; it is a path which gives a maximum net diffusivity and can change direction at any point which is an absolute maximum in any direction. Thus the primitive path is a random walk of step length a = $(3/\rho)^{1/2}$. I assume that when the system is disordered one will still get this law, a fact borne out by computer studies with a comparable if not identical definition and by alternative analysis (3).   The polymer has a step length ℓ and will have (a/ℓ) times as many step lengths as its primitive path, for we can now identify the primitive path as the topological skeleton of the polymer, a simplified version of the polymer which retains the correct topology

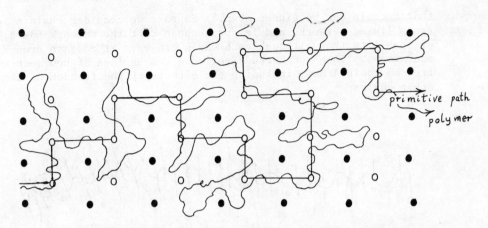

After the fashion of plasma physics, one can decompose the polymer coordinates into $R_\perp$ and R_{11} where $R_\perp$ is perpendicular to the tangent to the primitive path, and R_{11} parallel. There will be a sensible mean diffusion coefficient for R_{11} but $R_\perp$ has to have a functional dependence in its diffusion coefficient. This coefficient vanishes at any neighbour and in some average sense can be thought of as vanishing on the boundary of a pipe of radius a whose axis is the primitive path, $\mathcal{R}$ say. Thus one can write

$$\left(\frac{\partial}{\partial t} - \int \frac{\partial}{\partial R} \; D \; \left(\frac{\partial}{\partial R} + \frac{3R''}{\ell}\right) \right.$$

$$\left. - \int \frac{\partial}{\partial R_\perp} \; D_\perp(R_\perp - \mathcal{R}) \; \left(\frac{\partial}{\partial R_\perp} + \frac{3R_\perp''}{\ell}\right)\right)P = 0 \qquad (3.49)$$

as the analogue of (2.25), and then extend the analysis to the earlier work of this lecture. A difficulty seems to be that any simple averaging will have $D_\perp$ decreasing to zero say exponentially, but that is not good enough to <u>trap</u> the polymer. It must actually go to zero as our model (3.31) has shown, and of course it does along the polymer which is how the physical trapping takes place. There clearly is much more work to do to follow the detail of the complex but intuitively obvious problem. However to get visco-elastic behaviour one can take the tube picture on a purely intuitive basis and work out the consequences which we do in the next lecture.

Viscoelasticity

When the density of polymers is high, we have argued that there is a topological skeleton which causes a great reduction in the mobility of chains, roughly speaking the chain looking as if it is confined to a pipe by the cessation of diffusion processes which would cause chains to cut each other. Permanent cross linking is another way

that a chain can be pinned down.   Suppose we consider chain A in a
cross linked network, and let us suppose that the network has a level
of stability that although each chain can move it does so around some
mean locus.   Then the hatched region gives an idea of how much
freedom chain A has, including the effects of the freedom of the
other chains

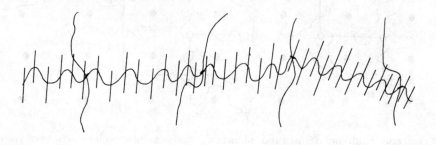

Roughly speaking we have again one primitive path picture with the
step length a_0 given by

$$a_0 = (\frac{L\ell}{N_c})^{1/2}$$ L total polymer length (ρV)

N_c no. of cross links = number steps in p.path

(4.1)

The picture is of course too simple for in addition to cross links,
the chains must inhibit one anothers motion by entanglement.   Suppose
for the moment this effect is small, perhaps in a very dilute gel.
Then the simple theory of rubber elasticity gives an entropy

$$\frac{kT}{2} \Sigma \lambda^2$$ per cross link   (4.2)

where $\lambda_1, \lambda_2, \lambda_3$ are the elastic deformations.   A more elaborate theory
can be constructed which allows all the cross links to be free
simultaneously and will clearly have a lower entropy.   This improvement
has only the minor effect of halving the entropy without altering
its structure.   If one extends this standard model to allow for
entanglements a variety of methods are available.   A simple model is
to allow slipping links as at X.   Suppose that there is a ring

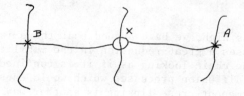

which can slide a certain
distance along either
chain between A & B.

Let η be a dimensionless measure of the slip

$$= (\ell\sigma)^{-1} \left[L\ell/N_c \right] \tag{4.3}$$

where σ is the distance the link can slip, and $L\ell$ the path between the bounding cross links. Note that the primitive path will now no longer be $L\ell/N_c$ but rather the spacing between slip links $L\ell/N_s$ where cross links are now just hard versions of slip links. Then it can be shown that each slip link contributes (6)

$$\frac{kT}{2} \sum_i \frac{\lambda_i^2(1+\eta)}{1 + \eta\lambda_i^2} + \frac{kT}{2} \sum_i \log(1 + \eta\lambda_i^2) \tag{4.4}$$

This is an interesting expression because the limiting cases are clear: for η zero one gets a standard cross link contribution. If η is very large the big term is $kT \log \lambda_1\lambda_2\lambda_3$ which is the loss of one degree of freedom in the perfect gas sense because $\lambda_1\lambda_2\lambda_3$ V original is the new volume. In the large limit the cross link slides over the whole length of polymer present and it can be shown that this is equivalent to precisely a log V addition in the entropy, just like one perfect gas degree of freedom. This behaviour is also characteristic of very large λ when the characteristic form is again a perfect gas term. Of course this formula does not allow for many other processes which come into play at large extensions or compressions. One process one can observe has been omitted is that if the primitive path increases by a length $(\Sigma\lambda^2/3)^{1/2}$ it will exhaust the polymer when

$$L = L_{\rho\rho}(\Sigma\lambda^2/3)^{1/2} \quad \text{i.e.} \tag{4.5}$$

$$(\frac{a}{\ell})^2 = \frac{1}{3} \Sigma\lambda^2. \tag{4.6}$$

This is very much shorter than the simple exhaustion, which would have the number of steps between cross links as $(L/N_c\ell)$ hence

$$\frac{\Sigma\lambda^2}{3} = \frac{L}{\ell N_c} = (\frac{a_0}{\ell})^2. \tag{4.7}$$

In a melt c may be say 5ℓ where as there may be 20ℓ or more between cross links. Although one can describe this effect in detail, we have not space to do it here. For small extensions it is customary to plot the ratio of the stress $\partial F/\partial\lambda$ against the simple cross link value. For incompressible material

$$\Sigma\lambda^2 = \lambda^2 + \frac{2}{\lambda} \tag{4.8}$$

when λ is the extension in the draw direction so that

$$\frac{\partial}{\partial \lambda} \Sigma \lambda^2 = 2(\lambda - \frac{1}{\lambda^2}) = \partial F_0 / \partial \lambda. \tag{4.9}$$

Since our expression is a sum of cartesian extensions an expansion in $\lambda_i^2 - 1$ must give

$$\frac{kT}{2} (\Sigma (\lambda_i^2 - 1) + \eta \sum_i (\lambda_1^2 - 1)^2 + \ldots) \tag{4.10}$$

so that if $i\lambda = 1 + \epsilon$

$$(\frac{\partial F / \partial \lambda}{\partial F_0 / \partial \lambda}) - 1 = (4\eta\epsilon \frac{(1 + 6\eta + 2\eta^2)}{(1+\eta)^3}) \qquad \text{per link} \tag{4.11}$$

Near $\lambda = 1$ the derivative is linear in $(\lambda - 1)$ and is in the sense found experimentally. So without attempting any detailed fit, the slip link model does change the simple result in the right direction.

The results above are quoted rather than derived for it is a fairly formidable calculation to do in spite of the simple conclusion. To get some idea of a simple derivation one can note that the slip links can be regarded as making up the tube and deformation will lengthen the tube by $(\frac{1}{3} \Sigma \lambda)^{1/2}$ since the tube is itself a random walk whose elements are extended by $|(\underset{\sim}{\lambda} \cdot \underset{\sim}{n})^2|^{1/2}$ where $\underset{\sim}{n}$ is the direction of the tube throughout that particular link. $(\Sigma \lambda^2)^{1/2}$ is an approximate average of this quantity. (The difference of $\sqrt{(\lambda \cdot n)^2}$ and $|\overline{\lambda \cdot n}|$ is very small.) So that the simplest level one can label the tube $\mathcal{R}(s)$ and the polymer $\underset{\sim}{r}(s)$ when $r(s) \overset{\sim}{=} \mathcal{R}(\frac{s\ell}{a})$.

Thus although the real effect of the entanglement comes through diffusion one can model it by a constraint like

$$\exp - \int_0^{L_c} (r(s) - \mathcal{R}(\frac{s\ell}{a}))^2 \frac{ds\ell}{a^4} \tag{4.12}$$

along a piece of tube length L_c between two cross links.

The coefficient ℓ/a^4 is chosen to make the mean value of $(r - \mathcal{R})^2$ to be a.

This model has $\underset{\sim}{r}(s)$ close to $\mathcal{R}\frac{(s\ell)}{a}$ in all dimensions. One can however make a and η independent since the Gaussian constraint has two parameters in it. However choose the assignment above for the

moment.   It is a straightforward matter to use it to find the same
result that the contribution per slip link is increased by a factor
i.e. $N_c \rightarrow N_s + N_c$.

If one studies the modulus at a high enough frequency that the polymer
has no time to slip down the tube one finds the modulus proportional
to N_s which is equivalent to saying that each piece of polymer along
the step length a of the primitive path behaves as if it were between
two cross links

$$F \propto \frac{kT}{2} \Sigma \lambda^2 \left(\frac{L}{a}\right) \tag{4.13}$$

$$F_{plateau} \propto (kT)(\Sigma \lambda^2)\rho^2. \tag{4.14}$$

In fact the power law is slightly higher. For swollen rubbers or
gels this may be due to the onset of semi dilute laws i.e. a scaling
regime, but it is usually studied in regions remote from the law
density, long length limit, and occurs in melts. It may be due to a
small but density dependent numbers of stars formed either covalently
or by very tight entanglements giving a higher functionality.

It will be noted that this discussion is at a cruder level than the
set piece formula (4.4) and clearly more work needs to be done to
harmonize the formal but accurate calculation with the intuitively
attractive but mathematically undeveloped subject of tube confinement.

Finally we turn to uncrosslinked materials. Our picture is dominated
by reputation, the creeping of chains down tubes, creating and
annihilating tube at their ends.   For deformation faster than the
time it takes to get out of a tube called the disengagement time by
de Gennes, the melt or concentrated solution behaves like a rubber.
For longer times a constant strain will create a stress (as of a
rubber) which will decay as the slow Brownian motion of the chain
regains the entropy lost in the straining.   One can then study the
dynamics of the stress strain relationship by translating the effect
of strain affinely onto the primitive path which carried that part of
the earlier history of the sample which has not been lost by the
stress relaxation of the reptative Brownian motion.

The detail of this picture has been worked out elsewhere (7) and,
even with the very crudest model outlined above, it gives a rich
rheology.

One can envisage two processes: A strain on the system with a
primitive path as in figure (a) produces figure (b)

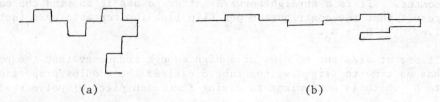

(a)   (b)

the polymer is now extended since any deformation will lengthen the primitive path and it shrinks reducing the number of links

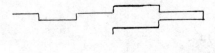

(c)

which then are eroded by the polymer diffusing and creating normal links fig (d)

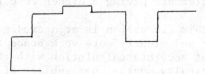

until finally a normal primitive path is evolved which will have no reference to the original fig (e).

The simplest argument will take b → c as almost instanteous (as it is for normal melts; but it can be made very slow near the glass transition) and employ the simple rubber elasticity arguments to calculate the stress due to the number of steps of the primitive path in existence. The formula for the stress at time t in terms of a strain tensor $\underline{\underline{E}}$ (our $\lambda_1\lambda_2\lambda_3$ in the earlier simple geometry) is a

stress tensor

$$\sigma_{\alpha\beta} = G_0 \, \mu(t) \, \mathcal{Q}_{\alpha\beta} + P \, \sigma_{\alpha\beta} \qquad (4.15)$$

where P is the pressure (not discussed for reasons given in lecture 1) G_0 is $3\rho kT(L/a)$

$$\mu(t) = \sum_{p \text{ odd}} \frac{8}{p^2\pi^2} \exp(-\, tp^2/Td) \qquad (4.16)$$

Where Td is the disengagement time, shown by de Gennes to be $\dfrac{L^2}{D\pi^2}$

(D the mean diffusion constant is however a complex thing to get as we have seen; but it doesn't affect the present argument)

$$\mathcal{Q}_{\alpha\beta} = <\frac{(E.n)_\alpha (E.n)_\beta}{|E.n|}> \ <|E.n|>^{-1} - \frac{1}{3}\delta_{\alpha\beta} \tag{4.17}$$

where $\underset{\sim}{n}$ is the vector lying along the primitive path.

The two terms $< >$ in $\mathcal{Q}$ come from the entropy and the number of links remaining after the first shrinkage. These results are quoted here simply to show that considering the complexity of the problem they are reasonably simple. The slithering back on tube stretching, the unreasonable perfection attributed to the tube, the absence of star like tight entanglements and so on all add complexity to the picture and have been studied already by several authors (7).   The general difficulty in improving the picture has in the fact that every improvement introduces a new parameter and one is left with more and more things to fit.   It will usually improve the fit with experiment, but is rather daunting in the sense of knowing whether some new physical insight of crucial importance has been unearthed.

References   Books covering the theory are

For single chains
Flory P.J. (1969) Statistical Mechanics of Chain, Molecules, Interscience, New York.

For solutions up to about 1970
Yamakawa H. (1971) Modern Theory of Polymer Solutions, Harper Row, New York.

For scaling theory and a discussion of modern developments
de Gennes P.G. (1979) Scaling Concepts in Polymer Physics, Cornell U.P. Ithaca N.Y.

Some recent or forthcoming references are

1) Graessley W.W. and Edwards S.F. (1981) Polymer 22, 1329
2) Muthukumar M. and Edwards S.F. (1982) J.Chem.Phys. to be published.
3) Evans K.E. and Edwards S.F. (1981) J.Chem. Soc. (Farad Trans)77, 1891; 1913; 1929.
4) Freed K.F. reviews in (1978) Progress in Liquid Physics Ed. C.A. Croxton Wiley, New York.
5) Zimm B.H. gives a modern assessment in (1980) Macromolecules 13, 592
6) Ball R.C., Doi M, Edwards S.F., Warner M. (1981) Polymer 22, 1010 see also Marrucci G (1981) Macromolecules 14, 434.
7) Doi M and Edwards S.F. (1978) J.Chem.Soc. (Farad. Trans II)74 1789; 1802; 1818 (1979) 75, 38.

The theory is assessed by
Graessley W.W. (1979) J.Pol.Sci.
and extended by
Curtiss C.F. and Bird R. Byron (1980) Kinetic Theory of Polymer
Melts, Rheology Research Center Report.   Univ. of Wisconsin.

STRUCTURAL STUDIES ON CRYSTALLINE POLYMERS

D.M.Sadler

University of Bristol, Physics Department

What are the shapes of molecules inside lamellar crystals, and
how do the crystals grow that way ? These questions have received renewed interest since the advent of small angle neutron
scattering. The aim here is to set the scene in terms of ideas
on crystal growth, and to review the results of a range of techniques with an accent on the contribution of neutron scattering
work. Some of the conclusions are as follows. Molecules in solution grown crystals fold preferentially along the growth facets.
Some at least of the surface "amorphous" layer is in fact due to
fluctuations in the position of the folds with respect to the
lamellar surface. The folding is largely adjacently re-entrant,
but for fast crystal growth competitive nucleation leads to some
longer folds and also to superfolding. The conformation in fast
grown crystals from the melt is very different. The molecule as
it is incorporated in the crystal contracts locally to form short
sequences of adjacent or "nearby" folds, even though on a large
scale the molecule retains the distribution in space it had in
the melt. Melt grown crystals which grow slowly have yet a different pattern of folding.

SOME IDEAS ON CRYSTAL GROWTH

Can we invent a perfect polymer crystal ? This is a useful
question to consider before we decide that nature's answers to it
are a little bizarre ! The introduction will take this question
as its theme in order to outline some of the basic ideas of crystallization in polymers.

The repeat unit in a crystal often consists of one molecule,

81

R. A. Pethrick and R. W. Richards (eds.), Static and Dynamic Properties of the Polymeric Solid State, 81–108.

or a small number of them, in those cases where the molecule is
small and fairly rigid. This is rarely true for polymer crystals,
although globular proteins are exceptional. In contrast to
synthetic polymers each molecule of a particular protein is identical, and "programmed" to form itself into a specific shape by
a precise sequence of different monomer units (1). It is easy
to see why synthetic polymers should not adopt this mode of crystallization. A second mode of crystallization, shown by paraffins, has all the chains packed together in an extended array.
Two problems occur if this were to be the pattern for the crystallization by polymers such as the longer chains homologue of paraffins (polyethylene). Firstly, since the molecules are not all
the same length, the ends of the molecules could no longer match,
and the resulting disorder would normally be prohibitive. The
second difficulty is more fundamental and deserves more detailed
attention.

It is not sufficient for a state of matter to be the most
thermodynamically favoured in order for it to exist. It is also
necessary for a series of states to be available which are intermediate between the state in which the material starts (e.g.a gas)
and the final state (e.g. a crystal). Figure 1 illustrates this

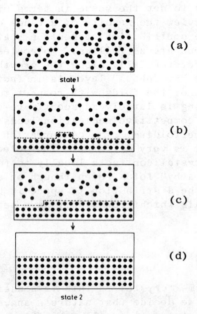

Figure 1. Schematic description of some of the possible nucleation steps (2) involved in proceeding from a gas (top) to a crystal (bottom).

schematically. Each crystal plane cannot form instantaneously
from the gas: this would require too high an activation energy.
It would be less "costly" for a so-called secondary nucleus to
form first (Figure 1(b)), after which no additional free surfaces
are incurred as each new molecule crystallized at the corners or
"niches" created at the edge of the growing nucleus (Figure 1(c)).
(In parenthesis, it may be remarked that non polymer crystals can
find an even less "costly" way of crystallizing: at the price of
one screw dislocation a spiral structure is formed whose axis is
along the dislocation. The spiral can grow along its length continuously by addition of molecules to a niche which will always
exist (2)). The formation of the hypothetical extended chain
polymer crystal from a state where the molecules were coiled must
also be considered in terms of a sequence of nucleation events.
However, the simple application of the arguments above,with each
molecule being an entire macromolecule would mean that each additional unit being added to the crystal would require the complete
straightening out of a coil. A two dimensional nucleus containing several molecules would require prohibitive activation energies.

     It can now be appreciated that it is not a simple matter to
invent a near perfect crystal which is capable of growth.It can
be asserted with confidence that almost none of the wide variety
of crystal types which have been observed are in the most favoured state thermodynamically. (Micrographs of some examples are
given in the lecture of Professor Wegner). Rather,the polymer is
obliged to crystallize into a metastable form. In order to explain
the existence and structure of any given metastable state it is
necessary to consider a sequence of intermediate states, analogous
to the nucleation events described above, which precede the final
one. Many of the crystal forms which might be conceivable can be
eliminated on the sole criterion that viable sequences of intermediate state do not exist. Incidentally, crystals of polyethylene can be grown where the degree of chain extension is very
high (cf. the second hypothetical state considered above). The
transitional existence under pressure of a liquid crystal-like
hexagonal phase is known to be a prerequisite (3),i.e. the growth
of extended chain crystals involves more than the transition between the normal coiled state and the crystal.

     The current status of the theories concerning the growth of
lamellar crystals will now be outlined. Many of the different
crystal types referred to above are various types of lamellae.
As was discovered in 1957 (4) the chains traverse these lamellae
many times. The data with which the crystal growth theories are
compared are lamellar thicknesses and growth rates which vary in
characteristic ways with the supercooling.

     The only theory which is available was pioneered and extended

largely by Hoffman and Lauritzen and their colleagues (5,6). The
key process is the creation of a secondary nucleus on a crystallographically smooth edge of an existing lamella (Figure 2).

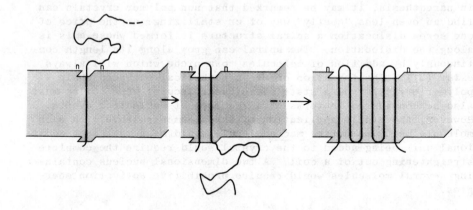

Figure 2. An artists impression of the steps involved in the
secondary nucleation of a molecule on to a smooth growth face.
The letters n denote niches.

That lamella is favoured which optimises this nucleation rate
and hence the crystallization rate. The calculation in its
simplest form of the resulting lamellar thickness is carried
out in the following steps:

1. A sequence of intermediate states is proposed. The
first consists of just one "stem" (a straight chain sequence)
being laid down on the edge. Each successive state adds another
parallel stem, with a fold joining adjoining stems (Figure 2).
The parts of the molecule not in the crystal are assumed to be
equivalent to a fully dissolved molecule.

2. Energies are assigned to each state, involving surface
and fold energies. An activation energy is also required for
each transition which normally depends in a simple algebraic
way on the energies of the two states.

3. Fluxes in both the forward and backward directions are
calculated using the activation energies, each one being contained in a Boltzman term.

4. The steady state condition is applied so that there is

a constant flux along the sequence of states.

5. The steps 1-4 assume one value for the layer thickness ℓ. The equation for the flux is now used to find that value of ℓ ($\ell*$) for which nucleation dominates. The result is

$$\ell^* = \text{constant}/\Delta T + \delta\ell \qquad\qquad (1)$$

where ΔT is the supercooling. The first term is the minimum value of ℓ which is thermodynamically stable. The second term $\delta\ell$ reflects the need for crystal growth to take place when the lamellae are more stable than the polymer in the coiled state; $\delta\ell$ increases with supercooling.

This basic method has been extended in a number of ways:

ℓ can be allowed to fluctuate as successive stems are added (7).

Multiple nucleation can be considered, which relaxes the condition that the "filling in" of the growth face is much faster than the formation of a secondary nucleus on a smooth growth face (8).

Two other modifications to the theory are designed to extend eqn.(1) to high supercoolings where the simplest theory fails since it predicts that $\delta\ell$ increases "catastrophically" as ΔT increases. The first (6) adjusts the method used to calculate the activation energies (step 2 above). The second (9) considers the constituent stages involved in the laying down of one stem. Instead of a coiled chain changing directly to a stem of a given length, the existence of a given stem length requires that a slightly shorter stem should exist already.

A more detailed description of the theories is not appropriate here. It is however worth emphasising some of the initial assumptions, and some of the conclusions which are relevant to structure.

The model assumes that the relaxation rate of the coiled part of the molecule in the partly crystallized state is fast compared with the crystallization rate (this follows from the way step 2 above is performed). This is the precise inverse of what has been proposed (10,11) for crystallization from quenched melts. In these other models the existence of lamellae is not explained, but the conformation of the molecule is predicted on the presumption of lamellae existing.

It is inherently difficult to obtain independent values for the surface free energies which are needed in order to predict

ℓ^* quantitatively. Much of the existing literature is necessarily concerned with this problem rather than with basic features of the theory.

If ℓ values are allowed to fluctuate from one stem to
another, a finite distribution of ℓ values is predicted of the
order of $\pm10\overset{o}{A}$ (7). In other words if the folds were crystallographically smooth the theory could not explain this smoothness
without invoking further assumptions (fold regularization).

Lastly, it is often not appreciated that steps 1 and 2 in
the theory do not depend crucially on the type of fold re-entry
(Lauritzen, personal communication). The existing theories do
indeed assume that all successive stems belong to the same molecule ("adjacent re-entry"), but this is desirable simply in order
to reduce the calculation to a manageable form. A more limited
degree of adjacency of folding (e.g. see the models suggested
below) would almost certainly not change the theories fundamentally (though this is not likely to be true for highly random
re-entry). There are many reasons why, at high supercoolings
especially, a sequence of stems being laid down from one molecule should be interrupted before all the molecule has crystallized. For example, if the crystallization is in regime II(where
multiple nucleation is important (8)) a second molecule could
crystallize so as to block the further laying down of the first
one. Another example is in regime I (where the filling in of a
crystal plane is fast compared with nucleation of a new plane).
It has been shown by measurements of crystal growth from solutions of different concentrations (12) that solvated molecules
tend to form a region of locally high concentrations around the
growth face. Under these circumstances two or more molecules
will be simultaneously in the vicinity of a niche as it passes
along the growth face. These molecules will "compete" so that
the final sequence of crystalline stems could consist, for
example, of a short sequence of stems from molecule A, followed
by another sequence from molecule B, then another from A etcetera.
Some crystallizing molecules may be left partly in the solution,
and these would be likely to continue crystallizing later in the
next layer which forms on the growth face. These events are
shown schematically in Figure 3. Fold sequences like this have
indeed been found by neutron scattering (see below).

ELECTRON MICROSCOPY

This section will extend what Professor Wegner has said on the
subject to those observations directly relevant to chain conformations in solution grown (lamellar) crystals. The methods involve mostly diffraction contrast; although they are unfamiliar
to many in the polymer field, they are central to other well

developed areas of material science.

The technique demonstrates unequivocally that the chains folding back into lamellae do so along preferred crystallographic directions. The crystal retains a memory of the growth face, and several physical properties can distinguish this direction from others which would be crystallographically identical on the basis of the unit cell. For example, a diamond shaped crystal of of polyethylene (PE) as shown in Figure 4 has two (110) and two (1$\bar{1}$0) growth faces. The crystal grows so that within each

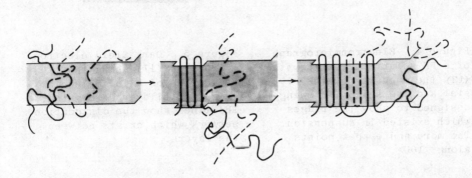

Figure 3. Diagrammatic representation of two molecules nucleating alternately on a growth face.

quadrant (or "sector") the <110> direction is no longer equivalent to <1$\bar{1}$0> even though the unit cell is orthorhombic. Many reviews exist (e.g. reference (14)); here the evidence which distinguishes sectors is simply listed.

1. The surfaces of the crystals are oblique to the crystal lattice, and this obliquity is different in each sector. The crystals as they grow in solution are in fact ridged or tent-shaped, with a common chain axis but with the lamellar surfaces sloping in different directions depending on the sector (13).

2. When the lamellae are fractured, the resultant faces show fibrils when the crack opens up perpendicular to the inferred fold direction, but not when the crack is parallel to the folds (15).

3. The crystal lattice is slightly distorted - e.g. the orthorhombic PE unit cell becomes slightly rhomboidal(14).

Figure 4.   Electronmicrograph
of a PE solution grown crystal
(13) the mounting of the crystal and the shadowing being
designed to show the ridges
which existed in suspension.
The more acute apex points
along <100>.

Figure 5.   Dark field electron
micrograph (17) using the 110
reflection of a low molecular
weight multilayer PE crystal.
The lines show the dislocation
network which exists between
the layers.

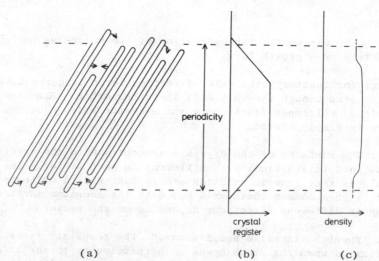

(a)   (b)   (c)

Figure 6.   Schematic view, approximately to scale, of the fluctuating fold model together with hypothetical density and crystal
register profiles. Features which are not shown here include the
collapse of longer folds (see arrows), departures from adjacent
re-entry (cf. Figures 3 and 12), and superficial amorphous material due to, for example, excluded "tails" of molecules.

   4. Sectors in PE crystals which grow with (010) faces melt
at lower temperatures than those on {110} faces (16).

     It should be added that there is no proof that the surface
obliquities of the crystals follow low index planes as would be
expected from arrays of folds which are smooth on an atomic scale,
although this idea did feature in the early literature. It was
found (13) that if the surfaces were crystallographic the most
likely surface planes were {312} and {314} (in PE). However, subsequently it was found that the interpretation was not unique.
An analysis of a large number of different crystals using a
number of preparations (17) showed a variability of fold surface
obliquities both within and between crystals which was best
explained by a non-crystallographic fold surface.

     I was myself involved with another electron microscope study
(18, 19, 20) which, although it was carried out on preparations
which were "special" in that the molecular weights were only
enough to fold twice on average, showed that the fold surface
could have some very unexpected properties. Figure 5 shows a
dark field micrograph (so far unpublished) which illustrates how
the crystal lattice can to some extent be continuous between
separate crystal layers. If the crystal layers were independent
of each other, moiré fringes would be seen, with an intensity
profile which depends on the cosine square of the distance along
the crystal. Figure 5 shows quite different sorts of lines, which
result from crystal distortion in dislocations. This distortion
changes the diffraction condition and hence the diffraction contrast. Dislocations by their very nature can only occur in a
crystal matrix (not in an amorphous layer !) so it might be
supposed that a multilayer stack is effectively one single crystal
with little scope for disorder. Yet the crystallinities of these
preparations were only about 80% by the conventional tests described by Professor Wegner. Putting these and other pieces of
evidence together it was found that the only possible surface
structure consisted of molecular ends and folds of different
heights above the crystal core. Two such rough surfaces could
mesh together in order to provide a degree of crystal continuity.
If this model is generalized then some if not all the disordered
material which is normally considered "amorphous" could in fact
be the result of folds of different heights (Figure 6).

SMALL ANGLE X-RAY SCATTERING

Professor Wegner has already introduced this topic; here a brief
resumé of the possibilities of the technique is as follows. A
stack of lamellae produces a very simple one dimensional "crystal".
Many studies simply use Bragg's law in order to find the period-

icity of this lattice. More detail may be obtained by measuring
(a) an integrated intensity (the "invariant") giving the mean
square density fluctuation   (b) the dependence of the peak intensities on the order of diffraction which in effect measures
the form factor for the disordered layer (c) the detailed dependence of the intensity function which is sensitive to both the
form factor for the disordered layer and the type of disorder in
the stack.

    It is incontrovertible using method (b) (21) that there
exists a disordered layer of about the width and density expected for amorphous polyethylene at room temperature (some uncertainty exists since the value of the density has to be extrapolated from molten polyethylene). It would be desirable to test
how sharp the boundary is between the crystal core and the disordered surface, and several attempts have been made to do this
using methods (a) - (c) (e.g. (22)). However, the published
conclusions are at variance, and it is probably fair to conclude
that the technique is rather insensitive to whether there are
transition zones of the order of 10Å thickness (23). It is worth
mentioning that method (a) has been applied to crystals with dislocation networks, as a function of temperature (Fischer and
Kloos, personal communication). I understand that similar results were obtained to other crystals of PE, although the temperature dependence was different. This is an interesting result
since crystals with dislocation networks are known not to have
true amorphous layers.

    A technique which gives information closely analogous to
small angle X-ray scattering is analysis of the shape of diffraction lines: it is in fact sensitive to a profile which describes
the probability of finding lattice planes in crystallographic
register. It has been concluded (24) that this profile along
the lamellar normal is trapezoidal for PE solution grown crystals,
with a decay of crystal register over 30Å. This of course agrees
with fluctuating folds (Figure 6). A density profile is more
difficult to predict from the model, but the transition must
occur over a distance shorter than 30Å. Although
crystal register may decay smoothly from unity to zero, the density cannot do so. If we consider a layer containing the 15%
innermost folds, the surface layers further out are very unlikely
to sustain a density lower than that of amorphous chains; presumably the long folds collapse to allow an approximately homogeneous filling of space. Figure 6 (b) and 6(c) show a hypothetical
crystal register profile and density profile.

    Polymers can give additional information because of a large
unit cell: Nylon 6.6 contains only four unit cells in each
lamella and consequently shows subsidiary diffraction maxima
which enable the number of unit cells to be counted (25).

DEGRADATION TECHNIQUES

The method relies on the detection by gel permeation chromato-
graphy of molecular fragments consequent to the removal of
surface layers by etchants (26). Early in the application of
the technique (27) it was found that some folds could be deep
in the crystal. The extension of this idea has enabled maps to
be made (28) of the number of folds as a function of depth in
the crystal (Figure 6(a) has been constructed so as to conform
to these data). The analysis is based on the etchant degrading
the chains completely as it penetrates into the surface. This
"slicing" of the surface removes the long folds first and the
short ones last, and the length of the molecular fragments
(e.g. once or twice the layer thickness) decreases continuously
during this process. Figure 7 shows a series of chromatographs.

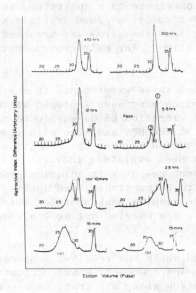

Figure 7. GPC traces of different stages of degradation with
ozone (28). Column of traces under (a) crystals grown at 70°C;
column of traces under (b) crystals grown at 90° C. The labels
1. and 2. on one of the chromatograms refer to fragments which
have traversed the crystals once or twice respectively. The peak
with the asterisks is from decane used for calibration. The pulse
numbers are marked on the individual traces.

The peaks correspond to integer numbers of traverses of the
lamellae, the right hand peak in each case being the single
traverse. It can be seen that the positions of the peaks moves
to lower molecular weight (to the right) as the degradation

proceeds. It has not been possible to explain the positions
areas and widths of the chromatograph peaks as a function of
etching with any other model for the fold type and etching
process.

SPECTROSCOPY

The first experiments which exploited the isotope effect effectively to "label" one chain were due to Krimm (e.g. (29)). This
exploits the fact that PE has two chains per unit cell, so that
in the pure material the CH_2 bending and rocking vibrations observed by infra red are split by about 10 cm^{-1}. The deutero PE
(DPE) has a vibration frequency about $\sqrt{2}$ lower than normal PE
(HPE). In an isotopic blend only those stems with the same
isotopes (i.e. with the same frequencies) and separated by
$<\frac{1}{2}\frac{1}{2}0>$ will interact so as to contribute to a splitting. In
essence, the results are that for solution grown crystals a
doublet is observed, indicating that the stems are arranged
adjacently along the direction <110>. For melt grown crystals
this is not the case.

Recent work (30) has extended these experiments in conjunction with neutron scattering, which also uses isotopic mixtures
(see below). The profile of the vibration bands was found to
vary with the molecular weight of the DPE and also with its
concentration. The general situation is that the vibration is
not a simple doublet. This has been explained (30) in
terms of lack of complete adjacent re-entry in solution grown
crystals. Recent work (Krimm, to be published, and Spells, to
be published) relies on the more complex peak profiles which are
obtainable when the measurements are carried out at low temperatures.

A second spectroscopic method which has received considerable
attention is observation of a low frequency Raman excitation,
which corresponds to standing waves along all trans sequences
of PE (reviewed in reference (31)). This has proved useful in
measurements of lamellar thickness, in cases where the lamellar
stacking is not regular enough for small angle X-ray scattering
to be useful. In principle the technique should also be sensitive to the surface structure. The difficulty has been to some
extent the experimental difficulties such as uncertainties in
modulus   but also the lack of a fundamental understanding of
what happens at the antinodes of the vibration i.e. at the end
of an all trans sequence where there will be one or more gauche
bonds. The debate is summarized in a recent review (31);
it does seem very likely that the all trans sequence is longer
than would be expected on the basis of a simple two phase model.

NEUTRON SCATTERING

The one simple message that is coming from the technique is
that there is no simple message. Expressed another way, the
detailed information that is now becoming available is breaking
new ground, and is not simply testing who was "correct" over the
last decades. By way of introduction it is important for the
interested onlooker to distinguish in this area of work (a) the
polymer (b) the crystallization conditions (c) the region of
scattering angle used (d) the terminology used to describe models.
Under the heading of terminology the use of the word "fold" can be
cited. In this article it means a chain returning to the crystal
from which it has emerged. If anything more specific is intended
adverbs such as "adjacently re-entrant" or "nearby re-entrant"
or "randomly re-entrant" will be used.

The basic theory for elastic coherent scattering is the same
whatever radiation is used, and reference should be made to
companion articles (e.g. by Professor Stein). Neutrons are scattered by nuclei, and since a proton (H) scatters very differently
from a deuteron (D) it is possible to study "solid solutions" of
an isotopically labelled molecule in a chemically very similar
matrix. (The scattering length is a scalar, like the X-ray case
but unlike light).

A recurrent theme in several lectures is that the measured
scattering function I(q) is the Fourier transform of a correlation function p(r). We are therefore able in principle to
measure p(r) for a single molecule, although in practice our
information is limited by the finite range of q over which measurements are possible ($q = 4\pi \sin \theta/\lambda$ where 2θ is the scattering angle and λ the wavelength). There is a more important limitation to the information available because more than one distribution of scattering matter in space can give the same correlation function. This is simply the equivalent of the crystallo-
graphers' phase problem which arises since intensities not amplitudes of radiation are measured. However,this lack of information
should be seen in perspective. Just as in crystallographic studies, as long as there is sufficient other information available
(e.g. that stems exist that are parallel within one lamella) the
information from scattering can usually select a small number of
closely related models.

It is convenient to separate the analysis according to the
range in q: Figure 8 shows a scattering curve in very schematic form. The small q values are sensitive to large scale structure, whereas local structure is more important as q approaches
the value corresponding to atomic lattice spacings.

The intensity I(q) given in the equation below is in units

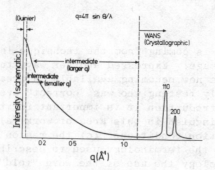

Figure 8. An idealized intensity I(q) showing the q ranges.
corresponding to

$$I(q) = (d\Sigma/d\Omega)/\{(b_D - b_H)^2 c_D c_H N\}$$

where $d\Sigma/d\Omega$ is the differential cross section per volume of
sample, b are scattering lengths, c are volume fractions, and
N the number of scattering units (e.g. hydrogen nuclei) per volume of sample.

A typical way to approach a scattering study is as follows:

1. Measurements in the so-called "Guinier Range" can be
used to derive radii of gyration for the whole molecule using:

$$I(0)/I(q) = 1 + R_g^2 q^2/3 \tag{2}$$

Heterogeneous mixing of isotopes also gives signals in this range.

2. At larger q the intensity is tested against general laws
of the type $I(q) \sim q_\nu$ which can often indicate likely models.

3. More detailed analysis then requires a model for which
calculations can be made to test against measurements, usually
suggested by a priori considerations and steps 1 and 2.

Methods of calculation

The basic equation is due to Debye;

$$I(q) = \sum_i \sum_j \sin(q\, r_{ij})/(q\, r_{ij}) \tag{3}$$

The correlation function p(r) is a histogram of the number of
distances r_{ij} as a function of r.   Equation (3) · can be seen
as the Fourier transform of p(r) averaged over all orientations of the molecule with respect to the incoming radiation.
For a large number of scatterers it is common to replace

the summation with an integral:

$$I(q) = \int p(r)\sin(qr)/(qr)dr \qquad (4)$$

Some standard results using these formulae are particularly useful as long as care is taken to choose a suitable range in q:

for rods: $I(q) = n_L/(\pi q)\, C_1(q)$ $\qquad\qquad$ (5)

for sheets: : $I(q) = 2\pi n_A/q^2\, C_2(q)$ $\qquad\qquad$ (6)

where n_L and n_A are the number of scatterers per unit length and per unit area respectively, and $C_1(q)$ and $C_2(q)$ are due to the finite thickness of the rods (or sheets) and are commonly near to unity. $C_1(q)$ and $C_2(q)$ are often approximated by gaussian functions. Equations (5) and (6) correspond to $\nu = 1$ and 2 respectively (see above).

A two dimensional version of equation (3) is valid for the problem of an assembly of rods with separations R_{ij}, which can be a useful approximation for the stems in a chain folded mole-cule:

$$I(q) = \{n_L/(\pi q)C_1(q)\} \sum_i \sum_j J_o\ (q\,R_{ij}) \qquad (7)$$

The first part of this expression will be recognized as the scattering for individual rods (equation (5)). The second part consisting of a double summation, arises from the mutual arrange-ment of the different stems and will have two types of terms. Those with i = j are usefully considered as the "single stem" contribution which would be present even for randomly arranged stems. Those with i $\neq$ j are the so called "interference terms", the sum of which is normally large as q decreases towards the Guinier range. A convenient way of displaying data is to plot

$$(Iq^2)_c = I(q)q^2/C_1(q)\ \text{against q.}$$

This should give a straight line through the origin for the sin-gle stem terms.

Occasionally it is possible to calculate the intensity analy-tically. However, for those (interesting) cases where there is a limited degree of disorder it is usually necessary to resort to numerical methods using computers. One approach is to use equation (3) where it is necessary for computational reasons to choose a group of monomers as the scattering unit, and this has so far limited the method to fairly small q. This method has the merit of including all the parts of the molecule including amorphous loops, but this is sometimes a mixed blessing since it is necessary then to specify the conformation in complete detail.

Many of the uncertainties concerning local details of the conformation in the surface have to be resolved, and it is not always
clear what are the primary physical quantities which are important in deciding whether a good fit is obtained with the data.
Another approach is to use the approximate equation (7), which
can be used to survey a very wide range of models with relatively small amounts of computer resources. A desirable procedure
would be to use this second method to narrow down the choice of
models and then use the complete method (equation (3)) to test
the most promising models.

Inhomogeneous Mixing Of Isotopic Species

Before proceeding further it is useful to mention fractionation
by isotope, which restricts experiments on polyethylene to crystals grown at high supercoolings. The way that slight chemical
differences between H and D polyethylene lead to extraneous scattering signals is not central to the study of chain conformation.
It has however gained some prominence in the literature, and it
is worth explaining briefly how these signals are understood and
that they do not prevent experiments when the fractionation is
minimized. Fractionation according to molecular weight is a well
documented phenomenon (32, 33); it results in inhomogeneous distribution of different species, with the more slowly crystallizing
polymer being concentrated on the perimeter of the crystals. There
is no evidence that fractionation by isotope is essentially different from this, and, of course, if there are fluctuations in isotope concentration over the crystals there will be an additional
intensity contribution at small q (34). This contribution goes
approximately as q^{-4} so that it can be neglected as q increases
(as long as the fractionation is not extreme).

Results For Solution Grown Crystals

It was found that for PE the intensity at intermediate q could be
rationalized by equation (6), i.e. the local structure was
sheet like (35, 34) even though the values of n_A and sheet
thicknesses (from $C_2(q)$) could not be interpreted simply in
terms of sheets corresponding to rows of stems (Figures 9(a) and
9(b)).

   R_g was found to be insensitive to molecular weight (Figure
10 (35)), which is in contrast not only to the gaussian coil
prediction of an increase as the square root of the molecular
weight, but also to each molecule being in one uninterrupted
sheet. "Superfolding" was invoked, whereby the folded ribbons
themselves fold into multiple ribbons (Figure 11); as the molecular weight increases so the number of ribbons increases. The
physical origin of this effect is almost certainly competition
between molecules during crystallization, as was explained in

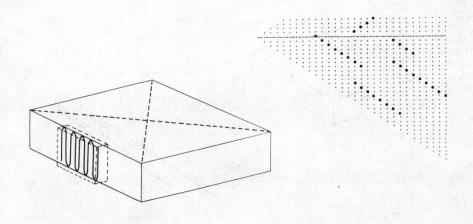

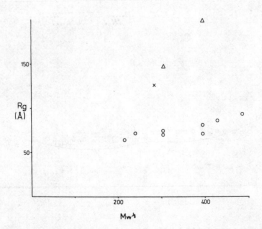

(a)   (b)

Figure 9.   Representations of adjacently re-entrant folding
(a) perspective view   (b) plan view, with parts of molecules shown
as rows of large dots. The broken line is a sector boundary divi-
ding regions of the crystal where the folds are along <110> from
where they are along <1$\bar{1}$0>.

Figure 10. Values of radius of gyration obtained from solution
grown crystals of polyethylene grown at 70°C in xylene ("high"
supercooling) (36). Measurements on oriented and unoriented mats
enables information to be obtained on dimensions in the plane and
perpendicular to the plane of the lamellae.

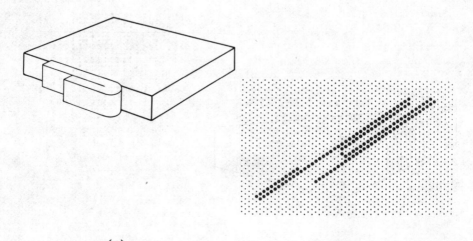

(a)   (b)

Figure 11. Folding of a sheet ("superfolding") shown in perspective and plan form. In (b) the stems are shown fully adjacent,
but recent results suggest a "dilution" along the rows of stems
(see Figure 12 inset).

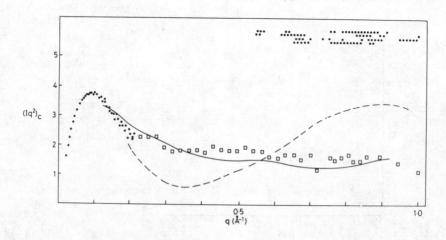

Figure 12. Results at intermediate q for solution grown crystals
(see also reference (40)). The q range was extended as far as
possible so as to test for the broad peak predicted for a 75%
preference for next-to-adjacent folding (broken line). The solid
line shows the results of calculations for a 75% preference for
adjacent re-entry (a representative stem arrangement is shown in
the inset).

connection with Figure 3. If a molecule loses its position on
the growth face in favour of another, it is likely to be incorporated into the next crystal plane the next time a niche
passes by.

It has already been mentioned in the spectroscopy section
that the adjacency of folding is not likely to be complete, at
least at high supercoolings. This has also been deduced from
the neutron scattering (37-39). It now remains to decide how
to arrange the stems in sheets while not making all the folds
adjacently re-entrant. If a label molecule leaves gaps in a
sheet, these gaps must be occupied by the stems of a matrix
molecule and we can assume that these two molecules are equivalent. Hence it is necessary to construct models in which the
pattern made by the stems is of the same type as that made by
the gaps. For example, the occupancy of the plane by one mole-
cule can only be $\frac{1}{2}$ $\frac{1}{3}$ $\frac{1}{4}$ etcetera. Models have been made (39)
in which a statistical probability is defined for the occupancy
of a new site in the sheet which is conditional on the occupancy
of the preceding site. For example, if the preceding site contains a label stem, the probability of the current stem belonging
to the same molecule could be, say, 25%. It follows that, if
the total occupancy is $\frac{1}{2}$, a gap has a 75% chance of being followed
by a label stem. A model with approximately these probabilities
has been proposed by Yoon and Flory (37). This model generates
a preferred label stem separation of 8.8Å, and hence a broad
peak at the equivalent Bragg spacing ($q = 2\pi/8.8\text{Å}^{-1}$) is predicted.
The measurements (38-40) do not show such a peak. Reasonable
agreement over the complete range of q values can be obtained (39)
by a model where the probability of adjacency is not 25% but 75%
(Figure 12). This is also reasonable from a consideration of
events on the growth face (Figure 3). Since the molecule whose
stem is already attached is likely to be nearer the niche than
any other molecule, it is reasonable that there is an increased
probability of that molecule providing the next stem.

The models of Yoon and Flory (37) differ from the one presented here only in the detail of the way one molecule is "diluted"
by another. A preference for a stem separation of 8.8Å is explained by them on a thermodynamic argument, rather than kinetic
as is outlined above. The model of Stamm et. al. ("stacked sheets")
does not differ from that shown in Figure 12 to any significant
degree.The only publication on PE solution grown crystals using
neutron scattering not mentioned so far (41) also favours sheet
type conformations. Hence, in contrast to the common conception,
there is virtual unanimity on the conformation, in this particular system at least .

Solution grown crystals of iPS have also been studied (42).
In contrast to the reputation of this polymer as one in which

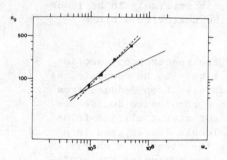

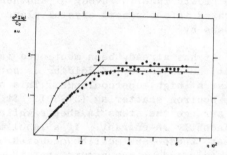

Figure 13. Double log plot of
R_g against molecular weight
(42) for (●) solution grown
crystals of isotactic polystyrene, (0) amorphous sample
moulded at $250°C$. The broken
line corresponds to adjacent
re-entry.

Figure 14. Plot of $I(q)q^2$
against q for:(●) solution
grown crystals, (o) amorphous
samples. The former obey
equation (5) for q < q* and
equation (6) for q > q*.

the crystallization is complex, the results are straight-forward
to interpret in terms of each molecule forming single uninterrupted sheets. The radii of gyration increase as molecular weight
to the power 0.91, (Figure 13) which is a higher power than for
the amorphous state (cf. PE where the dependence is less). In
Figure 14 is shown a plot of Iq^2 versus q (in a range of q where
$C_2(q) \simeq 1.0$. Below a value q* the intensities obey equation (5)(for
rods) whereas above this value they obey equation (6) (for sheets).
We would indeed expect the folded strip to appear as a rod at
low q but as a sheet for higher q.

Results On Melt Grown Crystals

The three polymers (PE, iPS, and polypropylene) have all been
studied under conditions where the rate of deposition on the
crystal is relatively fast compared with the relaxation times
expected for the whole molecule. It should be stressed that these
conditions are far from the ultimate rates of crystallization:
Barham (private communication) has achieved rates many orders
of magnitude faster than used in the neutron scattering studies.
In all three cases R_g is the same after crystallization as in
the melt (43, 44, 45) for high molecular weights. For low molecular weights this is not the case (44). It has been suggested
(40) that in fact the conformation of the molecule undergoes only
minor adjustments as it goes from the molten to the crystalline
state. This model has been shown not to be consistent with the

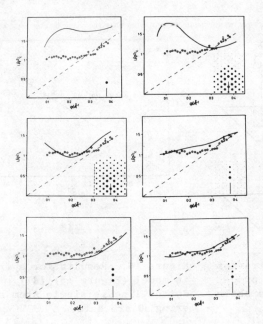

Figure 17. Calculations for subunit models, as explained in the text.

other 30% of the folds are randomly re-entrant in the more conventional sense, with a high chance that the molecule enters a different lamella from the one it leaves. The picture based on
these three models is that the majority of the folds are either
adjacently re-entrant or "nearby adjacent" (the latter indicating
about one to three lattice displacements of stems that are adjacent along the molecule). There are as a result groups or "subunits" of stems near together, but separated by longer loops.
Figure (17) shows predictions for various types of folding within
subunits whose size varies from five to ten stems. A stem is
imagined to be at the top of the short vertical line in the insets
to the diagrams. The molecule then folds so that the next stem
occurs at a neighbouring lattice site, the probability of which
is proportional to the areas of the large dots. In this way
Figure 17 (top left) shows short adjacently folded sheets. The
next two (reading top to bottom, left to right) are two dimensional random walks. The last three are the models which are
favoured; they correspond to various combinations of stem separations in the range 4-13Å, all leading to linear groups of stems.
Figure 18 shows schematically the distribution of subunits within
one lamella (the molecule will in fact change lamellae on occasion).
There is no evidence from the neutron scattering that there is a
preferred direction for the short rows in each subunit (cf. Figure 18 (a)) so Figure 18(b) shows a version of this model which
would be equally likely from these data.

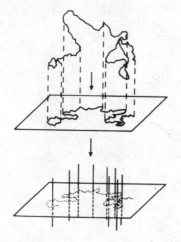

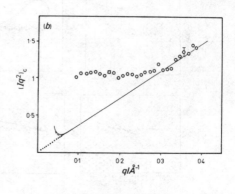

Figure 15. An illustration of the "freezing in" procedure which minimises chain rearrangements (47).

Figure 16. Results for a freezing in model (Figure 15) (47). The dotted line is single stem scattering. 0 are data.

intensities at intermediate q, at least if it means literally a "freezing in" model with an absolute minimum of chain movement during crystallization (as shown schematically in Figure (15)). An analytic calculation is possible for a two dimensional arrangement of stems which follows the projection of the original three dimensional distribution of the coil on to a lamella. The result is shown in a plot of $(Iq^2)_c$ versus q (Figure (16)). The dotted line refers to the "single stem scattering" (see above), so that the calculation shows far less constructive interference between stems than is observed i.e. the stems are too randomly arranged. A more commonly held view, based on intensities in the q range as shown in Figure (16) and also at smaller q, (46) is that there is a substantial rearrangement on a local scale. There is still some divergence of views concerning the exact nature of these rearrangements, but many of the differences are exaggerated by the way in which the models are described. In the broadest terms, it is necessary to disperse each PE molecule over a large region of the crystals in order to explain the R_g values, while at the same time the stems must be fairly compact locally in order to explain the intensity at larger q. This dichotomy is expressed in several ways: as a "subunit" model (47) or a "garland" model (45). In the case of the model of Yoon and Flory (48) the dichotomy is still present but less explicit: the setting of an "escape parameter" at 0.3 means that there is a 70% chance of a fold re-entry which gives a displacement of the new stem only a few lattice spacings away from its neighbour. The

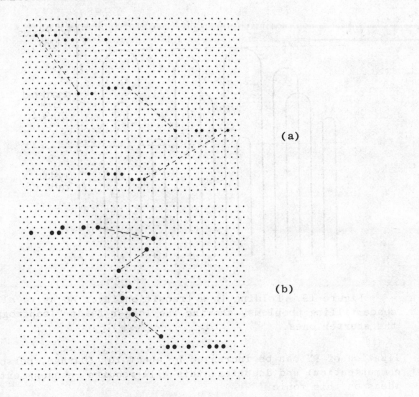

(a)

(b)

Figure 18. Diagrammatic representation in two dimensions
of the stem arrangements of one molecule (one lamella only is
shown, but the molecule will in fact change lamellae on occasion).
(a) and (b) refer to different possible directions for the subunits (see text).

A discussion of the results described above would not be
complete without mentioning two physical constraints which have
been cited as reasons a priori why certain models are not possible.
One of these is mobility in the crystallizing melt, which received
extensive attention in the Faraday Discussions (1979). Yoon and
Flory contended that the rates of molecular disentanglement were
not high enough to allow adjacently re-entrant folding. Two other
contributions, by Klein and diMarzio et. al., pointed out that
reptation increases the expected rates of rearrangements
so that at least short sequences of adjacently re-entrant folds
should be possible. The issue is probably not as clear cut as this
however since all models except the strictest "freezing in" model
(Figure 16) require significant rearrangement on a local scale.
A contribution by de Gennes pointed out that the process of reptation could itself influence the type of folding which occurs.
Since the Discussion it has been found that the rate of crystal-

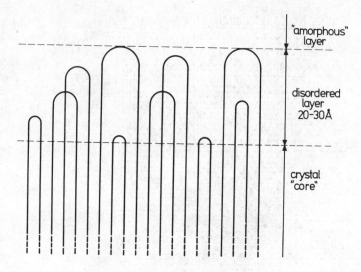

Figure 19. Folding in a transition layer so as to alleviate
space filling problems, showing the longer folds "leapfrogging"
the shorter ones.

lization of PE can be increased by about 10^4 (Barham, personal
communication) and doubtless this will require reassessment of
ideas on this topic.

The second constraint discussed was spacefilling at the
lamellar surfaces. A set of parallel crystalline PE chains cannot all extend into an amorphous layer since not only is the amorphous less dense, but also the chains are no longer straight (49).
These ideas were applied to models with high degrees of random
re-entry by Gutman et. al. (50) who showed that impossibly high
densities were being predicted. However, it was pointed out by
myself (51) that these considerations were less restrictive if the
condition of a sharp boundary between crystal and amorphous was
relaxed. Figure 19 shows the idea schematically: a degree of non-
adjacent but nearby folding is possible in principle if a transition zone of about 10-30Å thickness is allowed in which some
folds can "leapfrog" others. The idea of a transition zone is an
integral part of the fluctuating fold model discussed in some
detail earlier in this review (see Figure 6) and features extensively elsewhere (e.g. (52)). With an abrupt crystal to amorphous transition 70% of folds must be adjacently re-entrant (50).
With a transition layer 70% could be nearby re-entrant.

It will be noted that the ideas on crystal growth in the
first part of this review have not featured as extensively in the
description of melt grown crystals as they did in the case of

solution growth. This is not unreasonable since even solution
growth at high supercoolings can involve several complications
(superfolding and so on). Several questions remain unanswered:
for example, if indeed the subunits are straight, is this because
of preferred fold plane(s) associated with growth facets ? Perhaps
the straightness could be associated with the need to minimize
chain rearrangements, while at the same time obeying the restric-
tions imposed by spacefilling at the surfaces.

On the two polymers other than PE, measurements have been
made for melt crystallization at slower crystallization rates
than those leading to models of the subunit type (53, 54). In
both cases R_g values were no longer the same as in the melt. In
the case of polypropylene (54) departures from the melt values
are associated with stems passing through two adjacent lamellae.
For iPS both the R_g values and plots of the type shown in Figure
14 are explained in terms of one adjacently folded sequence,
with amorphous sequences at either end.

Preliminary Results on Fibres

The measurement of chain extension in order to supplement existing measures of chain orientation has always been an exciting
possibility. Results are now starting to become available on
polypropylene (starting material melt crystallized) (55) and
polyethylene (starting material both melt crystallized and solution crystallized) (56). Figure 20 (a) shows an intensity
contour plot (the main beam arrives in the centre of the detector)
showing the very high degree of anisotropy which is possible.
Figure 20(b) shows a plot of intensity taken from the strip marked
in Figure 20(a). From the width of the maximum an estimate of the
molecular length can be made. From the rate of decrease of intensity in the equatorial direction (horizontal) the width of the
molecule can readily be measured. The preliminary results show
that for room temperature drawing of melt crystallized PE the
molecular dimensions change approximately affinely with the sample
dimensions. For drawing at higher temperatures of solution grown
crystals the molecular deformation is far from affine.

CONCLUDING REMARKS

The aim of this review is to describe molecular shapes and some
ideas on the molecular processes which might lead to them. A
striking conclusion is the variety which has been found. For
example, the preferred fold directions, which can be taken as
proved for crystallization from solution, do not appear to be
general for melt crystallization. Equally, the initial result of
neutron scattering, that for melt quenched polyethylene the
molecular dimensions are the same as for the melt, has not been

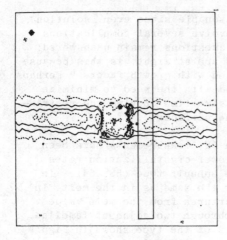 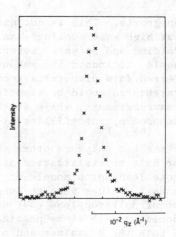

Figure 20(a) Contour plot of   Figure 20(b) Intensity profile
difference intensities for a   taken from the data in
drawn sheet of polyethylene   Figure 20(a).
(specimen to detector 9.14m,
wavelength 8Å)

found for either slowly crystallized polymers or solution grown
crystals. The results, then, suggest how versatile a crystalli-
zable molecule can be. Each metastable state is reached according to a series of intermediate states,and since these are influenced by the conditions prevailing, each final state is specific
to those conditions.

     If care is not taken to avoid generalizing from one crystal-
lization condition to another,an alarmist view can result which
mistakes lively scientific debate for chaos. I hope that this
review has shown that in fact the areas of actual disagreement on
well specified issues are now quite limited: for example: exactly
to what degree does the local conformation in melt quenched crystals reflect the disorder of the melt ? Even in this case there
are signs that the terminologies that are used tend to exaggerate
the real differences between models that are proposed. For solution crystals grown at high and at low supercoolings there is now
a satisfactory concordance between experimental results and ideas
on the nucleation processes during crystallization. Progress has
been made which tends to shift the emphasis away from simple two
phase models to ones where folds have fluctuating positions in
transition layers between the crystal core and any more truly
amorphous layer.

REFERENCES

1. Lehninger A.L., Biochemistry, Worth, N.Y. (1975).
2. Burton, W.K., Cabrera, N. and Frank, F.C., Phil. Trans. Roy.
    Soc. 243A, 299 (1951).
3. Bassett, D.C., Polymer, 17, 461 (1976).
4. Keller, A. Phil. Mag. 2, 1171 (1957).
5. Lauritzen, J.I. and Hoffman, J.D., J. Res. NBS 64A, 73 (1960)
6. Hoffman, J.D., Davis, G.T. and Lauritzen, J.I. in "Treatise
    on Solid State Chemistry" ed. N.B. Hannay, 3 Plenum Press
    N.Y. (1976).
7. Hoffman, J.D., Lauritzen, J.I., Passaglia, E., Ross, G.S.,
    Frohlen, L.J., and Weeks, J.J., Kolloid Z.u.Z. fur Polymere
    231, 564 (1969).
8. Hoffman, J.D., Frohlen, L.J., Ross, G.S., and Lauritzen,J.I.,
    J.Res. NBS 79A, 671 (1975).
9. Point, J.J., Macromolecules, 12, 770 (1979).
10. Ballard, D.G.H., Longman, G.W., Crowley, T.L., and
    Cunningham, A., Polymer, 20, 399 (1979).
11. Dettenmaier, M., Fischer, E.W. and Stamm, M., Colloid and
    Polymer Sci. 258, 343 (1980).
12. Keller, A. and Pedemonte, E., J. Crystal Growth 18, 111 (1973).
13. Bassett, D.C., Frank, F.C. and Keller, A., Phil. Mag. 8, 1739
    and 1753 (1963).
14. Keller, A., Rep. Prog. Phys. 31, 323 (1968).
15. Lindenmeyer, P.H., J. Polym. Sci. C,1, 5 (1963).
16. Keller, A. and Bassett, D.C., Proc. R. Microsc. Soc. 79,
    243 (1960).
17. Sadler, D.M., Thesis (1969).
18. Holland, V.F., Lindenmeyer, P.H., Trivedi, R. and Amelinckx,
    S., Phys. Status Solidi, 10, 543 (1965).
19. Sadler, D.M. and Keller, A., Kolloid Z.u.Z. fur Polymere,
    239, 641 (1970).
20. Sadler, D.M. and Keller, A., ibid. 242, 1081 (1970).
21. Fischer, E.W., Goddar, H. and Schmidt, G.F., J. Polym. Sci.,
    B5, 619 (1969).
22. Strobl, G.R., J. Appl. Cryst. 6, 365 (1973).
23. Windle, A., Faraday Discussions, 68, 466 (1979).
24. Windle, A., J. Mat. Sci.10, 252 (1975).
25. Atkins, E.D.T., Keller, A., and Sadler, D.M., J. Polym. Sci.
    A2, 10, 863 (1972).
26. Blundell, D.J., Keller, A., Ward, I.M., and Grant, I.J.,
    J. Polym. Sci. B4, 781 (1966).
27. Sadler, D.M., Williams, T., Keller, A., and Ward, I.M.,
    J. Polym. Sci. A2, 7, 1819 (1969).
28. Patel, G.N. and Keller, A., J. Polym. Sci. (Phys. Ed.) 13,
    2259 (1975).
29. Bank, M.I. and Krimm, S., J. Pol. Sci. A2, 7, 1785 (1969).
30. Spells, S., Sadler, D.M. and Keller, A., Polymer 21, 1121
    (1980).

31.   Keller, A., "Structural Order in Polymers", Int. Union of
       Pure and Appl. Chem. eds. F. Ciardelli and P. Giusti,
       Pergamon (1981).
32.   Sadler, D.M., J. Pol. Sci. A2, 9, 779 (1971).
33.   Wunderlich, B., Macromolecular Physics, Ac. Pr. N.Y. and
       London vol.1 (1973) vol.2 (1976).
34.   Sadler, D.M. and Keller, A.,Macromolecules, 10,1128 (1977).
35.   Sadler,D.M. and Keller, A., Polymer, 17, 37 (1976).
36.   Sadler,D.M. and Keller, A., Science, 203, 263 (1979).
37.   Yoon, D.Y. and Flory, P.J., Faraday Discussions 68, 288,
       (1979).
38.   Sadler, D.M., Faraday Discussions 68, 435 (1979).
39.   Sadler, D.M. and Spells, S. in preparation (presented at
       the U.K. Polymer Physics Meeting 1981).
40.   Stamm, M., Fischer, E.W., and Dettenmaier, M., Faraday
       Discussion 68, 263 (1979).
41.   Summerfield,G.C., King,J.S., and Ullman, R., J. Appl.
       Crystallogr. 11, 548 (1978).
42.   Guenet,J.M., Macromolecules, 13, 387 (1980).
43.   Schelten,J., Ballard, D.G.H., and Wignall,G.D., Polymer,
       15, 685 (1974).
44.   Ballard,D.G.H., Cheshire,P., Longman, G.W. and Schelten,
       J., Polymer, 19, 379 (1978).
45.   Guenet,J.M., Polymer, 21, 1385 (1980).
46.   Ballard,D.G.H., Schelten, J., Wignall,G.D., Longman,G.W.,
       and Schmatz, W., Polymer 17, 751 (1976).
47.   Sadler,D.M., Faraday Discussions 68, 429 (1979) and
       Sadler, D.M. and Harris,R., J. Polym. Sci. (Phys. Ed.)
       (in press).
48.   Yoon, D.Y. and Flory, P.J., Polymer, 18, 509 (1977).
49.   Frank, F.C., in Growth and Perfection in Crystals (Proc.
       Int. Conf. Crystal Growth)pp. 529 and 53, eds Doremus
       Roberts and Turnbull, John Wiley, N.Y. (1958).
50.   Gutman, C.M., Hoffman,J.D. and DiMarzio,E., Faraday
       Discussions 68, 297 (1979).
51.   Sadler,D.M., Faraday Discussion, 68, 106 (1979).
52.   Mandelkern,L., Faraday Discussions, 68, 310 (1979).
53.   Guenet, J.M., Polymer 22, 313 (1981).
54.   Ballard, D.G.H., Burgess, A.N., Crowley,T.L., and Longman,
       G.W., Faraday Discussions 68, 279 (1979).
55.   Ballard,D.G.H.,personal communication, and presented at
       the U.K. Polymer Physics Meeting 1981.
56.   Sadler, D.M. and Barham, P.J. to be published.

SMALL ANGLE LIGHT SCATTERING FROM THE POLYMERIC SOLID STATE

Richard S. Stein

Polymer Research Institute
University of Massachusetts
Amherst, Massachusetts   01003

ABSTRACT

The Rayleigh scattering of visible light from a polymeric
solid provides information about its morphology. Scattering
arises because of fluctuations in average refractive index
(density), anisotropy and orientation of optic axes. These contributions may be distinguished from observations of scattering
polarization. Scattering theory may be developed based upon
calculations from model structures or else upon statistical
theory. In the former, model parameters (radii of spheres,
lengths of rods, etc.) are calculated, whereas in the latter,
correlation functions are obtained by Fourier inversion. Applications are discussed for phase separation of amorphous polymers,
spherulite growth and deformation.

INTRODUCTION

A homogeneous medium does not scatter as a consequence of
destructive interference between rays scattered from the various
volume elements. Real media scatter because of incomplete
interference arising from fluctuations in scattering power from
point-to-point within the media. That is, the total scattered
amplitude is given by

$$E_s = K_1 \int \rho(\underset{\sim}{r}) \exp[k(\underset{\sim}{r} \cdot \underset{\sim}{s})]d^3r \qquad (1)$$

where $\rho(\underset{\sim}{r})$ is the scattering power of a volume element located at
position $\underset{\sim}{r}$, $k=2\pi/\lambda$ (where λ is the wavelength within the media)
and s is the scattering vector $\underset{\sim}{s}=\underset{\sim}{s}_0-\underset{\sim}{s}_1$ where $\underset{\sim}{s}_0$ and $\underset{\sim}{s}_1$ are unit

R. A. Pethrick and R. W. Richards (eds.), Static and Dynamic Properties of the Polymeric Solid State, 109–125.
Copyright © 1982 by D. Reidel Publishing Company.

vectors in the directions of the incident and scattering rays. The exponential term accounts for the phase difference arising from path length differences for the various rays.

The value of $\rho(\underset{\sim}{r})$ is dependent upon the polarizability tensor $|\alpha(\underset{\sim}{r})|$ of the volume element. That is

$$\rho(\underset{\sim}{r}) = |\alpha(\underset{\sim}{r})| \underset{\sim}{E} \cdot \underset{\sim}{O} \tag{2}$$

where $\underset{\sim}{E}$ is the field of the light ray at the volume element and $\underset{\sim}{O}$ is a unit vector in the direction of polarization of the analyzer in the scattered ray. The field E may differ from the external field of the incident ray because of the internal field arising from the surrounding volume elements. In the Rayleigh-Gans-Debye approximation, which is usually adequate for most of the appications discussed in this article, this perturbing effect of the internal field is neglected. This is adequate providing the refractive index difference between the scattering particles is not too big. A more rigorous treatment due to Mie can be applied to specific structures (spheres, cylinders) in cases where the approximation is not adequate. The theoretical calculations discussed here involve specific solutions of Equations (1) and (2).

The experimental measurements of scattering involves the determination of the variation of scattered intensity with scattering angle θ and azimuthal angle μ (Fig. 1). While earlier scattering measurements were made using a mercury arc light source (1), lasers are now universally used for providing an intense, monochromatic and parallel source. The incident and scattered light beams may be polarized in a direction parallel to each other ($I_{||}$ or I_{V_V}) or perpendicular ($I_\perp$ or I_{H_V}). (I_{H_V} refers to the incident beam being vertically polarized and the scattered beam horizontally polarized. The "horizontal" plane is that in which θ is measured. Generally, $I_{H_V} = I_{V_H}$).

The scattered intensity may be measured photographically (1). (Polaroid film is convenient). Alternatively, for quantitative intensity measurements, photometric scans are desirable (2). Recently, the use of optical multichannel analyzers (OMA) have been proposed providing scans of the scattered intensity in one (3) or two (4) dimensions. These utilize vidicon detectors which may be computer interfaced so as to provide scans of the variation of scattered intensity with θ or even a 2-dimensional intensity contour plot. A schematic diagram of a 2-dimensional OMA apparatus is shown in Fig. 2 and a typical H_V contour diagram for a spherulitic polyethylene terephthalate film is provided in Fig. 3. Such results can currently be obtained in about 1 sec.

Scattered intensities should be corrected for refraction,

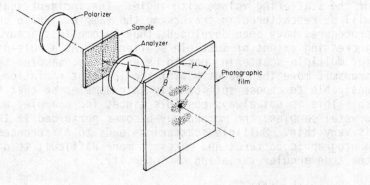

Figure 1.   The Light Scattering Angles θ and μ.

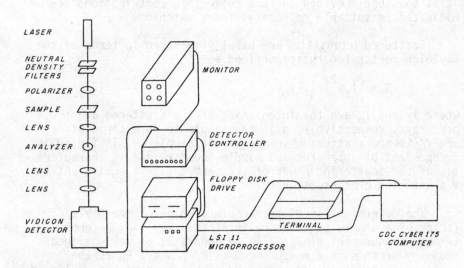

Figure 2.   A Block Diagram of the Two-Dimensional
            Optical Multichannel Analyzer.

reflection and multiple scattering (5,6). The light beam is refracted in passing through the film sample-air interface so that the scattering angle within the film differs from the measured value in air. Furthermore, this leads to a variation of the scattering volume with angle. The incident scattered ray will be rescattered in traversing the sample. While correction procedures have been developed, they become less accurate with increasing extent of multiple scattering. As a rule-of-thumb, for multiple scattering not to be excessive, samples should transmit more than about 80% of the incident radiation. It is desirable to choose sufficiently thin samples so that this is so. This is not always possible since, for example, with liquid crystal samples, the morphology becomes perturbed if the sample is very thin. Multiple scattering leads to differences in photographic patterns and makes it more difficult to discern the true angular variation of intensity.

With thin film samples, the scattering by surface irregularities can be significant. This contribution can be minimized by immersing the sample with a matching refractive index fluid or melting it between flat glass plates. Scattering from other geometry samples than films (e.g. fibers) is more difficult, but it is possible provided surface reflection contributions can be minimized by suitable refractive index matching.

Scattered intensities are usually measured in terms of the Rayleigh factor (or ratio) defined as

$$R = (I_s p^2)/(I_0 V_0) \tag{3}$$

where I_s and I_0 are the intensities of the scattered and incident rays, respectively, p is the sample to detector distance and V_0 is the scattering volume of the sample. This factor is independent of apparatus and sample geometry and is characteristic of the scattering power of the sample (it is equivalent to a scattering cross-section).

The determination of the absolute value of the Rayleigh factor of a sample requires calibration. This may be done through measurement of I_s/I_0 in an apparatus of well-defined geometry or through a measurement of R/R_0 where R_0 is the Rayleigh factor from some standard sample such as a pure liquid (e.g. benzene). When comparing I_s with I_0, intensities differ considerably so it is necessary to reduce the more intense beam through use of calibrated neutral filters.

Figure 3.   An H_v contour for PET obtained with the OMA2.

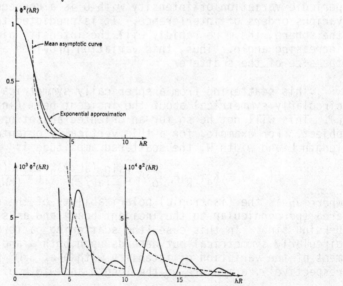

Figure 4.   Calculated Scattering Functions for an Isotropic Sphere (from A. Guinier and G. Fournet,
1955, Small Angle Scattering of X-Rays, John
Wiley and Sons, New York).

THEORY

The Model Approach

For specific model structures, Equations (1) and (2) may be integrated to provide closed solutions. For example, for isotropic spherically symmetrical systems with a radically dependent scaler polarizability $\alpha(r)$, Eqn. (1) becomes

$$E(h) = K_2 \int_{r=0}^{\infty} \alpha(r)\frac{\sin(hr)}{hr} r^2 dr \qquad (4)$$

where $h=(4\pi/\lambda)\sin(\theta/2)$. (This quantity h is known as the magnitude of the scattering vector ks and is often designated by q or k in physics literature.) The function $\alpha(r)$ can be obtained by Fourier inversion of $E(h)$. For the particular case of a uniform sphere of polarizability α_s immersed in a medium of polarizability α_0, the solution is

$$E(h) = K_3 V_s (\alpha_s - \alpha_0) \left[\frac{3}{U^3} (\sin U - U \cos U)\right] \qquad (5)$$

where V_s is the volume of the sphere $(4/3\pi R_s^3)$ and $U=hR_s$. The scattered intensity is, of course, proportional to $E(h)^2$ and depends upon the squre of the sphere volume and upon $(\alpha_s-\alpha_0)^2$, the "contrast". A plot of this result is given in Figure 4. The periodic variation of intensity with U is a consequence of various orders of interference. It is predicted that the larger the sphere, the more rapidly will the intensity diminish with increasing angle. Thus, this variation provides a measure of the size of the scatterer.

This scattering from a spherically symmetrical object is circularly symmetrical about the incident beam (independent of μ). This will not be so for an oriented anisotropically shaped object. For example, for a thin vertically oriented rod of length L and width W, the scattered amplitude is

$$E(\theta,\mu) = K_4(\alpha_R - \alpha_0)A \left[\frac{\sin (aL/2)}{(aL/2)}\right] \left[\frac{\sin (bW/2)}{(bW/2)}\right] \qquad (6)$$

where α_R is the (isotropic) polarizability of the rod, A is its area (perpendicular to the incident beam) and $a=ks\sin\theta \cos\mu$ while $b=ks\sin\theta \sin\mu$. In this case, the scattering pattern is no longer circularly symmetrical but depends upon both θ and μ. Measurement of the variation of intensity with θ at $\mu=0°$ provides, respectively, a measure of the length and width of the rod.

If the rods are completely randomly oriented, the pattern again becomes circularly symmetrical, in which case it becomes difficult to distinguish the scattering from a collection of rods from that from spheres. The intensity, for example, from a 3-dimensionally randomly oriented collection of infinitesimally thin rods of length L is given by

$$I(h) = K_5[1/W \, Si(2W) - (\sin W/W)^2] \qquad (7)$$

where $W = hL/2$ and $Si(2W) = \int_0^{2W} (\sin x/x) \cdot dx$.

For a partially oriented collection of rods, assuming that there is no phase coherence among them, the intensity may be calculated by summing over the distribution function of orientations

$$I(\theta,\mu) = \Sigma N(\psi,\phi) I(\theta,\mu,\psi,\phi) \qquad (8)$$

where $N(\psi,\phi)$ is the number of rods oriented at angles ψ and ϕ to the vertical direction, and $I(\theta,\mu,\psi,\phi)$ is the scattering intensity at θ and μ contributed by these rods. These equations presume intensity additivity which would only be true if the rods scattered incoherently. In general, the amplitude of scattering from an assembly of particles can be given by

$$E = \sum_i E_i e^{ik(R_i \cdot s)} \qquad (9)$$

where E_i is the scattered field strength from the i-th particle and R_s is the vector specifying the location of this particle. The intensity is then

$$I = (c/4\pi)EE^* = (c/4\pi)[\Sigma E^2 + \underset{i \neq j}{\Sigma\Sigma} E_i E_j e^{ik(R_{ij} \cdot s)}] \qquad (10)$$

where $R_{ij} = R_i - R_j$ and c is the velocity of light. The incoherent case leading to intensity additivity when values of the inter-particle separation occur randomly so that the second term of Eq. (10) averages to zero. Regularity in R_{ij} leads to inter-particle interference effect, ultimately resulting in diffraction.

Spherulitic polymers consist of spherically symmetrical aggregates of crystals within which a particular crystal axis is preferentially oriented at a particular angle with respect to the radius. The scattering from such aggregates can be approx-imated quite well by the scattering from an anisotropic sphere having polarizabilities α_r and α_t in the radial and tangential direction. The calculation for scattering from such an aniso-tropic object can be evaluated using Eqs. (11) and (2) where

$$\rho(r_i) = [(\alpha_1 - \alpha_2)_i (E \cdot a_i) a_i + (\alpha_2)_i E] \cdot 0 \qquad (11)$$

in which $\underset{\sim}{a_i}$ is a unit vector in the direction of the optic axis and α_i and α_2 are the principal polarizabilities parallel and perpendicular to the optic axis (assumed azimuthal). a_i may be assumed parallel to, perpendicular to or at some definite angle to the spherulite radius. For an isolated sphere with the optic axis along the radius, the interpretation of Eq. (1) leads to (1)

$$I_{H_v} = K_6 V s^2 \ [(3/U^3)(\alpha_t - \alpha_r) \cos^2 (\theta/2) \sin\mu \cos\mu$$
$$(4 \sin U - U \cos U - SiU)]^2 \tag{12}$$

and

$$I_{V_v} = K_6 V_s^2 \ \{(3/U^3)[(\alpha_t - \alpha_o) \ (2\sin U - U\cos U - SiU)$$
$$+ (4\sin U - U\cos U - 3SiU)]\} \tag{13}$$

It is assumed that the spheres are imbedded in a uniform iso-tropic medium of polarizability α_o.

A plot of Eq. (12) is given in Fig. 5 which may be compared with a series of experimental photographic patterns obtained along the crystallization of polyethylene terephthalate (PET), in Figure (6). The two are in reasonable correspondence.

It is evident that there are scattering maxima at odd multiples of $\mu=45°$ leading to a four-leaf clover shaped pattern. This occurs because of the $\sin\mu \cos\mu$ term in Eq. (12). [In some cases (e.g polybutylene terephthalate) the maxima occur at multiples of $\mu=90°$, which is interpreted in terms of the optic axis being oriented at an angle close to 45° to the radius]. A maximum occurs in the U direction at U=4.09. This corresponds to a θ_{max} given by

$$U_{max} = 4.09 = (4\pi R_s/\lambda)\sin(\theta_{max}/2) \tag{14}$$

Thus, the value of θ at which the maximum occurs decreases with increasing spherulite radius, as is evident in Fig. 6 as the PET spherulites grow. This shift in the position of the maximum provides a convenient measure of spherulite size and for following spherulite growth rates.

The V_v pattern exhibits two-fold symmetry as seen in the theoretical and experimental plots of Fig. 7 which is associated with the $\cos^2\mu$ factor in the third term of Eq. (13). It is noted that the first two terms of Eq. (13) are independent of the angle μ and are hence circularly symmetrical. These terms depend upon the polarizability of the surroundings, α_o and arise from the interference between the spherulite and the surroundings.

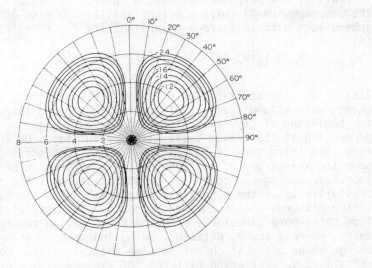

Figure 5.   A Theoretically Calculated H_V Scattering
Pattern for an Ideal Spherulite

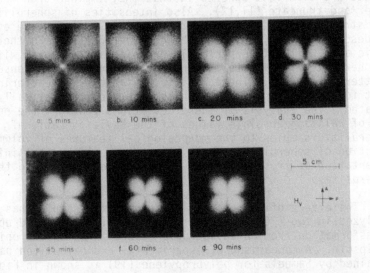

Figure 6.   A Series of H_V Scattering Patterns Obtained
in the Course of Crystallization of a Polyethylene Terephthalate Sample.

In reality, the surroundings of the spherulite may consist of
the amorphous phase (with a polarizability α_a) plus other
spherulites with an average polarizability $(\alpha_r + 2\alpha_t)/3$ so that
(9)

$$\alpha_0 = \phi_s(\alpha_r + 2\alpha_t)/3 + (1 - \phi_s)\alpha_a \qquad (15)$$

Thus α_0 and the contribution of these isotropic terms of Eq. (13)
change with volume fraction of spherulites, being largest at
the beginning of crystallization when ϕ_s is low and $\alpha_0 \approx \alpha_a$ and
decreasing as the medium fills with spherulites. On the other
hand, the third term of Eq. (13) which is μ dependent depends
upon the anisotropy of the spherulite $(\alpha_t-\alpha_a)$ which often
increased during the course of crystallization as the degree of
crystallinity of the spherulite increases. Thus the V_v scattering patterns change during the course of crystallization
from being more circularly symmetrical to having two-fold symmetry, whereas the H_v patterns, being independent of α_0 do not
change shape but just change in size and intensity. Computer
calculated V_v scattering patterns for various values of α_r, α_t
and α_0 have been published by Samuels (10).

The above theory of spherulites is idealized. Real spherulites are imperfect spheres in that they impinge upon each
other and truncate (11,12). Also intensities of spherulites are
not strictly additive and interspherulitic interference effects
as described in Eq. (9), must be considered (13). Furthermore,
optic axes will not be perfectly or uniformly oriented within
a spherulite, leading to internal heterogeniety with consequent
scattering (14,15). These lead to deviations in scattering from
that for perfect spherulites which have been theoretically
treated. The analysis of these deviations provides the possibility of relating absolute values of the Rayleigh ratio of
spherulitic samples to the morphological features of volume
fraction of spherulites, their size, degree of crystallinity and
orientation of crystalline and amorphous regions within the
spherulites.

The effect of deformation of spherulitic samples has been
analyzed in two (16) and three dimensions (17,18) based upon the
transformation from an anisotropic sphere to an anisotropic
ellipsoid. A comparison of experimental and calculated patterns
obtained by Samuels for polypropylene (19) is shown in Fig. (8).
From such comparisons, the axial ratio of the deformed spherulite
may be determined and compared with change in external sample
dimensions. These measurements may be made rapidly using cinema-
graphic techniques (20,21). The two-dimensional OMA presents the
possibility of carrying out such measurements more conveniently.
Measurements of the variation of light scattering under vibrational strain as a function of frequency and temperature have

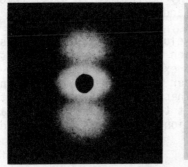
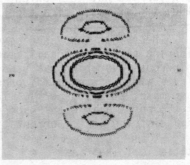

$$\bar{n} = 1.5261$$

Figure 7.   Theoretical and Experimental V_V Light Scat-
             tering Patterns (from R.J. Samuels: 1971,
             J. Polym. Sci., A2, p. 2165-2246, Fig. 15).

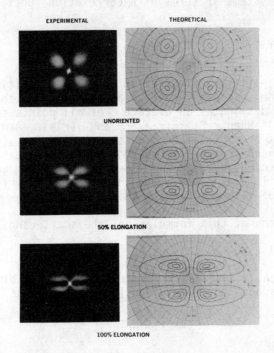

EXPERIMENTAL   THEORETICAL

UNORIENTED

50% ELONGATION

100% ELONGATION

Figure 8.   Experimental and Theoretical H_V Light Scat-
             tering Patterns for Vertically Drawn Poly-
             propylene Samples (Courtesy Hercules Incor-
             porated from R.J. Samuels: 1960, Hercules
             Chemist, No. 56, pp. 19-26, Fig. 5).

been carried out (22) which may be related to other rheo-optical studies of x-ray diffraction and birefringence (23).

Theory has also been developed for anisotropic rods, considering first the idealization of anisotropic infinitesmally thin rods (24) which has been extended to cases of rods of finite width, internal disorder and inter-rod interference, by Hashimoto, et al. (25-27). The effect of variation of the angle of the optic axis with respect to the rod axis as well as that of a distribution of orientation of rods has been considered.

Statistical Theories

In the above section the scattering from objects having a defined geometry has been considered. In many cases, systems are sufficiently complex so that geometry is ill-defined and a statistical description serves better. Such an approach has been pioneered by Debye and Bueche (28) who describe scattering from isotropic systems in terms of a correlation function $\gamma(\underset{\sim}{r})$ defined as

$$\gamma(\underset{\sim}{r}) = \frac{<\eta_1\eta_2>_r}{\overline{\eta^2}} \tag{16}$$

where $\eta_1 = \alpha_1 - \overline{\alpha}$ is the fluctuation in polarizability at point 1 and α is the spacial average polarizability for the medium. The symbol $<>_r$ designates an average over all points separated by distance r. When r=0, $<\eta_1\eta_2>_r = \overline{\eta^2}$ and $\gamma(r)=1$. At large r, η_1 and η_2 will not be correlated so that $<\eta_1\eta_2>_r = 0$ and $\gamma(r)=0$. Thus $\gamma(r)$ decreases from unity toward zero with increasing r in a manner dependent upon the geometry of the system and represents the probability of correlation in polarizability fluctuations for pairs of volume elements separated by r.

In terms of this formulation, the scattering is given by

$$I = K_7\overline{\eta^2}\int \gamma(\underset{\sim}{r}) \cos[k(\underset{\sim}{r} \cdot \underset{\sim}{s})]d^3r \tag{17}$$

For spherically symmetrical systems, this three-dimensional integral reduces to a one-dimensional integral as

$$I = 4\pi K_7\overline{\eta^2}\int_{r=0}^{\infty} \gamma(r) \frac{\sin(hr)}{hr} r^2dr \tag{18}$$

Here $\gamma(r)$ is a scalar correlation function which may be obtained from a Fourier inversion of the scattered intensity

$$\gamma(r) = K_8 \int_{r=0}^{\infty} I(h)\frac{\sin(hr)}{hr}h^2dh \tag{19}$$

It is often found experimentally and may be shown to theoretically follow (29) for a random two phase system that $\gamma(r)$ is exponential

$$\gamma(r) = \exp(-r/a_c) \tag{20}$$

where a_c is a correlation distance serving as a measure of the spacial size of the fluctuation. Generally

$$a_c = \int_{r=0}^{\infty} \gamma(r)dr \tag{21}$$

For a two phase system with sharp boundaries, the correlation distance may be related to the mean chord lengths through the two phases ℓ_1 and ℓ_2 by the Porod relationships (30)

$$\ell_1 = a_c/\phi_2 \quad \text{and} \quad \ell_2 = a_c/\phi_1 \tag{22}$$

where ϕ_1 and ϕ_2 are the volume fractions of the two phases. For such systems, the mean squared polarizability fluctuation is given by

$$\overline{n^2} = \phi_1\phi_2(\alpha_1 - \alpha_2)^2 \tag{23}$$

where α_1 and α_2 are the polarizabilities of the two phases. This predicts that $\overline{n^2}$ (and scattered intensity) should be a maximum when $\phi_1=\phi_2=0.5$.

The correlation function approach can be used to treat particle scattering. For example, for a sphere of radius R_s

$$\gamma(r) = 1 - 3/4(r/R_s) + 1/16(r/R_s)^3 \tag{24}$$

for $r<2R_s$ and is zero for larger r whereas for a randomly orienated assembly of thin rods of length L

$$\gamma(r) = C_9(L-r)/r^2 \tag{25}$$

for $r<L$ and is zero for larger r.

The Debye-Bueche formulation applies to isotropic systems for which there is no depolarized (H_V or V_H) scattering. Goldstein and Michalek (31) and Stein and Wilson (32) have generalized the treatment to anisotropic systems for which a simplified form of the latter treatment gives

$$I_{H_V} \qquad I_{H_V} = 4\pi K_7 \cdot 1/15\delta^2 \int_{r=0}^{\infty} f(r) \frac{\sin(hr)}{hr}r^2 dr \tag{26}$$

and

$$I_{V_V} = 4\pi K_7 \overline{n}^2 \int_{r=0}^{\infty} \gamma(r) \frac{\sin(hr)}{hr} r^2 dr + 4/45$$

$$\delta^2 \int_{r=0}^{\infty} f(r) \frac{\sin(hr)}{hr} r^2 dr \}$$

(27)

where δ is the anisotropy of the (assured uniaxial) volume element defined as the difference in its principal polarizabilities (assumed the same for all volume elements in this simplified version). $f(r)$ is a correlation function for optic axis orientation defined as

$$f(r) = [3<\cos^2 \theta_{12}>_r - 1]/2$$

(28)

where θ_{12} is the angle between optic axes for volume elements separated by distance r. $f(r) - 1$ for r=0 when optic axes are parallel and reduces to zero for large r where there is random correlation and $<\cos^2 \theta_{12}>_r = 1/3$.

Thus I_{H_V} arises from anisotropy and comes from orientation correlations while I_{V_V} arises from both orientation and density correlations. For $\delta=0$, $I_{H_V}=0$ and I_{V_V} of Eq. (27) reduces to the Debye-Bueche result of Eq. (18). $f(r)$ can be obtained from Fourier inversion of $(I_{V_V}-4/3I_{H_V})$. Often, both correlation functions are found to be exponential with characteristic correlation distances for both types of fluctuations. The relative importance of orientation and density fluctuation contribution to the scattering can be assessed by comparing I_{H_V} with I_{V_V}. For orientation scattering, these two contributions are comparable, whereas if there is appreciable density contribution to scattering, I_{V_V} is much greater than I_{H_V}. The former is true for a crystalline polymer in which crystals are uniformly distributed, whereas the latter is the case for a phase separated amorphous polymer blend.

The above treatment of orientation fluctuation scattering for unoriented systems leads to the prediction that the scattered intensities circularly symmetrical about the incident beam [μ does not appear in Eqs. (26) and (27)]. This occurs because of what is called the "random orientation fluctuation assumption". That is, it is assumed that the probability of optic axes being parallel depends only upon their separation and not upon the angle which the optic axes make with respect to the interconnecting vector r. This implies that correlated regions are spherically symmetrical and leads to the cylindrically symmetrical scattering. Many real systems do not follow this simplifying assumption and parallelness of optic axes may be more probable if they are parallel to r (rod-like correlation) or perpendicular to r (disc-like). To describe this, the correlation function must be described as a function of both r and the angles that the optic axis makes with r. For example, in a two-dimensional treatment (33) where a single angle, β suffices one may expand the correlation function in a Fourier series to give

$$f(r,\beta) = f_0(r) + f_2(r) \cos(2\beta) + f_4(r) \cos(4\beta) + \ldots$$

(29)

While this is in principle an infinite expansion, one finds that only the above terms are necessary in the scattering theory. The Stein-Wilson case corresponds to $f_2(r)=f_4(r)=0$ so that $f(r,\beta)=f_0(r)$. The $f_2(r)$ and $f_4(r)$ terms describe the shape of the correlated regions and lead to terms in the scattering results which are dependent upon μ. the theory has been generalized to three dimensions by van Aartsen (34) where the correlation function is expanded in spherical harmonies. While such generalized theores can generally describe the scattering from anisotropic systems such as those composed of spherulites or assemblies of anisotropic rods, they involve large numbers of parameters which are difficult to evaluate and therefore have limited usefulness.

The statistical theories of scattering can, in principal, be generalized to describe oriented systems (35) in which the correlation function becomes a vector function dependent upon direction within the sample. The scattering from oriented isotropic samples must then be described using Eq. (17) or its generalization for anisotropic systems. Scattering phenomena are then quite complex and depend not only upon the angles θ and μ but upon the polarization direction of the incident and scattered light. Under such conditions one must be concerned with the effect of sample birefringence on the state of polarization of the incident and scattered rays (36).

CONCLUSIONS

The study of the scattering of light from solid polymers leads to information about their morphology. Information may be in terms of the size, shape, polarizability and arrangement of contained particles, or else in terms of correlation functions providing a statistical description of the system. For anisotropic systems, additional information is provided by the observation of polarization of scattering.

Scattering may be observed during phase separation from amorphous systems, crystallization and deformation and may provide information characterizing morphology changes accompanying these phenomena.

ACKNOWLEDGEMENT

This paper represents a culmination of the work of many students, too numerous to mention here but referred to in appropriate references. The support for this preparation is acknowledged to the National Science Foundation, the Army Research Office (Durham), the Petroleum Research Fund of the American Chemical Society and the Materials Research Laboratory of the University of Massachusetts.

REFERENCES

1. Stein, R.S. and Rhodes, M.B.: 1960, J. Appl. Phys., 31, p. 1873.
2. Stein, R.S. and Plaza, A.: 1959, J. Polym. Sci., 40, p. 267.
3. Wasiak, A., Peiffer, D. and Stein, R.S.: 1976, J. Polym. Sci., Polym. Letters Ed. 14, p. 381.
4. Tabar, R., Stein, R.S. and Long, M.B., J. Polym. Sci., Polym. Phys. Ed., submitted for publication.
5. Stein, R.S. and Keane, J.J.: 1955, J. Polym. Sci., 17, p. 21.
6. Prud'homme, R.E., Bourland, L., Natarajan, R.T. and Stein, R.S.: 1974, J. Polym. Sci., Polym. Phys. Ed., 12, p. 1955.
7. Natarajan, R.T., Prud'homme, R.E., Bourland, L. and Stein, R.S.: 1967, J. Polym. Sci., Polym. Phys. Ed. 14, p. 1541.
8. Stein, R.S.: 1973, in Structure and Properties of Polymer Films, ed. by R.W. Lenz and R.S. Stein, Plenum, New York, p.1.
9. Yoon, D.Y. and Stein, R.S.: 1974, J. Polym. Sci., Polym. Phys. Ed., 12, p. 735.
10. Samuels, R.J.: 1971, J. Polym. Sci., A2, 9, pp. 2165-2246.
11. Picot, C. and Stein, R.S.: 1970, J. Polym. Sci., A2, 8, p. 2127.
12. Prud'homme, R.E. and Stein, R.S.: 1973, J. Polym. Sci., Polym. Phys. Ed. 11, p. 1683.
13. Picot, C., Stein, R.S., Marchessault, R.H., Borch, J. and Sarko, A.: 1971, Macromolecules, 4, p. 467.
14. Stein, R.S. and Chu, W.: 1970, J. Polym. Sci., 8, p. 1137.
15. Yoon, D.Y. and Stein, R.S.: 1974, J. Polym. Sci., Polym. Phys. Ed., 12, p. 763.
16. Stein, R.S., Clough, S. and van Aartsen, J.J.: 1962, J. Appl. Phys., p. 3027.
17. van Aartsen, J.J. and Stein, R.S.: 1971, J. Polym. Sci., A2, 9, p. 295.
18. Samuels, R.J.: 1966, J. Polym. Sci., C13, p. 37.

19.   Samuels, R.S.: 1968, Hercules Chemist, 56, p. 19.
20.   Erhardt, P.R. and Stein, R.S.: 1965, Polymer Letters 3,
      p. 553.
21.   Erhardt, P.F. and Stein, R.S.: 1967, Appl. Polym. Symp.,
      High Speed Testing, Vol. VI: The Rheology of Solids 5,
      p. 113.
22.   Hashimoto, T., Prud'homme, R.E., Keedy, D.A. and Stein,
      R.S.: 1973, J. Phys. Ed., 11, pp. 693, 709.

23.   Stein, R.S.: 1966, J. Polym. Sci., C15, p. 185.
24.   Rhodes, M.B. and Stein, R.S.: 1969, J. Polym. Sci., A2, 7,
      p. 1539.
25.   Moritani, M., Hayeshi, N., Utsuo, A. and Kawai, H.: 1971,
      Polym. Journal, 2, p. 74.
26.   Hashimoto, T., Morakami, Y. and Kawai, H.: 1975, J. Polym.
      Sci., Polym. Phys. Ed., 13, p. 1613-1631.
27.   Hashimoto, T., Yamaguchi, K. and Kawai, H.: 1977, Polym.
      J., 9, pp. 405-414
28.   Debye, P. and Bueche, A.M.: 1949, J. Appl. Phys., 20, p.
      518.
29.   Debye, P., Anderson, H.R. and Brumberger, H.R.: 1957, J.
      Appl. Phys., 28, p. 679.
30.   Kratky, O.: 1966, Pure Appl. Chem., 12, p. 483.
31.   Goldstein, M. and Michalek, E.R.: 1955, J. Appl. Phys.,
      26, p. 1450.
32.   Stein, R.S. and Wilson, P.R.: 1962, J. Appl. Phys., 33,
      p. 1914.
33.   Stein, R.S., Erhardt, P.F., Clough, P., and Adams, G.:
      1966, J. Appl. Phys., 37, p. 3980.
34.   van Aartsen, J.J.: 1971, in Polymer Networks, Structural
      and Mechanical Properties, ed. by A.J. Chompf and S. New-
      man, Plenum Press, New York, p. 307.
35.   Stein, R.S. and Hotta, T.: 1964, J. Appl. Phys., 35,
      p. 2237.
36.   Stein, R.S. and Stidham, S.N.: 1966, J. Polym. Sci., A2, 4,
      p. 89.

SMALL ANGLE NEUTRON SCATTERING (SANS) BY AMORPHOUS POLYMERS

Cl. PICOT

Centre de Recherches sur les Macromolécules, CNRS
6, rue Boussingault
67083 Strasbourg-Cedex, France

1. INTRODUCTION

Small Angle Neutron Scattering (SANS) is an experimental technique introduced since about ten years for the investigation of
polymer conformation in all the concentration range from dilute
solution to the melt (1-4).

The neutron scattering method presents some special advantages
that distinguish it from X ray scattering and light scattering.

- Neutron Scattering is the result of neutron nucleus interaction
and hence different isotopic species may interact with neutrons
quite differently. In particular, Hydrogen and Deuterium have
scattering power (Scattering length) that differ in both sign
and magnitude. As a consequence a deuterated molecule in a
protonated matrix will be visible to neutrons and vice versa.

- The other advantage results from the wavelength of thermal
neutrons $(2<\lambda(\text{Å})<20)$ and low angle spectrometers which makes
possible to reach scattering vectors $(q = 4\pi/\lambda \sin \theta/2, \theta = \text{scat-}$
tering angle$)$ ranging from 10^{-3} Å^{-1} to few Å^{-1}. This range of q
matches very well to the characteristic polymer dimensions.
Compared values of q ranges corresponding to Neutron, X ray and
light scattering are represented on figure 1.

In this lecture concerning SANS by amorphous polymers we will
devote a first part to a brief description of the experimental
methods. Some elements of SANS theory will be then presented
from the scattering by identical atoms to end by the scattering
laws by polymer blends.

R. A. Pethrick and R. W. Richards (eds.), Static and Dynamic Properties of the Polymeric Solid State, 127–172.
Copyright © 1982 by D. Reidel Publishing Company.

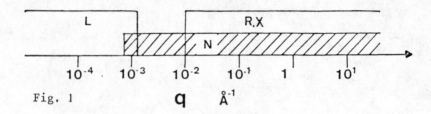

Fig. 1

Finally, a last section will be devoted to the presentation of
experimental results on some specific topics :
    - the amorphous polymers at equilibrium and in the stretched
state
    - the rubbery networks

More detailed informations about neutron scattering theory and
its application to polymer can be found in a number of excellent
books and review articles. See for example, Marshall and Lovesey
(5), Turchin (6), Bacon (7), Allen and Higgins (8), Maconnachie
and Richards (9) and Ullman (10).

2. INSTRUMENTATION AND EXPERIMENTAL TECHNIQUES

A basic neutron scattering experiment requires a source of
radiation, a device for selecting the energy or wavelength and
a method for detecting the scattered radiation. Detection
involves measuring the intensity as function of the scattering
vector for elastic scattering and as function of energy transfer
for inelastic scattering. In the investigations on polymer chain
conformation that we will be described we will be only concerned
by elastic scattering and then limit our description to the
spectrometer built for performing this type of measurements.
In her lecture, J. Higgins gives details on inelastic scattering
spectrometers which allow studies on polymer chain dynamic.
All the instruments use a reactor as neutron source. The uranium
fission in the core of the reactor produced highly energetic
neutrons. This energy is reduced by surrounding the core with
D_2O. By using a "cold source" it is possible to shift the
wavelength distribution to produce a greater proportion of long
wavelength neutrons. In this way a factor about 10 is gained
in the flux of long wavelength neutrons (cold neutrons) as seen
on figure 2.

For simplicity, we will limit our description of experimental
device to that of D11 system of ILL which has been widely used
by different teams working in the field of SANS by polymers.
D11 system has been described by Ibel (11), others ILL experimental devices description are given in reference (12). A recent
review by Schelten and Hendriks (13) surveys a number of

instrumental developments.

The D11 system presented on figure 3 operates in the following way.

The "cold neutrons" arrive through a slightly curved guide tube which filters γ rays and fast neutrons from the beam. The beam passes through a helical monochromator made of a large number of slotted discs, each disc bening slightly rotated with respect to the previous one.

The rotation of this helical monochromator at a defined speed defines a helical path and transmits only the neutrons that do not touch the helical guide, the others being absorbed. As elastic neutron scattering studies on polymer consist in diffuse scattering measurements, the experiment can tolerate a beam with a certain wavelength distribution. On D11 two monochromators are available : one with $\delta\lambda/\lambda = 0.1$, the other with $\delta\lambda/\lambda = 0.45$ calculated with full width at half maximum.

After the monochromator, there are several collimator diaphragms (which are adjusted according experimental requirements) and the sample target. The detector (position sensitive BF_3 detector) is a grid of 64x64 electrodes giving 4096 cell (each 1 cm^2) and is placed after the sample at distances which may be varied from 1.68 to 40 m to provide a wide range of scattering vectors. Scheme (4) shows the q range which can be covered as function of the wavelength and sample-detector distance. Finally the detector is connected to a data acquisition computer.

The data are then averaged at different values of r over the whole detector for isotropic scattering or over preferential directions for anisotropic scattering measurements.

The rough data are then used to extract the scattering signal of interest by introducing background and attenuation corrections.

Detailed data treatment programs very easy to handle have been written by R. Ghosh in a booklet available ILL (Grenoble) (14).

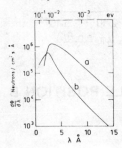

Fig. 2 : Plot of the logarithm of the neutron flux in Å vs. the wavelength λ measured at the exit of the neutron guide B I of the Saclay reactor EL 3. The curves (a) obtained with the cold source and (b) without the cold source. (from Ref. 32).

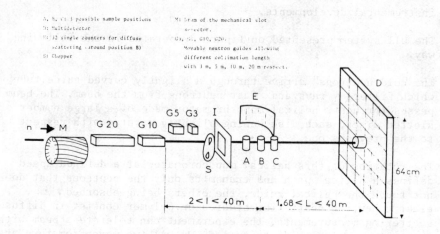

A, B, C: 3 possible sample positions
D: Multidetector
E: 32 single counters for diffuse
   scattering (around position B)
S: Chopper

M: Drum of the mechanical slot
   selector.
G5, G3, G10, G20:
   Movable neutron guides allowing
   different collimation length
   with 1 m, 5 m, 10 m, 20 m respect.

SCHEMATIC VIEW OF THE SMALL ANGLE SCATTERING
CAMERA D11

Fig. 3   (from Ref. 12)

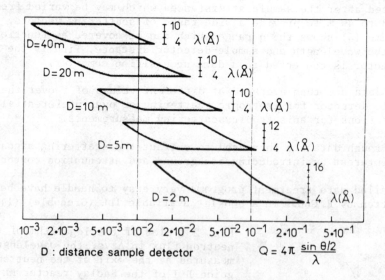

$$Q = 4\pi \frac{\sin \theta/2}{\lambda}$$

D : distance sample detector

Q–RANGE ACCESSIBLE FOR SAMPLE POSITION B

Fig. 4   (from Ref. 12)

3. ELEMENTS OF SMALL ANGLE NEUTRON SCATTERING THEORY

It is not our purpose in this introduction to give a detailed
derivation of the theory of thermal neutron scattering but
rather to introduce the basic equations and concepts which will
be used throughout this lecture. Several good books and reviews
already exist which describe these theories in details ; see for
example : Marshall and Lovesey (5), Turchin (6) and Allen and
Higgins (8).

Neutron interacts with matter in two ways. The main interaction,
by which we will be concerned, is that between neutron and
nucleus and is called nuclear scattering. The second interaction
the magnetic scattering occurs when unparred electrons are
present, then the neutron interacts with the associated magnetic
moment.

The scattering laws that we will describe are only concerned
with the first type of scattering by low energy neutrons (thermal neutron scattering) which concerns essentially the investigations which can be applied to the study of polymeric systems.
Indeed, as we are concerned with neutron nucleus interactions,
different isotopic species may interact with neutrons quite
differently in particular hydrogen (H) and deuterium (D) have
that is called "scattering length" that differ in both magnitude and sign. Thus a deuterated molecule is "visible" to
neutron in a protonated matrix of the identical chemical composition and vice versa.

In a scattering experiment, the scattered intensity is measured
as function of energy and/or scattering angle. One define the
scattering vector $\vec{q}$ by $\vec{q} = \vec{k} - \vec{k}_o$

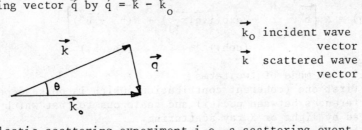

$\vec{k}_o$ incident wave
          vector
$\vec{k}$ scattered wave
          vector

In an elastic scattering experiment i.e. a scattering event
which occurs without energy exchange (*) $|k| = |k_o|$, $|q|$ reduces
to the classical expression

$$|q| = \frac{4\pi}{\lambda} \sin \frac{\theta}{2} \qquad \text{since} \quad |k_o| = \frac{2\pi}{\lambda}$$

* (case which will be only considered, for inelastic scattering
c.f. J. Higgins)

The amplitude of a scattered wave is proportionnal to -b where
b is called the scattered length. b is complex, the real part
can be positive or negative and the imaginary part is a measure
of neutron absorption which can occur.

3.1. General scattering relation by an assembly of identical nuclei - coherent and incoherent scattering

Let us consider an assembly of N identical nuclei denoted by
indices i and j with scattering lengths b_i and b_j. The intensity
scattered is described by the general relation :

$$\tilde{S}(q) = A \left\langle \sum_{i,j=1}^{N} b_i b_j \exp (i\vec{q}.\vec{r}_{ij}) \right\rangle \tag{1}$$

where $\vec{r}_{ij} = \vec{r}_i - \vec{r}_j$

A is a experimental constant proportional to the intensity of
incident beam. Neutrons are particles of spin 1/2 and the scattering length b^+ is generally different to b^- ; b^+ occurs with
relative frequence w^+ and b^- with relative frequence w^-. Thus
the average values $\overline{b}$ and $\overline{b^2}$ are given by :

$$\overline{b} = w^+ b^+ + w^- b^- \tag{2a}$$

$$\overline{b^2} = w^+ (b^+)^2 + w^- (b^-)^2 \tag{2b}$$

If one separates in $\tilde{S}(q)$ the terms in which i=j from those in
which i≠j and if one averages over spin states, $<b_i b_j> = \overline{b}^2$ for
i≠j and $\overline{b^2}$ for i=j, (1) becomes :

$$\tilde{S}(q) = A \left[\overline{b}^2 \sum_{i,j=1}^{N} \exp(i.\vec{q}.\vec{r}_{ij}) + N(\overline{b^2} - \overline{b}^2) \right] \tag{3}$$

$$\quad\quad\quad\quad\text{(coh.)}\quad\quad\quad\quad\quad\quad\text{(incoh.)}$$

$\tilde{S}(q)$ is then made of two terms :
- The first one (coherent contribution) which is phase dependent
(interferences between nuclei) and analogous to that which is
measured by light or X ray scattering.
- The second term (incoherent contribution) which leads to isotropic scattering and which does not depend on the phase.

For sake of simplicity $\overline{b^2} - \overline{b}^2$ is often denoted $b^2_{inc.}$ and the
cross-sections σ defined by :

$$\sigma_{coh.} = 4\pi \ \overline{b}^2 = 4\pi \ b^2_{coh.} \tag{4a}$$

and $$\sigma_{incoh.} = 4\pi \ b^2_{inc.} \tag{4b}$$

are often used.

It must be noticed that the incoherent term is unique to neutron. In particular the incoherent cross-section σ_{inc} of hydrogen is very high and, accordingly , polymers which usually contain important amounts of hydrogen, will exhibit a large incoherent background independent of angle.

Table I gives some coherent scattering lengths, incoherent and absorption cross-sections of some common elements. Cadmium and Boron, which absorb strongly neutrons are used as neutrons absorbers on experimental devices.

Element	$b_{coh} \times 10^{12}$ cm	$\sigma_{inc} \times 10^{24}$ cm^2	$\sigma_A \times 10^{24}$ cm^2
H	−0.374	79.7	0.19
D	0.667	2	≈ 0
B	0.54 + 0.021 i	0.7	430.0
C	0.665	≈ 0	≈ 0
N	0.94	0.3	1.1
O	0.58	≈ 0	≈ 0
F	0.56	≈ 0	≈ 0
Na	0.36	1.8	0.28
Si	0.42	≈ 0	0.06
S	0.28	0.2	0.28
Cl	0.96	3.4	19.5
Cd	0.37 + 0.16 i		2650.0
V	−0.05	5.1	2.8

* σ_A is determined at a neutron wavelength of 1.08 Å.

Table I : Scattering lengths, incoherent cross sections, and absorption cross sections for some elements

3.2. The hypothesis of incompressibility

Let us now return to equation (3) and focus our attention on the first coherent scattering term which contains the interesting information of the measurement. The second term can be substrac- ted from the total scattering by using convenient background or introducing known numerical values.

Let us suppose that the scattering medium is still made of N identical atomes. Each atoms of coherent scattering length b occupies a volume v = V/N (V being the volume of the system).

At this stage it is useful to introduce the following operators :
- the local density of atoms $\rho(r)$
- the local density of scattering length b(r)
- the local number of atoms n(r) by using the Dirac function such that : $\int \delta(r) d_3 r = 1$

Then $\rho(r) = \sum\limits_{i} \delta(r-r_i)$ (5)

$n(r) = v\,\rho(r)$

$b(r) = b\,\rho(r)$

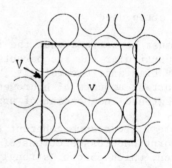

It can be shown that $\tilde{S}(q)$, the coherent part of $\tilde{S}(q)$ can be written as :

$$\tilde{S}(q) = <b(q)\ b(-q)> = b^2 < \rho(q)\ \rho(-q)> \qquad (6)$$

$S(q)$ can be then expressed as function of $b(q)$, Fourier transform of $b(r)$. Furthermore, $\tilde{S}(q)$ is directly proportional to the pair correlation function of the Fourier transform of the local density operator $\rho(r)$.

In a SANS experiment one measures correlation lengths greater that few atoms size. At that scale it is reasonable to admit that the local fluctuations of the number of atoms are negligibles : in other words that the system is incompressible. This is equivalent to say that all the position of the system are occupied by a atom $(n(r)=1)$. In these conditions $\rho(r)$ is a constant and its F.T. is equal to zero for $q\neq o$. The result is that the system does not exhibit central scattering.

3.3. Scattering by a binary mixture of particles. The contrast factor

The volume V contains now two types of particles denoted by this indices 1 and 2.
The preceeding relations (5) can be then written as :

$$\rho(r)= \sum\limits_{i\in 1} \delta(r-r_i)+ \sum\limits_{i\in 2} \delta(r-r_i) = \rho_1(r)+\rho_2(r)$$

$$n(r) = v_1\rho_1(r)+v_2\rho_2(r)$$

(7)

$$b(r) = b_1\rho_1(r)+b_2\rho_2(r)$$

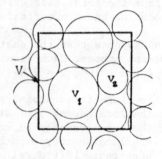

The hypothesis of incompressibility $n(r)=1$ implies now compensation of the fluctuations of concentrations of species 1 and 2. It can be written as :

$$V \delta(q) = v_1 \rho_1(q) + v_2 \rho_2(q) \tag{8}$$

at q=o it leads to :

$V = N_1 v_1 + N_2 v_2$ showing that v_1 and v_2 have to be considered as partial molar volume.
for q≠o the expression   of $\rho(q)$ as function of $\rho_2(q)$ leads to :

$$b(q) = (b_1 - b_2 \frac{v_1}{v_2}) \rho_1(q) \quad q \neq o \tag{9}$$

b(q) is then the F.T of the density of particules   uniquely, weighed by an apparent scattering length K (contrast)

$$K = (b_1 - b_2 v_1/v_2) \tag{10}$$

The coherent scattered intensity can be then written as :

$$\tilde{S}(q) = K^2 <\rho_1(q) \rho_1(-q)> \tag{11}$$

which takes into account only fluctuations of particles 1.

The hypothesis of incompressibility reduces the problem to the knowledge of spatial correlation of particules 1.

To summarize, $\tilde{S}(q)$ is the product of two terms :
- a technical term K^2 the scattering power due to the weight given by the radiation to the labelled particle in the medium.
- a physical term which is function of the position of scattering particles in the system only. All the previous relations have been established for the case of samples containing atoms. They can be generalized to the case of molecular systems provided that the dimensions of the molecules are smaller than the q^{-1} values required by the experiment. In other terms the approximation holds only in the limit that the term exp(iqr) of   expression (3) is equal to unity . In the specific case where the monomer size (in the case of a polymer system) is large and scattering angles not sufficiently low, the phase relationships between atoms of the monomer unit has to be taken into account.

When the approximation holds, the coherent scattering length for a monomer is the sum of the scattering lengths of the atoms composing this structural unit.

Table II, reported from Ullman's review (10) gives a list of values of coherent scattering lengths and incoherent scattering cross-sections of several usual monomer units.

Monomer	Chemical formula	$b_{coh} \times 10^{12}$ cm	$b_{coh}^* \times 10^{-10}$ cm^{-2}	$\sigma_{inc} \times 10^{24}$ cm^2
Ethylene	C_2H_4	-0.166	-0.342	318.8
Ethylene d$_4$	C_2D_4	3.998	8.24	8.0
Propylene	C_3H_6	-0.257	-0.331	478.2
Propylene d$_6$	C_3D_6	5.997	7.72	12.0
Ethylene oxide	C_2H_4O	0.414	—	318.8
Ethylene oxide d$_4$	C_2D_4O	4.578	—	8.0
Styrene	C_8H_8	2.328	1.420	637.6
Styrene d$_8$	C_8D_8	10.656	6.50	16.0
Methyl methacrylate	$C_5H_8O_2$	1.493	1.069	637.6
Methyl methacrylate d$_6$	$C_5D_8O_2$	9.821	7.029	16.0
Methacrylic acid	$C_4H_6O_2$	1.576	—	478.2
Methacrylic acid d$_6$	$C_4D_6O_2$	7.82	—	12.0
Tetrahydrofuran	C_4H_8O	0.248	—	637.6
Tetrahydrofuran d$_8$	C_4D_8O	8.576	—	16.0
Propylene oxide	C_3H_6O	0.323	—	478.2
Propylene oxide d$_3$	$C_3H_3D_3O$	3.446	—	245.1
Propylene oxide d$_6$	C_3D_6O	6.569	—	12.0
Sodium styrene sulfonate	$C_8H_7SO_3N_a$	5.106	—	559.9
Sodium styrene sulfonate d$_7$	$C_8D_7SO_3N_a$	12.379	—	16.0
Acrylonitrile	C_3H_3N	0.873	1.169	239.4
Acrylonitrile d$_3$	C_3D_3N	3.996	5.351	6.3
Dimethylsiloxane	$C_2H_6S_2O$	0.086	—	478.2
Dimethyl siloxane d$_6$	$C_2D_6S_2O$	6.332	—	12.0
Butadiene	C_4H_6	0.416	0.426	478.2
Butadiene d$_6$	C_4D_6	6.662	6.823	12.0

* Here, $b_{coh}^* = b_{coh} N_A \rho / M_0$, where ρ is density, N_A is Avogadro's number, and M_0 is the molecular weight of a monomer.

Table II : Coherent scattering lengths and incoherent cross-
sections of monomer units (from Ref. 10).

In equation (11), if the monomer (1) are linked together to form
a polymer chain, the correlation function will correspond to the
spatial position of all these monomers in a same chain or of
monomers belonging to different chains. In the limit of very
low concentrations of polymer, the single chain correlation
function is obtained. For practical use, we will discuss
briefly that last case which was the situation mostly encountered
in the experiments of scattering by polymer solutions.

3.4. Scattering by a dilute solution of polymer

Let us then consider n polymer chains made of N_1 scattering
segments (scattering length b$_1$) dispersed in a medium of small
molecules (b$_2$).

According to the relations given before, the coherent scattering
contribution of this chain can be written as :

$$\tilde{S}(q) = (b_1 - b_2 \frac{V_1}{V_2})^2 \, S_{1coh}(q) \tag{12}$$

with $$\tilde{S}_{1coh} = \sum_{i.j} \langle \exp(i \, q.r_{ij}) \rangle \sim \int \exp(i\underline{q}\underline{r}) P(r) dr \tag{13}$$

where $P(r)$ is the probability of having two monomers at the distance r (see Daoud)

The expression of $P(r)$ for an ideal gaussian chain is very well known and leads to the classical Debye function f_D.

Thus $$\tilde{S}_{1coh} = N_1^2 \, f_D \tag{14}$$

with $$f_D = \frac{2}{X^2} \left[\exp(-X) - 1 + X \right] \tag{15}$$

$$X = q^2 \, \overline{R_g^2}, \qquad \overline{R_g^2} \text{ mean square radius of gyration} = Nl^2/6 \text{ for a gaussian coil}$$

Finally $S(q)$ is written as :

$$S(q) = n \, N_1^2 \, K^2 \, f_D(q) \tag{16}$$

By introducing more experimental parameters such as :
- the molecular weight of the polymer M and of its monomer unit m
- the polymer concentration (w/vol)
The intensity scattered for unit volume $I(q)$ is :

$$I(q) = K^* \, c \, M \, f_D(q) \tag{17}$$

with $$K^* = \frac{K^2 N_A}{m^2} \, , \quad N_A \text{ avogadro's number}$$

Depending on the q values with respect to the molecular parameter of the polymer chain, four regions of interest can be distinguished :

A) The Guinier range ($q \, R_g < 1$) which corresponds to long-range correlations between monomers.

In that case f_D reduces to $$(1 - \frac{q^2 R_g^2}{3})$$

Thus the measurements leads to the value for the radius of gyration Rg which corresponds to the global size of the polymer.

B) The intermediate range $(R_g^{-1}<q<a^{-1})$ where a is the persistence length of the polymer.

In that range, also called submolecular range, the correlations at short distances between monomers inside a chain are observed, and for gaussian chains $f_D(q)$ behaves like q^{-2}.

C) The range of q corresponding to the persistence length (local rigidity) $a^{-1}<q<l^{-1}$ a persistence length
l statistical step length

In that range the chain is "seen" like a rod and presents a scattering behaviour proportional to q^{-1}.

D) Finally the very high q range $(l^{-1}<q)$ where the scattering is governed by the structure of the chain segments. It corresponds to discrete scattering by the atoms of the monomer units.

These different ranges of q can be in principle distinguished from each other by using the classical representation of Kratky. A schematic and idealized representation is given on figure 5

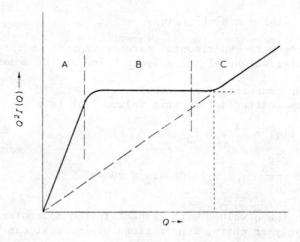

Fig. 5 : An idealized Kratky plot   (from Ref. 9)

The equation developped for I(q) is quite similar to that obtained for light scattering.

It can be shown that for non-ideal solutions of polymers the concentration effects can be taken into account in a form similar to the well known Zimm relation :

$$\frac{K^{*}c}{I(q)} = \frac{1}{M} (1 + \frac{q^2R_g^2}{3}) + 2A_2c + \ldots \tag{18}$$

Thus, by extrapolating a series of measurements at different c and q to c=o and q=o, Rg and M can be obtained. For polydisperse systems the z average is measured for R and weight average for M.

The contrast factor in polymer solutions

For polymer solutions, the relation (17) is valid for all radiations but K is a function of the nature of the radiation matter interaction.

For neutrons, K is an apparent coherent scattering length of the monomer in the medium (see rel. 10). So, by using a judicious pair solute-solvent, it is possible to have b_1 positive and b_2 negative and thus obtain a very large scattering power. For photons, K is given by a difference between the absolute values of the corresponding parameters (polarizability for light, electron density for X rays). These quantities are only dependent on the electron shells of the atoms. Then the values of the scattering power are small compared with those obtained with neutrons.

In neutron scattering, due to the difference in coherent scattering length between hydrogen ($b_H = -0.374.10^{-12}$ cm) and deuterium ($b_D = +0,667.10^{-12}$ cm), an important contrast can be obtained by dispersing small amount of fully deuterated polymers in a hydrogenous matrix or vice versa. Neutron contrasts factors for some classical polymer diluent systems are given in Table III together with X rays contrasts. Clearly deuteration offers larger possibilities, especially for dispersion of deuterated polymer in a hydrogeneous matrix. Furthermore, isotopic substitution does affect only slightly thermodynamical properties of polymeric

Polymer	Solvent	$10^{28}K$ (m^2)	$10^{28}K_x$
Hydrogenous polystyrene	Benzene, C_6H_6	0.13	6.3
	Benzene, C_6D_6	73.70	2.3
	Cyclohexane, C_6H_{12}	8.1	11.75
	Deuterated polystyrene	70.2	0
D–PMMA	H–PMMA	70.2	0

Table III : Contrast factors for neutron (K) and X-ray scattering (K_x) for polymer-diluent systems

systems as shown by Strazielle and Benoit (15).

Improvements of contrast in X ray scattering can be obtained by heavy   atoms labelling, but this labelling procedure alters the chemical integrity of the polymer molecules.

3.5 Scattering from a mixture of labelled and unlabelled chains and from partially labelled chains

In his lecture, Daoud gives general calculations of correlation functions of labelled chains in a melt. It is the purpose of this following part to present a more simple and physical approach according a method proposed by Leibler, Benoit and Koberstein (16,17).

a) Case of non interacting chains

- general relation
Let us assume that we are dealing with a homogenous system made of p different constituents. Each constituent i can be made of z_i scattering units with coherent scattering length b_i. For sake of simplicity we will assume that all scattering unit has the same volume, b_i is then proportional to the coherent scattering length per unit volume.

If we forget about incoherent scattering that can be substracted from experimental results, the coherent scattered intensity may be written by similarity to equation (3) in the general form :

$$I(q) \sim \sum_{ij} b_i b_j \exp{(i \cdot q \cdot (\vec{r}_i - \vec{r}_j))} \tag{19}$$

In order to simplify the calculations, let us assume that we have three components (a,b,c,) in the system and let us call A,B,C the scattering lengths of these components respectively.

In these conditions, and spliting I(q) into the different terms depending on i and j are a,b,c scatterers one obtains:

$$I(q) \sim A^2 \sum_{Na} \exp{iq(\vec{r}_{i_a} - \vec{r}_{j_a})} + B^2 \sum_{Nb} \exp{i\vec{q}(\vec{r}_{i_b} - \vec{r}_{j_b})} + C^2 \sum_{Nc} \cdots$$

$$+ 2AB \sum_{NaNb} \exp{i\,\vec{q}(\vec{r}_{i_a} - \vec{r}_{j_b})} + 2BC \sum_{NbNc} \cdots + 2CA \sum_{NcNa} \cdots \tag{20}$$

which can be written in the simplified form :

$$I(q) \sim A^2 S_{AA} + B^2 S_{BB} + C^2 S_{CC} + 2AB \, S_{AB} + 2BC \, S_{BC} + 2CA \, S_{AC} \tag{21}$$

Assuming the hypothesis of incompressibility we will show that we can reduce the number of scattering functions appearing in this expression.

Hypothesis of incompressibility

As shwon in the previous part, $S(q)$ is the Fourier Transform of the density of segments

$$S(q) = \rho(\vec{r_i}) \, \rho(\vec{r_j}) \, \exp - i q (\vec{r_i} - \vec{r_j}) \tag{22a}$$

$$= \tilde{\rho}(q) \cdot \tilde{\rho}(-q) \tag{22b}$$

by introducing the average density of segments ρ_o and the local fluctuation $\Delta\rho = \rho - \rho_o$

$$S(q) = \Delta\tilde{\rho}(q) \cdot \Delta\tilde{\rho}(-\vec{q}) \tag{23}$$

Thus
$$S_{AA} = \Delta\tilde{\rho}_A(q) \cdot \Delta\tilde{\rho}_B(-q)$$
$$S_{BB} = \Delta\tilde{\rho}_B(q) \cdot \Delta\tilde{\rho}_B(-q) \tag{24}$$
$$S_{CC} = \Delta\tilde{\rho}_C(q) \cdot \Delta\tilde{\rho}_C(-q)$$

The hypothesis of incompressibility means (no density fluctuations) :

$$\Delta\rho = \Delta\rho_A + \Delta\rho_B + \Delta\rho_C = 0 \tag{25}$$

By developping S_{CC} and using $\Delta\rho_C$ as function of $\Delta\rho_A$ and $\Delta\rho_B$ one obtains :

$$S_{CC} = \Delta\rho_C(q)\Delta\rho_C(-q) = \left(-\Delta\rho_A(q) - \Delta\rho_B(q)\right)\left(-\Delta\rho_A(-q) - \Delta\rho_B(-q)\right) \tag{26}$$

$$S_{CC} = S_{AA} + S_{BB} + 2S_{AB} \tag{27}$$

In the same way one obtains :

$$S_{AC} = -S_{AA} - S_{AB} \tag{28}$$
$$S_{BC} = -S_{BB} - S_{AB} \tag{29}$$

By eliminating C terms of S one is lead to :

$$I(q) = (A-C)^2 S_{AA} + (B-C)^2 S_{BB} + 2(A-C)(B-C) S_{AB} \tag{30}$$

Thus in $I(q)$ appears only the scattering response of molecules A and B, the terms corresponding to C having been eliminated by the hypothesis of incompressibility and leading to a contribution just by the contrast that they provide.

This type of calculation could be extended to a system of p constituents. In that case the number of S functions appearing in (21) would be $p(p+1)/2$ and could be reduced by the hypothesis of incompressibility to $p(p-1)/2$.

- Applications
  Two polymers in solution

Let us consider that a and b are the same polymer (Z monomer units) with different scattering length A and B and C is a solvent (scattering length S).

Assuming that they are N_a molecules of a and N_b molecules of B $N = N_a + N_b$ and $x = N_a/N$, $(1-x) = N_b/N$.

The functions $S_{\alpha,\beta}$ can be split in intra and inter molecular parts : (respectively $P(q)$ and $Q(q)$)

$$S_{AA} = Z^2 N_a \left[P(q) + N_b Q(q) \right] \tag{31}$$

$$S_{BB} = Z^2 N_b \left[P(q) + N_b Q(q) \right] \tag{32}$$

$$S_{AB} = Z^2 N_a N_b Q(q) \tag{33}$$

where $P(q) = \dfrac{1}{Z^2} \langle \sum_i \sum_j e^{iq(r_{i_a} - r_{j_a})} \rangle = \dfrac{1}{Z^2} \langle \sum_i \sum_j \dfrac{\sin q\, r_{ij}}{q r_{ij}} \rangle \tag{34}$

$$Q(q) = \dfrac{1}{Z^2} \langle \sum_i \sum_j e^{iq(r_{i_a} - r_{i_b})} \rangle = \dfrac{1}{Z^2} \int (g(r)-1) \dfrac{\sin q_r}{qr} \, d\vec{r} \tag{35}$$

where $g(r)$ is the pair correlation function between monomers of two different chains.
By using (30,31,32,33) one obtains :

$$I(q) = Z^2 \left\{ \left[N_a (A-S)^2 + N_b (B-S)^2 \right] P(q) + \left[N_a (A-S) + N_b (A-S) \right]^2 Q(q) \right\} \tag{36}$$

They are only two unknown functions $P(q)$ and $Q(q)$ to determine. These two functions P and Q can be determined by two methods.
i) by adjusting   S=B   (Williams method used for poly-
                                              electrolytes (18))

(36)   then becomes :

$$I(q) = N Z^2 (A-B) \left[x P(q) + N x^2 Q(q) \right] \qquad (37)$$

$$I(q) = \phi \left[x P(q) + N x^2 Q(q) \right] \qquad (38)$$ where ϕ is a technical
constant

Thus, by plotting $I(q)/\phi . x = P(q) + x N Q(q)$ (39), one obtains
a kind of Zimm plot which leads for $x = o$ (zero content of
labelled species for example) the function $P(q)$. The function
$Q(q)$ is obtained by the slope of this function at a given q
value.

ii) a second method consists to adjust the coherent scattering
length of the solvent to the mean value of the polymers
defined as :

$$\overline{B} = x A + (1-x)B = S \qquad (40)$$

Introducing this value in relation (36) leads to :

$$I(q) = N Z^2 (A-B)^2 x(1-x) P(q) \qquad (41)$$

which allows to determine directly $P(q)$.

This method preconized by Akcasu and Han (19) has been used for
instance for neutron scattering investigation on polymer chain
conformation in swollen networks. An example of application
will be given in the second experimental part of this lecture.

   Two polymers in bulk

Let us suppose that N_a molecules are deuterated. The medium
is then made only by a and b polymers, therefore we can write
as before for case (ii)

$$S = \overline{B}$$

and again :

$$I(q) = N Z^2 (A-B)^2 x(1-x) P(q) \qquad (42)$$

The q dependence of the signal is independent of x and intermolecular interferences do not play any role.

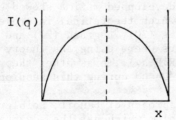

At this stage few practical remarks can be made :
i) when a scattering experiment is carried out on a blend of
deuterated and hydrogenated chains of the same degree of polymerisation (for which segregation effects are weak) the
scattering intensity is directly proportional to the single
chain scattering function. Then the use of high concentration
of deuterated chain will be a very convenient method since
stronger scattered intensity allows either the use of neutron
sources of medium flux either to do more precise studies of the
scattering functions. In the experimental part we will give
some experimental examples of this method and discuss its limitations especially when polymer samples are polydisperse.

ii) If (42) is rewritten in terms of light scattering one gets
at (q=o)

$$\frac{K'C}{I(q)} = \frac{1}{M} + x \frac{1}{M}$$
(43)

The second term corresponds to the second virial coefficient :
$A_2 \sim 1/M$, which is the classical result of the Flory theory for
the case of athermal solution (the polymer system becomes
"theta" when molecular weight tends to infinity).

iii) Relation (37) can be rewritten as :

$$I(q) = N \overset{2}{Z} (A-B)^2 x \left[P(q) + xNQ(q) \right]$$
(44)

using relation (42) on gets then :

$$NQ(q) = -P(q)$$
(45)

Qualitatively this result can be explained by the fact that the
term Q(q) corresponds to interferences on all points excepts
one molecule. It is therefore the contribution of an homogeneous
medium with a hole at the place of the considered molecule.

Partially labelled chains

During the last few years experiments have been carried out
with partially labelled polymers. These are kind of "copolymers"
H-D on which the labelled sequence D can be adjusted with
precision by chemistry. These systems provide interesting
scattering response with well developped maxima showed experimentally by Duplessix and for which theoretical interpretation
has been developped by De Gennes (20), Legrand (21) and
Leibler (22). These last authors where using the theory of
"random phase approximation" (R.P.A.). This rather theoretical
approach has been presented by Daoud during this session.

We would like to present the approach of Benoît, Leibler and
Koberstein (16,17) of this problem which has the advantage to

use only simple geometrical considerations and which leads in some special cases to the results of RPA.

i) Case of monodisperse systems

We assume that all the chains in the sample are identical, all of them being labelled. Using the considerations developped before for a two components system we can write :

$$I(q) \sim (A-B)^2 S_{AA}(q)$$

S_{AA} being the scattering function for the labelled part of the polymer.

By splitting S_{AA} in two terms like in equation (31) $S_{AA}(q)$ becomes :

$$S_{AA}(q) = N Z_A^2 \left[P_A(q) + NQ_A(q) \right] \tag{47}$$

The quantity $P_A(q)$ is the intramolecular scattering function of labelled part(Z_A labelled segments) which can be evaluated. $Q_A(q)$ is the intermolecular scattering function which is less easier to evaluate.

Let us now consider the case of half labelled molecules in which $Z_A = Z_B = Z/2$.

The relation (47) can be written :

$$S_{AA}(q) = N(\frac{Z}{2})^2 \left[P_{1/2}(q) + NQ_{1/2}(q) \right] \tag{48}$$

where the subscript 1/2 refers to half part of the molecule. Let us now demonstrate that $Q_{1/2}=Q$, Q being the intermolecular scattering function of a completely labelled chain.

If the molecules were completely labelled $Q(q)$ would take the form:

$$Q(q) = \frac{1}{Z^2} \sum_{i=1}^{Z} \sum_{j=1}^{Z} < \exp - i q r_{i_1 j_2} > \tag{49}$$

Then $Q(q)$ is the average value of the summ of coefficients $e^{iqr_{i_1 j_2}}$ of a square matrix of Z rows and Z columns.

It is evident that $\exp (-iqr_{i_1 j_2}) = \exp(- iq r_{i_1}(Z-j_2))$. This means that this matrix has two axis of symetry and from that considerations :

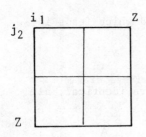

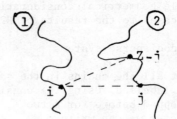

$$Q_{1/2} = \frac{4}{Z^2} \left(\frac{1}{4} \text{ matrix}\right) = \frac{1}{Z^2} \text{ matrix} = Q \qquad (50)$$

This leads finally to :

$$I(q) = (A-B)^2 N\left(\frac{Z}{2}\right)^2 \left[P_{1/2} - P\right] \qquad (51)$$

This equation would be valid for all kinds of symetrical linear molecules even if they were not gaussian.

Fig. 6 shows calculated variations for half labelled gaussian coil of $P_{1/2}$, P and $I(q)$ in arbitrary units.

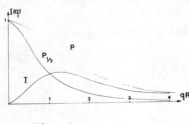

Fig. 6

In starts to zero and goes zero at high q values. It has therefore a maximum in the range $qR{\sim}1$, where R is the radius of gyration of the labelled part.

Such geometrical approach could be carried on molecules presenting a given degree of symetry, like rings and stars with some labelled branches. The details of these calculations are given in ref. (16).

ii) Effect of polydispersity

The effect of polydispersity has been calculated rigorously by Leibler and Benoît (17) using RPA and Daoud has presented in his lecture some aspects of its influence. We will limit us to present a physical approach if this problem which has been developped in ref. (16).

Let us consider the two type of polydispersity which can be present on a partially labelled polymer.

- Length polydispersity

If all the molecules have the same composition but differ by their length, relation (51) can be written :

$$I(q) \approx \left[\overline{P_{1/2}}(q) - \overline{P}(q) \right] \tag{52}$$

where $\overline{P_{1/2}}$ and $\overline{P_q}$ are the classical average values of $P_{1/2}$ and P for a polydisperse system.

- Polydispersity in composition

This case is more interesting because it results often from chemical synthesis.

Let us examine this effect on the simple system already presented by Daoud using RPA method. Let us then suppose a mixture of N_a half labelled molecules and N_b unlabelled molecules.

By calling $x = N_a/N_a+N_b$ and by similarity to previous expressions :

$$I(q) = (A-B)^2 S_{AA} = (A-B)^2 \left(\frac{Z}{2}\right)^2 N_a \left[P_{1/2}(q) + N_A Q_{1/2}(q) \right] \tag{53}$$

as according (45) and (50) $Q^{1/2} = Q = -\frac{P}{N}$

$$I(q) = N_a \left(\frac{Z}{2}\right)^2 (A-B)^2 \left[P_{1/2}(q) - x P(q) \right] \tag{54}$$

The important point is that the zero scattering angle is no more zero as soon as x is different from unity. The repartition of labelled is deuterated phase is no more homogeneous. The zero scattering intensity is given by :

$$I(o) = N_a \left(\frac{Z}{2}\right)^2 (A-B)^2 (1-x) \tag{55}$$

and the initial slope in the case of Gaussian chains is given by :

$$\frac{q^2}{3} R^2 \frac{1-2x}{1-x}$$

The maximum in the curve desappears for $x < \frac{1}{2}$

The figure (7) gives a graphic representation of the calculated effect that has been described, while figure (8) gives an experimental example of neutron scattering by to "copolymers" PSH-PSD one having a successful preparation and low polydispersity, the other one being rather polydisperse.

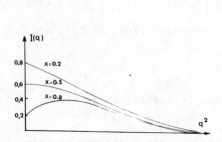

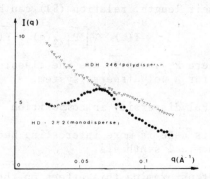

Fig. 7 : Influence of the
polydispersity in composition
for half-labelled chains.

Fig. 8 : Experimental results
on two-block polystyrene
polymers Hydrogen-deuterium-
hydrogen, one with low polydispersity, the other with
large polydispersity (from
Ref. 16)

As conclusion we have shown by simple geometrical considerations
and taking into account incompresssibility hypothesis that by
using high concentration of completely labelled chains, the
correlations corresponding to a single chain can be obtained
improving the accuracy of the measurements.

- The existence of a maximum in the scattering by partially
labelled chains does not result from a Bragg reflection which
could be attributed to local order in the system. As already
mentioned by Daoud, this result from the correlation hole which
is a measure of the repulsion between chains due to constant
density hypothesis.

- The influence of the polydispersity is very strong in the
same manner that it has been shown in the case of copolymer
in dilute solutions (23,24).

As shown by Daoud, RPA provides a more power method for calculating all the scattering functions for any mixtures of chains
with an arbitrary sequence of tagged molecules. These calculations hold for chains having no specific interactions. These
interactions which are present in the case of copolymers with
different chemical nature can be taken into account in the frame
of RPA theory. These calculations have been recently widely
developped by Leibler (22).

As example we will limit us to describe the case of a copolymer with two interacting sequences.

b) Case of interacting chains

By using the Random Phase Approximation method, Daoud has given
the general form of the correlation functions for mixtures of
labelled and unlabelled chains. These calculations were restricted to the case were labelled and unlabelled species do not
interact strongly. This is generally the case for hydrogenated
and deuterated species of a same polymer. In the case of
polymer blends of different chemical structure, the strong
incompatibility leads to macroscopic phase separation (segregation) except for a limited number of polymer couples which
can lead to homogeneous phases. In the case of copolymers (A,
B for instance) the effect of segregation leads to the formation
of microdomains whose structure have been extensively studied by
X Ray diffraction. In a very elaborated theory based on RPA
method, Leibler (22) has described the phase diagramm of copolymers and characterized the transition from homogeneous melt
to ordered mesophases. This author has also calculated the
scattering correlation functions for selectively labelled
species in the case of homogeneous melt.

We would like to show by presenting as a example results on
labelled diblock copolymers, how the scattering correlation
function is modified by the additional contribution of the
interaction between sequences.

Example of a diblock copolymer with interacting sequences .

Let us then consider a block copolymer A-B in which the sequence
A is deuterated for instance. Let us call f the fraction of A
segments ($f = N_A/N$, $N=N_A+N_B$).

Daoud has shown that the scattered intensity in the case of non
interacting chains can be written as :

$$I(q) = (A-B)^2 S_{AA}(q) \qquad (56)$$

with the general form for $S_{AA}(q)$:

$$S_{AA}(q) = \frac{S^\circ_{AA} S^\circ_{BB} - \left[S^\circ_{AB}\right]^2}{S^\circ_{AA} + S^\circ_{BB} + 2S^\circ_{AB}} \qquad (57)$$

$$= \frac{W(q)}{S(q)} \qquad \text{according Leibler's notations} \qquad (58)$$

S° are the scattering correlation functions of ideal independant copolymer chains.

It can be shown by straigthforward calculations that in the case of 50/50 copolymer (57) reduces to (51).

In the case of interacting chains characterized by the Flory interaction parameter χ, Leibler has shown that the scattering correlation function becomes :

$$S_{AA}(q) = \frac{W(q)}{S(q) - 2\chi W(q)} \qquad (58)$$

By supposing that the chain conformation remains Gaussian the different functions $S^{\circ}_{ij}(q)$ of $W(q)$ and $S(q)$ can be expressed in terms of Debye functions. S_{AA} can be then written in the reduced form :

$$S_{AA}(q) = \frac{N}{F(X) - 2\chi N} \qquad (59)$$

where $F(X)$ is a composition of Debye functions of the two sequences and can be easily calculated.

In (59) $X = q^2 R^2 = \frac{N l^2}{6}$ (1 length of statistical segment identical for both sequences)

- for $qR \gg 1$

$$S(q) \simeq 2 N f(1-f)/q^2 R^2 \qquad (60)$$

$S_{AA}(q) \to 0$ like q^{-2}

This means that at small scale of distances (submolecular) the fluctuations of monomers are like those of a Gaussian chain and then independent of the repulsion between monomers.

- for $qR \ll 1$

$$S(q) \simeq 2Nf^2(1-f) \frac{q^2 R^2}{3} \qquad (61)$$

at small q the scattered intensity vanishes (consequence of incompressibility hypothesis)

As q increases, $S(q)$ increases due to the correlation hole effect.

As for non interacting chains the correlation function S_{AA} presents a maximum.

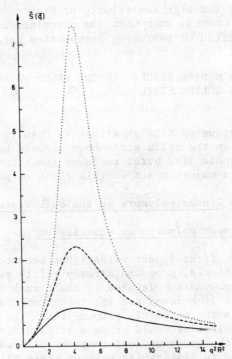

Scattering intensity (in arbitrary units) of the molten diblock copolymer with composition $f = 0.25$ as a function of $x = (qR)^2$ ($q = 4\pi[\sin(\theta/2)]/\lambda$) for three values of interaction parameter: $(\cdots)$ $\chi N = 17.5$; $(---)$ $\chi N = 16.0$; $(-)$ $\chi N = 12.5$. The interaction parameter for which $\bar{S}(q^*)$ diverges (spinodal point) is given by $\chi N = 18.2$. The MST is expected to occur for $\chi N = 17.6$.

Fig. 9 (from Ref. 22)

It must be stressed again that this maximum does not result from segregation effects. The shape of this maximum, whose position is not q dependent for a given copolymer, depends strongly on the interaction parameter χ. As shown in relation (59), the relevant parameter is rather χN than χ.

Figure 9 shows an example of scattering correlation function as function of q^2R^2 for $\neq$ values of χN. The peak becomes more narrow and its maximum increases with χN.

As mentionned by Leibler there is a certain critical values of $(\chi N) = (\chi N)_s$ for which $S_{AA}(q)$ diverges. It corresponds to the spinodal point. However the instability is masked by the earlier appearance of a first order transition to a mesophase for $(\chi N)_t < (\chi N)_s$.

Taking into account the high sensitivity of S(q) to interactions,
Leibler suggested to use the analyse of the correlation function
S_{AA} as a direct method of measuring interaction parameter χ.

4. SOME SANS EXPERIMENTAL RESULTS IN THE FIELD OF AMORPHOUS POLYMERS IN THE SOLIDE STATE

Exhaustive reviews on SANS experimental results concerning
amorphous polymer in the solid state have already been published
(9,10,25,26). We would like here, to focus the attention on
results which had a major impact in this field of polymer science.

4.1. Amorphous linear polymers in the bulk state

4.1.1. Linear polymers at equilibrium

Undoubtely, the first impact of SANS has been to ckeck expe-
rimentaly the hypothesis, proposed already thirty years by
Flory (27), that polymers molecules in their bulk amorphous
state should behave like ideal and interpenetrable Gaussian coil.
A number of other authors (28-31) proposed that the linear structure of polymer molecules would produce local ordering leading
to deviation from the Gaussian conformation. SANS was the only
experimental method to distinguish between these alternatives.
As mentionned us section 3.4 SANS results on single labelled
chain can be discussed in two characteristic q ranges : The
Guinier range and the submolecular range of q.

The Guinier range $(qR_g \ll 1)$

In this range of q, the variation of the reciprocal value of
the intensity leads to the determination of the mean dimensions
of the labelled molecule (R_g^2). For a chain obeying Gaussian
statistics $(R_g^2)^{1/2}$ must be proportional to $M^{1/2}$.

SANS experiments have been carried out on numerous polymeric
systems :
- Benoît et al. (32) have investigated deuterated Polystyrene in
matrices of protonated polystyrene of equivalent molecular
weight. There results have been compared to those obtained with
Polystyrene D in cyclohexane.

Figure 10 shows an example of Zimm representation obtained
for a bulk system. It must be noticed that the second virial
coefficient A_2 is equal to zero.
Figure 11 gives the variation of R_w^2/M_w as function of M_w.
The results obtained in bulk state were compared to those
obtained in a θ solvent (cyclohexane) and those obtained in CS_2

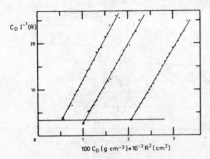

$C_D J^{-1}(R)$ is plotted $vs.$ $C_D + R^2$ where R is a function of the scattering angle $\theta = 2\pi R/D$. The slopes of the curves give the uncorrected radius of gyration and the slope of the points (X) obtained for each C_D by extrapolation at $R = 0$ gives the value A_2 of the second virial coefficient. The data are obtained from PSD 4 in its PSH matrix.

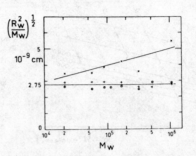

$(R_w{}^2/M_w)^{1/2}$ is plotted as a function of log M_w. The data are (X) for PSD in CS_2, (+) in θ solvent, and (⊙) in the bulk. The horizontal line, which is the mean for the data obtained in θ solvent and in the bulk, is consistent with a Gaussian configuration.

Fig. 10   (from Ref. 32)   Fig. 11 (from Ref. 32)

which is a good solvent. On the entire range of molecular weight investigated the dimensions of chains in bulk are quite comparable to those in the "θ" situation.

Ballard, Wignall and Schelten (33) obtained results in very good agreement with the previous one and showed in particular that, heating the sample throughout the glass transition, the overall chain dimension does not vary.

- Kirste, Kruse and Ibel (34) have carried out similar experiments with Polymethylmethacrylate in bulk but using H polymer as solute and deuterated species as the matrix. Their results indicate that the mean square dimensions are proportional to M_w but the constant of proportionality being slightly greater (10%) than that found for the same polymers in a theta solvent (butylchloride).

- Particularly interesting results have been obtained by Kirste and Lehnen (35) on polydimethylsiloxane (PDMS) in bulk state. The experiments have been carried out by using a 200,000 molecular weight protonated PDMS as solute (probe) in several deuterated matrices of molecular weight ranging from 180 to 250.000. As shown on Figure 12 the second virial coefficients were positive for lower molecular weight matrices and close to zero for higher molecular weight matrices. The experimental

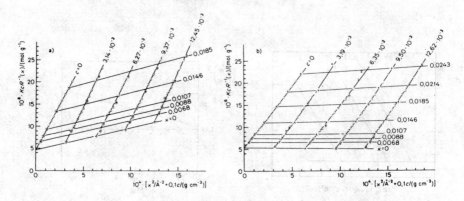

Fig. 12 : Zimm representation for blends of PDMS (M_w=200 000) in dPDMS. a) M_w(dPDMS)=3000, b) M_w(dPDMS=250 000) (from Ref. 35)

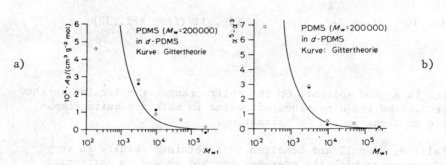

Fig. 13 : Variation of A_2 (a) and $\alpha^5-\alpha^3$ (b) for PDMS blends as function of molecular weight of the matrix. (from Ref. 35)

variation of A_2 as function of the molecular weight of the matrix is represented on Figure 13a and compared to the expected value from the Flory Huggins theory for athermal solutions ($A_2 \approx 1/M_{matrix}$). The expansion factor α of the chain as function M_{matrix} has also been studied and compared with the Flory prediction ($\alpha^5-\alpha^3 = c\ P_2^{1/2}/P_1$; P_1,P_2 polymerisation degrees of the matrix and of the solute polymer respectively)(see Figure 13b) The experimental results are in both cases in satisfactory agreement with the theoretical predictions.

- Lieser, Fisher and Ibel (36) have studied fractions of H polyethylene dissolved in D polyethylene in the molten state. Measurements confirm that the radius of gyration is proportional to $M^{0.5}$. No ordering effect is detectable in molten crystallize polymer.

- Finally it must be mentionned that identical behavior have been obtained for Polyethylene Oxyde in the molten state by Allen (37) and for Polyvinyl chloride in the amorphous state by Hershenroeder (38)

The Intermediate or submolecular range of q $(R_g^{-1} < q < a^{-1})$

In this range of q the scattering correlation function of a single chain reflects spatial distribution of chain segments inside the chain. For an ideal Gaussian chain I(q) should behave like q^{-2}. Thus a plot of $q^2 I(q)$ should reach a plateau in this q range.

Such plateau have been indeed observed for polystyrene (32,33) and molten polyethylene (39) and an example is given on Figure 14 for polystyrene (33). At the opposite, at high values

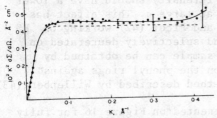

Fig. 14 : Kratky plot for PSD/PSH blend (from Ref. 33)

of q but still in the intermediate range, PMMA does not follow the Gaussian model. The results obtains by Kirste et al. (34) on atactic PMMA in bulk plotted in the Kratky representation exhibit a strong downward curvature comparable to the effect observed in dilute solutions on syndiotactic PMMA. Calculations by Yoon and Flory (40) using rotational isomeric state model have shown theoretically that for polymers presenting strong local correlations, the scattering at high q values can be strongly influenced by preferential local conformation. Figure 15 presents the comparison between experimental results and theoretical calculations which are in good qualitative agreement.

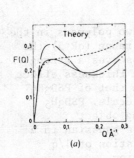
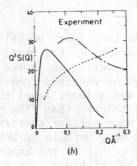

Fig. 15 : Kratky plot for PMMA chains (SANS and X Ray scattering)
—— atactic
—·—· syndiotactic
------ isotactic
(from Ref. 34 and 40)

Although the rotational isomeric state model is the more realistic approach to analyse the conformation of real polymer chains, the thickness of the chain is neglected in the calculation of the scattering function. By analysing scattering behaviour in the high q range, one reaches correlation distances comparable to the size of few monomers. At this level the cross section of the chain may have determinant effect. Calculations of Koyama (41) have clearly shown this contribution. Very crudely the problem can be formulated as follow. If $S_o(q)$ is the scattering function for an ideal thin polymer coil, the scattering function for a chain with finite cross-section (defined by its mean square radius of gyration R_ϕ^2) should be :

$$S(q) = S_o(q) \exp \left(- \frac{q^2 R_\phi^2}{2} \right)$$

Therefore the observed scattered intensity should have a lower value than the theoretical value at large q. This effect has been clearly evidenced by Rawiso and Picot (42) working on PS chains fully deuterated (PSD8) and selectively deuterated on the backbone (PSD$_3$H$_5$). This last sample can be obtained by selective exchange of deuterons on the phenyl rings against protons of benzene according a method described by Willenberg(43).

Experimental results are presented on Figure 16 for fully labelled and partially labelled PS of 140,000 molecular weight immersed in a matrix of protonated PS of equivalent M_w (c/w=20%).

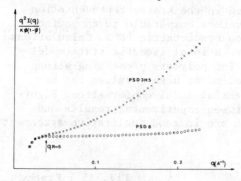

Fig. 16 : Kratky plot for PS labelled chains PSD8 : fully deuterated;PSD$_3$H$_5$: deuterated on the backbone (from Ref. 42)

The difference between the behaviours of this two polymers in the Kratky representation is striking. For small q range (qR<5) the two scattering functions converge, while one observes a strong divergence as q increases. As expected from the thickness effect relation, PSD$_8$ behaviour has a lower level than that of PSD$_3$H$_5$ which appears to neutrons thinner than PSD8 molecule. PSD$_3$H$_5$ behavior correspond to a much stiffer scattering object ($I(q) \sim 1/q$). The fact that PSD$_8$ appears to behave gaussian in a very wide range of q is the result of the compensation of 1/q behaviour convoluted by the thickness contribution

$$\left(e - \frac{q^2 R_\phi^2}{2} \right)$$

This effect raised up the precautions that have to be taken into account for the precise analysis of scattering behaviour in the high q range.
Experiments reported below have essentially been performed with small relative concentrations of labelled molecules and involved extrapolations to zero concentration in order to eliminate interchain interference effects. The recent developments in scattering theories, described in the theoretical section, have shown that the single chain scattering function does not require a regime of small concentration of labelled species. Thus, they have been recents experimental efforts to support these theoretical predictions.

In particular working on PSD/PSH blends Wignal et al.(44) and Boué et al. (45) have clearly checked these predictions.

Wignal et al. have worked on PSD/PSH mixtures of equivalent narrow molecular weight distributions (M_w=75,000) and have shown that the mean-square dimensions and the zero angle scattering extrapolation normalized to $\phi_D(1-\phi_D)$ are concentration independent

Boué et al. have examined in details the effect of the polydispersity of H and D species on the scattering functions. Using RPA calculations for data corrections they showed that the scattered intensities normalized to $\phi_D(1-\phi_D)$ are very well superposed in the intermediate range, while this is no longer the case for low angle results for which microvoids contribution has been invoqued particulary in the case of high deuterium content.

Recent SANS measurements have been carried out on PDMSH/PDMS mixtures by Beltzung et al. (46). In that case a very good superposition of the scattering intensities normalized to $\phi_D(1-\phi_D)$ have been obtained leading to constant values of determined mean-square dimensions. The results are still valid in the case of chain labelled networks. The variation of R^2 as function of M_z leads to :

$(R_z^2)^{1/2} = 0.23 \, M_z^{0.51}$ for the melt of linear chains

$(R_z^2)^{1/2} = 0.24 \, M_z^{0.50}$ for chains incorporated in a network

At the end of the survey of the results obtained on amorphous polymer at equilibrium, the main following conclusions can be drawn :
    - SANS experiments results show that a polymer molecule behaves like a Gaussian coil in the bulk amorphous state.

Excluded volume effects are found if the molecular weight of the
solvent polymer is considerably different from that of the
solute.

Gaussian behaviour is complementary supported by the results
obtained on partially labelled chains (see Daoud)

- The analysis of scattering functions at high q values must
be done with peculiar attention taking into account local
conformation on one hand and chain cross section on the second
hand.

- The use high concentration of labelled species is a very
promising procedure to get precise informations on chain conformations especially for deformed samples for which anisotropic
data acquisition requires long exposure time.

4.1.2. Oriented linear polymers

As seen before, SANS provides information about intrachain
pair correlation function over a wide range of q vectors allowing
the description of the chain conformation from distances ranging
from few statistical units to its overal dimension (Rg). It was
then tempting to use the possibility of this method to investigate the modification of this conformation when samples are
submitted to external constrains.

SANS experiments in this field have been carried out by
Picot et al. (47,48) and Boué et al. (49) on uniaxially hot
stretched mixtures of PSD/PSH.

In this type of experiments, samples are stretched at constant rate above the glass transition temperature Tg and then
rapidly quenched at room temperature. SANS measurements are then
performed on these frozen samples.

An example of two dimensional scattering pattern obtained is
represented on Figure 17 (49) where it can be noticed that

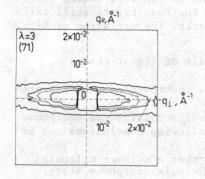

Fig. 17 : Anisotropic
scattering pattern from
uniaxially hot stretched PS.
(from Ref. 49)

isointensity lines present a strong anisotropy. A detailed analysis of the scattering function can be made in the directions parallel and transverse to the stretching direction ($q_{//}$ and $q_{\perp}$) leading to chain correlation functions in these two privileged directions.

In order to have some feeling about the chain response and the resulting scattering functions, it is necessary to have some insight in the flow process involved in this type of deformation This flow process is indeed very complex. The major factor governing the overall molecular deformation and motions of the chains in such dense viscous and elastic system is the effect of entanglements and Ferry's classical book (50) and Graessley review (51) give detailed descriptions of this problem.

For purpose of simplicity and to try to give schematic ideas about scattering response let us consider this very simplified picture : the chain system can be described as a physical network but, at the opposite of chemical networks for which crosslinks are permanent, polymer melts will present temporary physical crosslinks whose times of life and density will depend on temperature and time scale involved in the deformation process. Therefore, considering a labelled chain in the melt we can consider that the forces can be applied at different points 1 and j separated by distance $\vec{r}_{ij}$ characteristic of the experimental conditions of deformation.

During the deformation $(\vec{r}_{ij})_o$ at rest, becomes $|\lambda|(\vec{r}_{ij})_o$ where $|\lambda|$ is the tensor of deformation

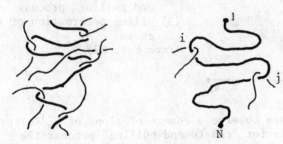

Two extreme case can be then considered :
  i) $\vec{r}_{ij}$ corresponds to the statistical segments and the chain will deform affinely at all distance scale
  ii) the forces are applied at the extremities of the chain (i=1, j=N) : the chain will deforme by a procedure called "end to end pulling". It is obvious that to this particular case will correspond a considerable loss of affinity in the chain deformation process resulting from conformational rearrangements of the chain segments.

Benoît and Jannink (52) have analyzed the scattered functions
resulting from such chain by calculating the scattering behaviour in the Guinier and the intermediate range of q
  . in the Guinier range :
    for a pure affine deformation (i) they obtain :

$$\overline{R}_{\!/\!/}^{2} = \lambda^2\, \overline{R}_{o}^{2} \quad \text{and} \quad \overline{R}_{\perp}^{2} = \overline{R}_{o}^{2}/\lambda$$

  . for "end to end pulling" process

$$\overline{R}_{\!/\!/}^{2} = \overline{R}_{o}^{2}\,(\frac{1+\lambda^2}{2}) \quad \text{and} \quad \overline{R}_{\perp}^{2} = R_{o}^{2}\,(\frac{\lambda+1}{2\lambda})$$

where $\overline{R}_{o}^{2}$, $\overline{R}_{\!/\!/}^{2}$, $\overline{R}_{\perp}^{2}$ are the radii of gyration at rest, in
the direction $/\!/$ and $\perp$ to stretching direction respectively.

  . in the intermediate range of q
    very differentiated behaviour are obtained depending on the
process of deformation considered as seen on Figure 18. For

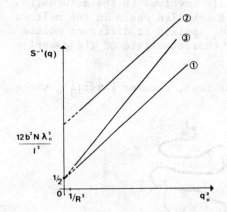

Fig. 18 : Representation of
$S^{-1}(q)$ as a function of q^2
(in the intermediate range
of q)
(1) undeformed chain
(2) chain deformed by "end to
    end pulling" process
(3) affine deformation of the
    chain
(from Ref. 52)

affine deformation one observe a change of slope of $S_{\!/\!/}^{-1}(q)$ proportional to λ^2 while for "end to end pulling" process the
slope is kept constant but the initial intercept is modified.

These two behaviors, characterized by strong differences in
the degree of affinity gives qualitative limits on the scattering functions that can be observed. The reality lies obviously
between these two extreme cases and the Figure 19 from the
work of Capelle (48) shows it clearly. This figure represent
reciprocal scattering intensity of labelled samples of PS
(M_w = 100,000) deformed at equivalent constant rate but at
different temperatures (110 and 115°C). The affinity is more
accentuated for deformations carried at low temperature as it
could be expected. In the paper of Picot et al. (48) and

Boué et al. (53) it is also shown that the actual deformational
changes tended to be intermediate between the two simplified
models resembling to affine deformation of segments if stretching took place close to the glass transition and following the
model of affine deformation of effective crosslinks for samples
oriented at elevated temperature. In this later case indeed,
the reduced orientation observed in the consequence of molecular
relaxation which occurs during stretching and quenching phases
which are not "sudden events".

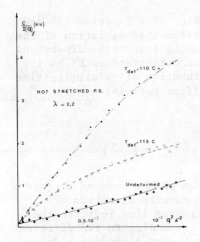

Fig. 19 : Variation of
$I^{-1}(q)$ as function of q^2
for hot stretched PS.
(from Ref. 48)

In order to investigate in detail this molecular relaxation
process Boué et al. (49) have studied in details the relaxation
of rapidly hot stretched PS samples after an extensional step
strain of a factor 3. The labelled polymer probe had a molecular
weight of 650,000. The relaxation of the scattering function
has been examined for times of relaxation ranging from 10sec
to 30 minutes. The temperature of deformation and subsequent
relaxation was above Tg and varied from 113°C to 134°C.

The results have been analysed in the frame of recents theories
(54,55,56) using the chain reptation model which gives a
description of the dynamical process of chain entanglement and
disentanglement. These theories have been largely developped by
using the tube   concept leading to a complete description of
the non linear relaxation of stress and molecular configuration
in the melt state (57,58,59).

The aim of the experiments of Boué et al. (49) experiments was
using time-temperature superposition to discuss the reptation
predictions for the form scattering factor proposed by Doi and
Edwards (60). On the base of tube model the molecular motion of
a single chain can be divided into two types : the small
wrilling motion of the chain in the tube (characterized by τeq)

which does not alter the topology of the entanglements and the large scale diffusive motion which changes the topology of the entanglements (characterized by τ_d). The time scale of these two motions are quite separated ($\tau_{eq} \sim M^2$, $\tau_d \sim M^3$).

The theory (60) shows that the interesting parameter to evidence these two characteristic times in a SANS experiment is the measure of the rleaxation of $R^2_\perp$ as function of time.

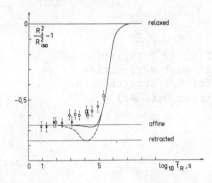

Fig. 20 : Experimental and calculated variation of the chain transverse dimensions of hot stretched PS as the function of relaxation time (from Ref. 49)

Figure 20 shows the representation of the reduced average square radius of gyration in the transverse direction at different relaxation time. The continuous line indicate the theoretical variations expected corresponding to two different ratios of τ_e/τ_{dis}). These theoretical curves predict a decrease of $R^2_\perp$ (corresponding to τ_e : retraction of the chain in the tube by wrilling motion) wich could not be evidenced experimentally.

This desagreement has been interpreted by these authors by the fact that the molecular weight of the polymer involved in the experiences was too low and consequently the retraction mode (τ_e) was masked by the diffusional motion (τ_d).

This type of experience shows the potential ability of SANS to provide informations about molecular motion in melt polymers.

4.2. Polymer Networks

SANS experiments provides an excellent opportunity to investigate dimensions and conformations of chains linked to a network as well at their deformation. These informations at the molecular scale are indeed of fundamental interest regarding the theories of rubber elasticity.

Most of the SANS experiments have been carried on model networks where the length of polymer chain between crosslinks can be controlled and is uniformly the same. This involves the preparation of monodisperse polymers and then end linking them. Two procedures of labelling have been adopted, one in which the

crosslinks are labelled (A) and another in which small percentage of the polymer chain in the network are labelled (B).

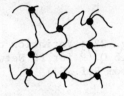　　　　　　　　　　　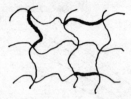

A Type Network   B Type Network

4.2.1. Networks with labelled crosslinks

SANS experiments carried out on this type of labelled networks should provide informations about the spatial correlations
between crosslinks.

Experiments on A type networks have been performed by
Duplessix (61) on Polystyrene networks obtained by anionic
copolymerization of Styrene and Divinyl-benzene.

The neutron scattering angular distribution exhibits a
single broad maximum. This maximum moves to lower angles as the
network is swollen as seen on Figure 21. From the angular
position of the maximum one can calculate a characteristic

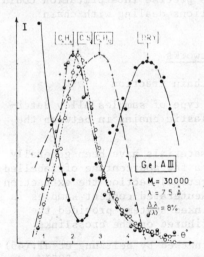

Fig. 21 : Neutron scatte-
ring intensity envelope
of type-A PS network at
different degrees of
swelling
(from Ref. 61)

distance h using simply a Bragg law which can be interpreted in
crude approximation as an average distance between crosslinks.
This average distance seems to follow the macroscopic deformation of the swollen network a shown on Figure 22 where h is
plotted against $Q^{1/3}$ (Q : swelling ratio). When type A networks

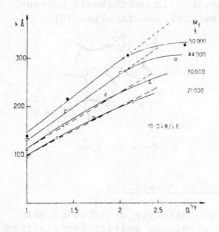

Fig. 22 : Variation of h
with $Q^{1/3}$ for PS Networks
(from Ref. 61)

are deformed uniaxially the peak position is shifted in opposite
directions when the intensity is analysed in the parallel and
transvers directions. A plot of h in these two directions shows
that the mean distance between crosslinks follows in first
approximation an affine deformation law.

The interpretation of this peak in terms of mean distances
has been very qualitative and more precise interpretation could
be certainly given by RPA calculations dealing with chain
labelled at their extremities.

4.2.2. Chain labelled networks

a) Influence of the labelled chain fraction

SANS experiments made on this type of samples allow determination of the conformation of elastic chains in between the
crosslinks.

Earliest experiments on such materials have been generally
carried out on samples containing a few pourcentage of labelled
chains. As mentionned in the theoretical section the extraction
of single chain signal on highly deuterated content samples is
still valid in the case of crosslinked system, provided the
labelled chains are randomly distributed on the crosslinks.

This point has been recently checked by Beltzung et al.(46)
on PDMS networks over a range of concentration up to 50%/W. as
well on dry networks as in networks swollen in the solvent of
appropriate coherent scattering length which leeds to cancela-
tion of intermolecular contributions.

Table IV gives the values of radii of gyration for mixtures of
PDMSH/PDMSD blends Mn = 10.500) and corresponding networks in the
dry state and swollen in cyclohexane.

ϕ_D	Rg(Å) Mixtures	Rg Å Networks	ϕ_D	Rg(Å) Mixture + cyclo	Rg(Å) swollen Netw.
0.020	40	36	0.08	48	58
0.077	41	38	0.20	51	62
0.193	41	39	0.50	50	58
0.493	41	38			

Table IV : PDMS Network   Swollen by cyclohexane
in the dry state $(c_{polM+D}=0.71/w)$

Identical values of radii of gyration are thus obtained for
each different system over all the range of labelled species
content ϕ_D investigated. This experimental check allows then the
use of high ϕ_D for network investigations. Similar results
have been obtained by Yu and Picot (62) working on polyisoprene
chain labelled networks. SANS measurements can be then performed
on networks samples with high deuterated chains content. The
single chain signal is obtained in the same manner than in the
melt, provided the labelled and non labelled chains are equivalent and randomly distributed on the crosslinks.

b) Conformation of the chains linked to a network

According Duplessix (61) experiments the chain remains
Gaussian in a polystyrene network. Figure 23 shows that the

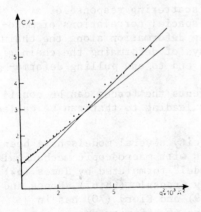

Fig. 23 : Scattering envelope
of type-B PS Network
(from Ref. 61)

scattering enveloppe of a labelled crosslinked chain fits very
well the Debye function.

The mean dimension are also quite comparable for the
crosslinked chain (Rg = 44 Å) and the chain in the bulk
(Rg = 45 Å). This result are likewise verified by the experiments of Beltzung (46) on PDMS (see Table IV) and by those of
Yu and Picot (62) on Polyisoprene (Rg = 24 Å in the network and
28 Å for the corresponding unperturbed dimension of the equivalent free chain).

One can then conclude that in the series of investigated
networks, the elastic chains are gaussian and their dimensions
comparable to the unperturbed dimensions of the equivalent
free chains. This result invalids the hypothesis of supercoiling
of the chains in dry networks often invoqued in the theories of
network swelling (63).

c) Networks deformation

Networks are very complicated topological objects composed
of linear polymer chains connected together by permanent
chemical crosslinks. The permanent crosslinks are not the only
constraints imposed to the chain when the sample is submitted to
external deformations. Entanglements acting as physical crosslinks can lead to suplementary constraints and have incidences
on the process of chain deformation.

In the last twenty years many theoretical calculations have
been devoted to the analysis of the problem of rubber elasticity
invoquing different molecular models. In order to try to
interpret SANS measurements on chain labelled networks it is
necessary to dispose of same simple models to analyse the
results. As mentioned in the previous section devoted to
stretched amorphous polymers, the scattering response of a
labelled chain will depend on the spacial correlations of monomers and on its modification during deformation along the chain.
We already mentioned two simple ways of deformaing the chain :
the pure affine deformationand the end to end pulling deformation.

In a network without entanglements the forces can be considered to act on chemical junctions leading to that can be called
a "junction affine deformation".

In the theory of rubber elasticity several models have been
developped to get better agreements with macroscopic mechanical
behaviour. The "phantom network model" formulated by James (64)
and James and Guth (65) extended by Graessley (66,67), Ronca and
Allegra (68), Deam and Edwards (69) and Flory (70) has in
particular allowed to introduce loss of affinity due to the
fluctuations of the crosslinks and topological constraints due
to the presence of neighbour chains. Some theoretical calculations of the scattering functions in the frame of this model
have been performed by several authors (Pearson (71), Edwards
and Warner (72), Ullman (73)). We will limit the discussion to

the comparison of some experimental results with these models.

•Uniaxial stretching

Let us at first consider results obtains for uniaxial stretching. The first experimental results have been obtained by C. Han (74,75) working on chain labelled polybutadiene networks. The results show that chain deformation process follow the "phantom network" prediction and is certainly not affine.

Such type of behavior have been recently confirmed by Clough and Macomachie (76) working on γ irradiated polystyrene networks stretched above Tg.

Recent experiments have been carried out by Beltzung et al. (46) on uniaxially stretched chain labelled networks.

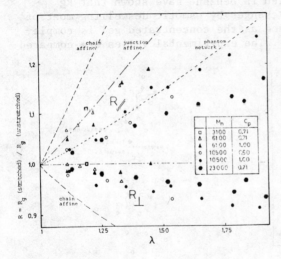

Fig. 24 : Molecular deformation for uni- axially stretched PDMS Networks. ∥ and ⊥ to stretching direction. (from Ref. 46)

Results are represented on Figure 24 and show clearly a depen- dence of chain deformation process on the molecular weight of the elastic chains and on the concentration during crosslinking process. To smaller molecular weights correspond higher affine deformation and the chain orientation is more pronounced when the crosslinking is achieved in bulk (Cp=1) than in concentrated solutions (Cp=0.6). This effects does certainly results from the entanglement density which is higher in the case of bulk crosslinking. For uniaxial stretching, the experimental results seems then to be scattered around the "phantom network model" exhibiting more pronounced affinity in the case of low molecular weight elastic chains and networks crosslinked in bulk and less affinity in the case of high molecular weight and crosslinking process achieved in solution.

In addition to the loss of affinity introduced the "phantom
network model" by fluctuations of the crosslkings around their
mean positions, another contribution to the loss of affinity
can be invoqued, the network unfolding which is more clearly
evidenced by the SANS obtained on swollen networks.

Swollen Networks

Experiments of Duplessix (61) on polystyrene chain labelled
networks have shown that the variation of R_g as function
of swelling degree falls well below the values predicted either
for affine deformation, end to "end pulling" or "phantom network"
models whatever the functionality of the network. In that kind
of experiments swelling was modified by using solvents of
different thermodynamical quality. More recents SANS measurements
carried out by Duplessix, Bastide and Picot (77), on the same type of networks highly swollen in benzene have shown that Rg
exhibits relatively small changes by osmotic deswelling, most of
the Rg decrease taking place as the concentrated gel is completely dried (see Figure 25). The experimental values are compared

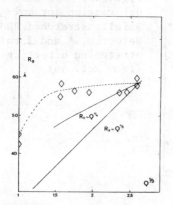

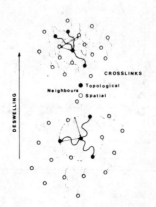

Fig.25 Osmotic deswelling of PS   Fig.26   Chain desinterspersion
type B Networks (Ref. 77)   process

to those one should obtain for affine deformation at the molecular level $Rg \sim Q^{1/3}$ and to the variation of Rg in equivalent
semi-dilute solution $Rg \sim Q^{1/4}$. In order to explain these
unexpected results, Bastide (78) has proposed that in the
network swelling appears an additional contribution to the loss
affinity which consists in a process of chain desinterspersion
in which the chain dimensions are not appreciably modified.

This process can be simply described on the scheme represented on Figure 26.

In the concentrated gel, the networks chains are highly
intertwined and the number of spatial neighbours crosslinks

(not directly connected by an elastic chains) is higher than
the topological neighbours (directly connected). Upon swelling
the network partially untolds but the mean dimensions of the
chains is not modified until the polymer is swollen at equilibrium in the pure solvent . In conclusion the chain desinterspertion contribution should be an additional parameter to introduce in the theories of network swelling, in addition to isotropic swelling of the coils resulting from excluded volume
effects and extension of the chain by network effect which are
generally only taken into account.

This desinterspertion contribution could be an explanation
to the loss of affinity observed in some cases for uniaxial
stretching of dry networks but in that case it must be notably
limited by the entanglement contribution.

5. CONCLUSION

In this lecture it is not possible to cover the entire
body of researchs done by SANS in the field on amorphous
polymers. The object was to focus us essentially on the field
of homogeneous polymers. Numerous promising works have been
undertaken these last years in the field of heterogeneous
systems like polymer blends and copolymers.

- In the domain of blends, the effort have been essentially
focused on the use of SANS for testing the compatibility at the
molecular level. In this frame, systems like PMMA-Poly α methyl
styrene have been investigated by Kruse et Kirste (79), Polysty-
rene-Poly α methylstyrene by Ballard (80), Polybutadiene-Poly-
styrene by Allen (81), PMMA-PVC by Jelenic (82), Polystyrene -
Polydimethylphenylene oxyde by Kambour (83) and Wignal (84),
PVC-Chlorinated PE by Walsh and Higgins (85) and recently PS-
Brominated PPO by Maconnachie (86). The measure of molecular
dimensions of labelled species and interaction parameters
deduced from A_2 allowed to appreciate the compatibility of these
systems.

- In the domain of copolymers SANS has been used by Richards
and Thomason (87) to determine the structure styrene-isoprene
copolymers. Due to the possibility of "playing" with the
contrast, the domains morphology, their size and the interfacial
layer thickness as well as the dimensions of the styrene blocks
have been determined .

- In a recent work of Hadziioannou, Skoulios and Picot (88)
the chain conformation in lamellar styrene-isoprene block copolymers oriented in large single crystals has been determined.
The mean size of the labelled PS blocks has been measured
around the normal to the interfaces. It has thus been shown that
the lateral interpenetration of the sequences is smaller than
than expected.

As final conclusion it can be assert that these last few years elastic SANS had a very important impact on polymer science especially in the field of amorphous polymers in bulk state. This technic has allowed to get information on molecular size and conformation. The use of this technic will certainly know these coming years important activity due to the developments of the theory and its very promising possibilities in the field of deformed polymers and heterogeneous polymer blends.

REFERENCES

1. Cotton J.P., B. Farnoux, G. Jannink, J. Mons and C. Picot, C.R.Acad.Sci., C275, 175 (1972)
2. Kirste R.G., Kruse W.A. and Schelten J., Makromol.Chem. 162 299 (1973)
3. Ballard D.G., Schelten J., Wignall G.D., Europ.Polym.J., 9, 965 (1973)
4. Cotton J.P. et al. Phys.Rev.Lett., 32, 109 (1974)
5. Marshall W. and Lovesey S.W., "Theory of Thermal Neutron Scattering" Clarendon Press, Oxford, 1971
6. Turchin V.E., "Slow Neutrons" Israel Program for Scientific Translations, Jerusalem, 1965
7. Bacon G.E., "Neutron Diffraction Clarendon Press , Oxford (1967)
8. Allen G. and Higgins J.S., Rep.Prog.Phys., 1973, 36, 1073
9. Maconnachie A. and Richards R.W., Polymer, 19, 739 (1978)
10. Ullman R., Ann.Rew.Mater.Sci., 10, 261 (1980)
11. Ibel K., J.Appl.Crystallogr., 9, 236 (1976)
12. ILL Neutron Beam Facilities, 1981
13. Schelten J., Hendricks R.B. J.App.Cryst., 11, 297 (1978)
14. Ghosh R., A Computing guide for Small Angle Scattering experiments ILL (Grenoble) 1978
15. Strazielle C. and Benoit H., Macromolecules, 8, 203 (1975)
16. Benoît H., Koberstein I. and Leibler L., Makromol.Chem.Suppl. 4, 85 (1981)
17. Leibler L. and Benoit H., Polymer, 22, 195 (1981)
18. Williams C. et al., J.Polym.Sci. (Letters Ed.) 17, 379 (1979)
19. Akcasu et al., J.Polym.Sci. (Polym.Phys.Ed.) 18, 863 (1980)
20. de Gennes P.G., J.Phys (Paris) 31, 235 (1970)
21. Legrand A.D. and Legrand D.G., Macromolecules, 12 (3) 450, (1979)
22. Leibler L., Macromolecules, 13, 1602 (1980)
23. Ionescu L. et al., J.Polym.Sci., (Phys.Ed.), 19, 1019 (1981)
24. Ionescu L. et al., J.Polylm.Sci. (Phys.Ed.) 19, 1033 (1981)
25. Higgins J.S. "Neutron Scattering in Material Science" Ed. G. Kostorz 1978
26. Richards R.W., Developments in Polymer Characterization (Ed. J.V. Dawkins) Applied Science, 1978
27. Flory P.J., 1953, Principles of Polymer Chemistry Ch. 14 Ithaca N.Y., Cornell Univ.Press

28. Kargin V., J.Pol.Sci., 30, 247 (1958)
29. Robertson R.E., J.Phys.Chem., 69, 1575 (1965)
30. Pechhold W., Kolloid Z.Z., Polym., 228, 1 (1968)
31. Yeh G.S., J.Macrom.Sci.Phys. B6, 451 (1972)
32. Benoît et al., Macromolecules, 6, 863 (1974)
33. Wignal G.D., Ballard D.G. and Schelten J., European Polym.
    Journal, 10, 861 (1974)
34. Kirste R.G., Kruse W.A., Ibel K., Polymer, 16, 120 (1975)
35. Kirste R.G., Lehnen B.R., Makromol.Chem., 177, 1137 (1976)
36. Lieser G., Fisher E.W., Ibel K., Polymer Letters Ed.13,39(1975)
37. Allen G. and Maconnachie, Unpublished results (1976)
38. Herchenroeder P. and Dettenmayer M. (Unpublished results 1977)
39. Fischer E.W. et al., Polymer Preprints,Am.Chem.Soc.Div.Polym.
    Chem., 15 (2), 8, (1974)
40. Yoon D.J. and Flory P.J., Polymer, 16, 645 (1975)
41. Koyama R., J.of Physical.Soc. of Japan, 36 (5) 1409 (1974)
42. Rawiso M. and C. Picot, (To be published)
43. Willenberg B., Makromol.Chem., 177, 3625 (1976)
44. Wignal G.D. et al., Polymer, 22, 886 (1981)
45. Boué F., Nierlich M., Leibler L. (To be published in Polymer)
46. Beltzung M. et al., IUPAC, Strasbourg 1981, p. 275
47. Picot C. et al., Macromolecules, 10, 436 (1977)
48. Picot C., Capelle, Froelich D. (Unpublished results)
49. Boué F., Nierlich M., G. Jannink, R. Ball (To appear in
    Journal de Physique)
50. Ferry J.D., Viscoelastic Properties of Polymers 2nd Ed.
    John Wiley and Sons, N.Y. 1970
51. Graessley W., Ad.Pol.Sci., 16 (1974)
52. Benoît et al. , Macromolecules, 8, 451 (1975)
53. Boué F., Jannink G., J.Phys., C2, 39, 183 (1978)
54. de Gennes P.G., J.Chem.Phys., 55, 572 (1971)
55. Edwards Proc.Phys.Soc., London, 9, 92 (1967)
56. Daoudi S., Journal de Physique (Paris), 38, 731 (1977)
57. de Gennes, Scaling Concepts in Polymer Physics, Ch. VIII,
    Cornell University Press (London) (1979)
58. Klein J., Macromolecules, 11, 852 (1978)
59. de Gennes, J.Pol.Sci., A2, 17, 1971 (1979)
60. Doi M., Edwards S.F., J.Chem.Soc., London, Faraday Trans.
    2 74, 1789, 1802, 1818 (1978)
61. Benoit H. et al., J.Pol.Sci.Phys., 14, 2119 (1978)
62. Yu M. and Picot C. (unpublished results)
63. Dusek K. and Prins W., Adv.Pol.Sci., 6, 1 (1969)
64. James H.M., J.Chem.Phys., 15, 651 (1947)
65. James H.M. and Guth E., J.Chem.Phys., 15, 669 (1947)
66. Graessley W., Macromol., 8, 186 (1975)
67. Graessley W., Macromol., 8, 865 (1975)
68. Ronca G., Allegra G., J.Chem.Phys., 63, 4990 (1975)
69. Deam R. and Edwards S.W.,Philos.Trans.R.Soc. London, Ser.
    A 280, 1296 (1976)
70. Flory P.J., Proc.R. Soc. London Ser. A 351,,(1976)

71. Pearson D.S., Macromolecules, 10(3), 701 (1977)
72. Warner M., Edwards S.F., J.Phys.A. Math.Gen. 11(8) 1649
                                                    (1978)
73. Ullman R., J.Chem.Phys., 71,(1), 436 (1979)
74. Hinkley J., Han C., Mozer B., Yu M., Macromolecules, 11,
                386 (1978)
75. Yu H., Han C. (Unpublished results)
76. Clough S.B., Maconnachie A. and Allen G., Macromolecules,
    13, 774 (1980)
77. Duplessix R., Bastide J., Picot C., (to appear in Macromo-
    lecules)
78. Bastide J. and Picot C., J.Macrom.Sci.Phys., B19(1), 13
                                                    (1981)
79. Kruse W.A. and Kirste R.G., Makromol.Chem., 177, 1145(1976)
80. Ballard D.G.H. et al., Polymer, 17, 640 (1976)
81. Allen G., Higgins J.P. and Yip W. (unpublished results)
82. Jelenic J., Kirste R.G. et al., Makromol.Chem., 80, 2057
                                                    (1979)
83. Kambour R.P. et al., Polymer communications, 21, 133 (1980)
84. Wignal G. et al., Polymer 21, 131 (1980)
85. Walsh D.J. et al., Polymer, 22, 168 (1981)
86. Maconnachie A. and Kambour R.P., IUPAC Strasbourg 1981,
                Reprints p. 1227
87. Richards R.W. and Thomasson J.L., Polymer, 22, 581 (1981)
88. Hadziioannou G. et al., Macromolecules (to appear Feb. 1982)

THE MECHANICAL PROPERTIES AND STRUCTURE OF ORIENTED POLYMERS

I.M. Ward

Department of Physics, University of Leeds,
Leeds LS2 9JT, UK.

First considerations of the mechanical properties of
oriented polymers require an understanding of the definition of
anisotropic behaviour at a macroscopic level.   This is obtained
by a generalization of Hooke's law to anisotropic materials.
From this starting point, experimental measurements can be
undertaken by a wide range of techniques from dead-loading creep
to dynamic mechanical and ultrasonic measurements.

The interpretation of mechanical anisotropy requires
consideration of the molecular structure in the first instance.
This leads to examination of crystal moduli, either from
theoretical calculations or experimental data.   Next, there is
the influence of orientation and morphology, both of which may
be important in particular systems.   Attempts to undertake
quantitative analyses of mechanical anisotropy have two simple
starting points.   These are the aggregate model, which deals
with molecular orientation and the Takayanagi model which
recognises the composite nature of crystalline polymers.
Developments based on these two approaches will be discussed.

The anisotropic mechanical behaviour also reflects
relaxation processes, which are related to orientation and
morphology.   Some definitive situations will be considered,
including both reversible relaxation behaviour and plastic flow.

INTRODUCTION

A valid starting point for studies of the mechanical
properties of oriented polymers is the recognition that at low
strains the behaviour of an anisotropic elastic solid can be

173

R. A. Pethrick and R. W. Richards (eds.), Static and Dynamic Properties of the Polymeric Solid State, 173–195.
Copyright © 1982 by D. Reidel Publishing Company.

represented by the generalised Hooke's law

$$\varepsilon_{ij} = S_{ijk\ell} \, \sigma_{k\ell}$$

$$\sigma_{ij} = C_{ijk\ell} \, \varepsilon_{k\ell}$$

where ε_{ij}, $\sigma_{k\ell}$ define the components of strain and stress respectively, $S_{ijk\ell}$ and $C_{ijk\ell}$ are the compliance and stiffness constants, respectively.

It is customary to adopt an abbreviated nomenclature

$$e_i = S_{ij} \sigma_j$$

$$\sigma_i = C_{ij} e_j$$

where e_i $(i = 1, \ldots .6)$ define the six independent engineering strain components

σ_j $(j = 1, \ldots .6)$ define the six independent stress components

S_{ij}, C_{ij} are the compliance and stiffness constants respectively.

From a practical viewpoint, it is easier to envisage the compliance constants because these define the strains developed in different directions for application of a simple stress e.g. a single normal stress or shear stress and hence relate to well-known Young's moduli, shear moduli, etc.

e.g. the Young's moduli $E_1 = \dfrac{1}{S_{11}}$, $E_2 = \dfrac{1}{S_{22}}$ etc

shear or torsional moduli $G_1 = \dfrac{1}{S_{44}}$, $G_2 = \dfrac{1}{S_{55}}$, etc

and Poisson's ratios $\nu_{13} = \dfrac{-S_{13}}{S_{33}}$, $\nu_{23} = \dfrac{\nu_{23}}{S_{33}}$,etc

For a solid with at least three orthogonal planes of symmetry such as a polymer film there are only nine independent elastic constants, and choosing the 1,2,3 axes to coincide with the axes of symmetry the compliance matrix is given by

$$
\begin{pmatrix}
S_{11} & S_{12} & S_{13} & 0 & 0 & 0 \\
S_{12} & S_{22} & S_{23} & 0 & 0 & 0 \\
S_{13} & S_{23} & S_{33} & 0 & 0 & 0 \\
0 & 0 & 0 & S_{44} & 0 & 0 \\
0 & 0 & 0 & 0 & S_{55} & 0 \\
0 & 0 & 0 & 0 & 0 & S_{66}
\end{pmatrix}
$$

The compliance constants for this system are shown in Figure 1.

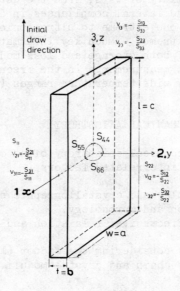

Figure 1.   The compliance constants for a polymer film

For a fibre or oriented film with fibre symmetry there are only five independent elastic constants, and choosing 3 as the fibre axis we have

$$
S_{23} = S_{13}, \quad S_{55} = S_{44} \text{ and } S_{66} = 2(S_{11} - S_{12})
$$

More comprehensive reviews of this aspect of mechanical anisotropy are given elsewhere (1,2,3,4).

Although oriented polymers are non-linear viscoelastic materials, it is a useful approximation to assume that they are linearly viscoelastic. The compliance constants for an elastic solid are then replaced by time dependent creep compliances and the stiffness constants by the corresponding stress relaxation moduli. An alternative equally acceptable representation is to express the results in terms of complex compliances or moduli usually obtained directly from dynamic mechanical measurements.

MEASUREMENT OF MECHANICAL ANISOTROPY

The mechanical behaviour of oriented polymers has been studied in a wide variety of ways, from the determination of extensional, shear and lateral compliances in dead loading creep experiments (5,6) at one extreme to ultra-sonic measurements (7,8) at the other extreme. Much key information has come from determination of both storage and loss moduli in conventional dynamic mechanical measurements in the frequency range $1-10^3$ Hz over comparatively wide temperature ranges (9,10).

INTERPRETATION OF MECHANICAL ANISOTROPY

The observed mechanical anisotropy of an oriented polymer will depend on three interrelated factors

(1) The molecular and crystallographic structure
(2) Orientation and morphology.
(3) Dynamic effects i.e. relaxation and creep.

We will start by considering the factor (1), which leads us to a discussion of chain and crystal moduli.

CHAIN AND CRYSTAL MODULI

The estimation of chain moduli was first made by theoretical calculations (11) for cellulose and, not surprisingly for a linear covalently bonded chain, the values obtained were high ~ 100 GPa, comparable to inorganic materials such as glass or aluminium. Such calculations were later undertaken by Lyons (12) and Treloar (13), and were recently for the polyaramids such as Kevlar and Nomex by Tadokoro and co-workers (14). It is of particular interest that the theoretical chain moduli are so high for many polymers, including polyethylene, polyethylene terephthalate and nylon 6 as well as the polyaramids, and, of course, in several cases this has now been realised in practice.

There have been rather few estimates of the other elastic constants. A notable paper by Odajima and Maeda (15) followed the Born method familiar to students (where it is applied to sodium chloride) to determine the complete matrix of elastic constants for polyethylene.   These are shown in Table 1.

$$\begin{pmatrix} 14.5 & -4.78 & -0.019 & 0 & 0 & 0 \\ -4.78 & 11.7 & -0.062 & 0 & 0 & 0 \\ -0.019 & -0.062 & 0.317 & 0 & 0 & 0 \\ 0 & 0 & 0 & 31.4 & 0 & 0 \\ 0 & 0 & 0 & 0 & 61.7 & 0 \\ 0 & 0 & 0 & 0 & 0 & 27.6 \end{pmatrix} (GPa)^{-1} \times 100$$

Table 1. Crystal compliance matrix for polyethylene

Again not surprisingly, all the constants, other than S_{33} and C_{33} corresponding to the chain modulus, indicate comparatively low stiffnesses in shear and in lateral deformation. Comparison with results for highly drawn ultra high modulus polyethylene shown in Table 2 indicate that this expectation, which reflects

ELASTIC CONSTANTS OF HIGH MODULUS POLYETHYLENE

	20°C	-196°C	Theoretical (Odajima & Maeda)
AXIAL MODULUS	70 GPa	160 GPa	250 GPa
TRANSVERSE MODULUS	1.3 GPa	-	5.6 GPa
SHEAR MODULUS	1.3 GPa	1.95 GPa	1.8 GPa
POISSON'S RATIO	0.4	-	0.5

Table 2. Elastic constants of high modulus polyethylene

the much weaker intermolecular forces between chains, is borne out by experimental data.

The crystal elastic constants can also be determined experimentally by three techniques.

(1)   The measurement of crystal strains by X-ray diffraction
(2)   The measurement of the chain modulus by low frequency Raman spectroscopy
(3)   The measurements of extensional and lateral stiffnesses by inelastic neutron scattering.

X-ray measurements of crystal moduli from the distortion of the crystal lattice under strain were first undertaken by Sakurada and his coworkers (16) and by Dulmage and Contois (17). The results have been evaluated on the basis that the stress applied to the crystalline regions is the macroscopic stress i.e. applied load divided by the cross-sectional area of the sample. This assumption of homogeneous stress has been called into question by recent work of Jakeways, Ward and co-workers (18). They have shown that there can be considerable increases in apparent crystal modulus as the temperature is reduced and the non-crystalline regions take more stress (Figure 2). X-ray crystal moduli

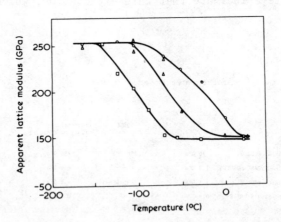

Figure 2.   The temperature dependence of the crystal moduli in polyethylene determined from X-ray crystal strain measurements.

should therefore be regarded as lower bounds to the true crystal moduli. Nevertheless, measured chain moduli are in the regions anticipated, being very high for extended molecular chains (Table 3 shows some selected results) and much lower for helical

Polymer	Crystal modulus (GPa)
Poly(p-phenylene terephthalamide)	153
Poly(p-benzamide)	182
Nomex	88
Polyethylene	255
Polyoxymethylene	53-57
Polypropylene	41-47

Table 3.   Experimental values for chain modulus from
           X-ray strain measurements.

chains such as polypropylene. The lateral stiffnesses are found
to be much lower, which is expected, as we have discussed above.

Crystal chain moduli can also be obtained from measurement
of the frequency of the longitudinal acoustic mode vibration by
Raman spectroscopy (19).

$$\nu_L = \frac{m}{2Lc} \sqrt{\frac{E}{\rho}}$$

where E is the chain modulus, ρ density
      L length of vibrating chain, c velocity of light
      m = mode number (1,3,5....)

The problem with this method is that the length of the
vibrating chain may not only be difficult to measure but that
there may be coupling between the chain in the crystal and the
non-crystalline regions or end groups so that L does not
correspond to the crystal thickness. Although the values
determined by this technique may therefore represent another
bound to the true values, they are also in the anticipated range.

Finally, chain moduli and lateral moduli can be determined
by inelastic neutron scattering (20). In this case the chain
length is much longer than the neutron wavelength so the dispersion curve for the longitudinal acoustic mode can be obtained
from the frequency shift as a function of neutron wavelength.
Somewhat higher values for the chain modulus have been obtained
by this technique compared with those obtained from X-ray crystal
strain and Raman spectroscopy data.

ORIENTATION AND MORPHOLOGY

There have been two very different starting points for

understanding the mechanical behaviour of polymers, both of which
can form the basis for discussion and further development.

(1) The aggregate model (21), which considers the partially
     oriented polymer as an aggregate of anisotropic units
     which are aligned by orientation processes such as
     drawing or rolling. The mechanical anisotropy then
     relates to the orientation of these units and their
     intrinsic nature.

(2) The Takayanagi model (9), which considers the
     oriented polymer to be a simple composite of completely
     aligned crystalline regions and isotropic amorphous
     regions. This leads to more sophisticated modelling
     in terms of the theory of composite materials.

We will consider these models and their developments in turn.

(1)   The Aggregate Model

     In its simplest form the aggregate model is a single phase
model and is therefore likely, a priori, to be most useful for
amorphous polymers. This expectation is realised in practice
and the aggregate model has been applied successfully to explain
mechanical anisotropy in polymethylmethacrylate (PMMA) and other
amorphous polymers.   It also has been shown to provide a useful basis for understanding mechanical anisotropy in polyethylene
terephthalate (PET) of low comparatively low crystallinity and
more surprisingly, low density polyethylene, for rather special
reasons, which will be explained.

     In its simplest form the aggregate model is applied to
systems showing fibre symmetry, and the anisotropic elastic units
are also considered to fibre symmetry.   The orientation of an
individual unit is then defined by the angle θ which its
symmetry axis makes with the fibre axis (or draw direction in
uniaxially oriented film).   The elastic properties of the
oriented polymer are then calculated by averaging procedures,
either averaging the compliance constants of the aggregate
(which assumes homogeneous stress) or the stiffness constants
(which assumes continuity of strain).   These represent lower
and upper bounds for the elastic constants of the polymer
(commonly termed the Reuss and Voigt bounds).

     Consider the compliance of an elastic unit with fibre
symmetry in a direction making an angle θ with the fibre axis.
This is given by

$$S_\theta = \cos^4\theta S_{33} + \sin^4\theta S_{11} + \sin^2\theta\cos^2\theta(2S_{13}+S_{44})\ldots \qquad (1)$$

where S_{33}, S_{11} etc are the compliance constants of the unit. For an aggregate of units the Reuss average is then

$$\bar{S}_{33} = \overline{\cos^4\theta}S_{33} + \overline{\sin^4\theta}S_{11} + \overline{\sin^2\theta\cos^2\theta}(2S_{13} + S_{44}) \qquad (2)$$

where $\overline{\cos^4\theta}$, $\overline{\sin^4\theta}$ $\overline{\sin^2\theta}$ $\overline{\cos^2\theta}$ are orientation functions which are the average values of $\cos^4\theta$, etc over the aggregate of units.
Similarly the Voigt average is given by

$$\bar{C}_{33} = \overline{\cos^4\theta}\,C_{33} + \overline{\sin^4\theta}\,C_{11} + \overline{2\sin^2\theta\cos^2\theta}\,(C_{13} + 2C_{44}) \qquad (3)$$

These are similar equations for the other elastic constants, and a full account is given in references 2 and 21.

To apply the aggregate model the orientation functions for the units are required. In the first instance, values of $\overline{\cos^2\theta}$ were estimated from birefringence data, and $\overline{\cos^4\theta}$ predicted on the basis of deformation schemes (either the affine rubber-like network deformation scheme or the pseudo-affine floating rod deformation scheme). The uncertainties of such procedures stimulated the attempt to develop methods for obtaining $\overline{\cos^4\theta}$, $\overline{\sin^4\theta}$ directly which has been done by broad line NMR, Raman spectroscopy, and polarized fluorescence spectroscopy (4,22). In PMMA and PET the measured mechanical moduli lay between the Reuss and Voigt bounds (Figure 3 and Table 4). In both cases the

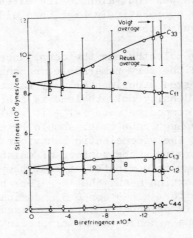

Figure 3. Comparison of measured elastic constants for polymethylmethacrylate with Reuss and Voigt averages calculated from the aggregate model.

**Comparison of Reuss and Voigt predicted values for extensional
moduli (in units of 10^{10} dyn cm^{-2}) with measured values for range of
natural draw ratios (NDR) and initial birefringences Δn_e**

NDR	$\Delta n_e \cdot 10^3$	E_R	E_V	$\frac{1}{2}(E_R + E_V)$	E_{33} (measured)	$\frac{2E_{33}}{E_R + E_V}$
1·25	67·8	1·81	6·76	4·29	4·94	1·15
1·59	46·0	1·93	7·58	4·76	5·35	1·12
1·92	28·0	1·98	7·94	4·96	5·58	1·12
2·36	13·07	2·14	8·48	5·31	6·28	1·18
2·52	9·52	2·22	8·62	5·42	6·09	1·12
2·64	5·43	2·24	8·70	5·47	6·27	1·15
3·19	3·52	2·68	9·62	6·15	7·02	1·14
3·99	1·37	3·31	10·55	6·93	7·85	1·13

Table 4.   Comparison of Reuss and Voigt predicted values
for extensional moduli (in units of 10^{10} dyn cm^{-2})
with measured values for range of natural draw
ratios (NDR) and initial birefringences Δn_e

In both cases the elastic constants of the unit were estimated
empirically by extrapolation to a high draw ratio or high
birefringence, which was equivalent. The extensional compliance
S_{33} of the unit was comparatively high, and did not correspond
to that for an extended chain.

In low density polyethylene the compliance averaging
procedure predicted the very remarkable mechanical anisotropy
which had been observed experimentally by Raumann and Saunders
(23). Because low density polyethylene is a highly crystalline
polymer - indeed a composite material - it would not be expected
that a single phase aggregate model should be appropriate.
Detailed studies of the viscoelastic behaviour (10) showed that
this simple model predicts the pattern of anisotropy correctly
because the behaviour is dominated at room temperature by the
c-shear relaxation, which means that S_{44} for the "unit" is very
large and hence the term $\sin^2\theta \cos^2\theta \; S_{44}$ dominates the behaviour.
The changes in c-axis orientation are expressed by changes in the
orientation functions - hence the model predicts the anisotropy
(Figure 4).

A recent application of the aggregate model, which is valid
in structural terms, has been to the influence of molecular
orientation on the Young's modulus of Kevlar. Because these

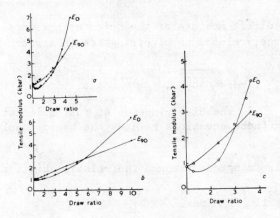

Figure 4. Mechanical anisotropy of low density polyethylene
(a) Measured Young's moduli E_0 and E_{90}, parallel
and perpendicular to the draw direction
respectively
(b) Predictions of aggregate model with the
pseudo-affine deformation scheme
(c) Aggregate model predictions with orientation
functions measured by broad line NMR

materials are essentially single phase it is justified to
correlate the modulus with the X-ray orientation functions, and
this has been done successfully quantitatively by Northolt and
Van Aartsen (24).

The extension of the aggregate model to crystalline polymers,
irrespective of their possible two phase nature, was essentially
made by Charch and Moseley (25) who proposed the use of sonic
modulus as a measure of orientation in polymers. As pointed out
many years ago (26) this follows from the aggregate model by
making several assumptions.   For comparatively low degrees of
orientation because S_{33} is small the term $\overline{\cos^4\theta}\, S_{33}$ is also
compared with the other terms in the compliance average equation.
We therefore have

$$S_{33} = \overline{\sin^4\theta}\, S_{11} + \overline{\sin^2\theta\cos^2\theta}\,(2S_{13} + S_{44}) \qquad (4)$$

S_{13} is also small, and if we assume $S_{44} = S_{11}$ it follows that

$$\overline{S_{33}} = \overline{(\sin^4\theta + \sin^2\theta \cos^2\theta)S_{11}} = \overline{\sin^2\theta} \; S_{11} \qquad (5)$$

Thus S_{33} is a direct measure of the orientation function $\overline{\sin^2\theta}$ which, of course, relates to birefringence Δn through the well-known formula

$$\Delta n = \Delta n_{max} \; (1 - 3/2 \; \overline{\sin^2\theta}), \qquad (6)$$

where Δn_{max} is the birefringence at full orientation. Combining these equations relates the sonic modulus $E = \dfrac{1}{\overline{S}_{33}}$

to the birefringence through the relationship

$$\overline{S_{33}} = \frac{2}{3} \; S_{11} \; (\Delta n - \Delta n_{max}) \qquad (7)$$

Although these simplifications do lead to an explanation of the observed correlations between modulus and birefringence (e.g. Figure 5), the basis is not very sound and the value

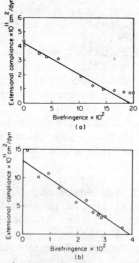

Figure 5.   Correlation between extensional compliance
             and birefringence for polyethylene
             terephthalate fibres

obtained by extrapolation to zero birefringence does not correspond to S_{11}, the lateral compliance.   The major point of the aggregate model, as we have seen, is that shear cannot be neglected, and we therefore cannot equate S_{11} and S_{44}.   With some reservations, therefore, we may proceed to Samuels'

development (27) of Charch and Moseley's approach. Samuels
assumes a two phase model, and assuming homogeneous stress i.e.
equivalent to the compliance averaging scheme with the Charch
and Moseley approximations, we have

$$\frac{1}{E} = \frac{\beta}{E^o_{t,c}} \quad (1-\overline{\cos^2\theta}_c) + \frac{1-\beta}{E^o_{t,am}} \quad (1-\overline{\cos^2\theta}_{am}) \tag{8}$$

where $E^o_{t,c}$, $E^o_{t,am}$ are the lateral moduli of the crystalline and
amorphous regions respectively; $\overline{\cos^2\theta}_c$ and $\overline{\cos^2\theta}_{am}$ are the
orientation averages for the crystalline and amorphous regions
respectively and β is the fraction of crystalline material.

For the isotropic polymer, $\overline{\cos^2\theta}_c = \overline{\cos^2\theta}_{am} = \frac{1}{3}$ and the
isotropic sonic modulus E_u is given by

$$\frac{3}{2E_u} = \frac{\beta}{E^o_{t,c}} + \frac{1-\beta}{E^o_{t,am}} \tag{9}$$

Samuels' procedure is to take isotropic samples of different
crystallinity and hence determine $E^o_{t,c}$ and $E^o_{t,am}$.

If orientation averages for the crystalline and amorphous
regions are defined as $f_c = \frac{1}{2}(\overline{3\cos^2\theta}_c - 1)$, $f_{am} = \frac{1}{2}(\overline{3\cos^2\theta}_{am} - 1)$
we have

$$\frac{3}{2}\left\{\frac{1}{E_u} - \frac{1}{E}\right\} = \frac{\beta f_c}{E^o_{t,c}} + \frac{(1-\beta)f_{am}}{E^o_{t,am}} \tag{10}$$

Then equation (10) can be used to determine f_{am}, provided that f_c
is measured by X-ray diffractometry. Samuels (Figure 6) showed
internal consistency for his scheme in uniaxially oriented polypropylene. The use of sonic modulus to determine amorphous
orientation does, however, require acceptance of the validity of
all the assumptions, which as we have seen require detailed
justification. Although in the case of polypropylene this
simplified theory may well be a good approximation, it does not
justify the use of sonic modulus to determine amorphous orientation for polymers in general.

(2)   The Takayanagi Model and Composite Models

In one sense, by discussing Samuels' extension of the
aggregate model, we have anticipated composite models and the
Takayanagi model.   Takayanagi, however, was the first to attempt

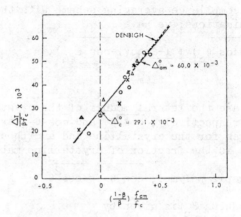

Figure 6. Correlation between $\dfrac{\Delta n}{\beta f_c}$ and $\left(\dfrac{1-\beta}{\beta}\right)\dfrac{f_{am}}{f_c}$ (after) (Samuels)

to deal with the composite crystalline and amorphous nature of polymers. In a highly oriented annealed polymer such as linear polyethylene or polypropylene the crystalline regions are close to full alignment in the draw direction or fibre axis, and the amorphous regions are relaxed. Often these polymers show a two-point small angle pattern which has been interpreted as indicating that the crystal blocks alternate with the amorphous regions in a regular repeating pattern, which I shall call a parallel lamellar structure.

The simple Takayanagi model is shown in Figure 7. It very neatly explains the observed cross over in extensional and transverse moduli at high temperatures, which Takayanagi observed for annealed drawn linear polyethylene (Figure 8).

The Takayanagi model, however is essentially one-dimensional and it succeeds in a sense because of this. If we think again about the response of a sequence of parallel lamellar stacks, we might conclude that even if the amorphous regions soften, the stiff crystalline sheets might constrain the structure (the Poisson's ratio effect) so that the amorphous regions could not deform under the application of a stress parallel to the lamellar normals. This leads to the recognition of the importance of shear deformations, and in particular interlamellar shear as a key deformation mechanism, and hence the orientation of the

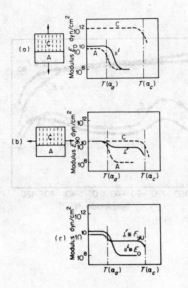

Figure 7.   The simple representation of a change of
            modulus E with temperature on the Takayanagi
            model for (a) the parallel and (b) the
            perpendicular situations corresponding to E_0
            and E_{90} respectively, C crystalline phase, A
            amorphous phase. Diagram (c) shows combined
            results.

lamellae as a controlling factor.   The Takayanagi model can
work for a parallel lamellae sheet because the lamellae are not
infinite in two dimensions, they are like planks rather than
plates so that the interlamellar material can deform in pure
shear when an axial stress is applied.

    The importance of interlamellar shear was shown by the
examination of the dynamic mechanical behaviour of specially
oriented sheets of low density polyethylene.   These studies
also provided a link with the previous aggregate modelling of
this polymer where the importance of chain axis orientation and
shear parallel to the chain axis direction, the c-shear mechanism

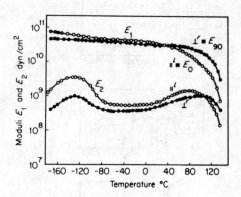

Figure 8.   Temperature dependence of the components of
the dynamic modulus E' and E" for the direction
parallel (||) and perpendicular (⊥) to the
draw direction for annealed samples of high
density polyethylene (after Takayanagi et al).

were the dominant factors.   By suitable drawing and hot rolling
techniques it is possible to make three key simple structures
(Figure 9).   One of these has been called 'b-c sheet', where
the c axis is in the initial draw direction and the b-axis in
the plane of sheet perpendicular to this direction.   The lamellar
texture shows lamellar planes inclined at about 45° to the draw
direction, the lamellae being arranged like the roof-tops of a
house.   A second structure is 'a-b sheet', where the b axis
remains in the plane of the sheet but the a axis is now in the
initial draw direction i.e. also in the plane of the sheet.
This sheet also shows a roof-top lamellar structure.   Finally
there is the parallel lamellae texture where the lamellar plane
normals are parallel to the initial draw direction.   The b-axes
remain in the plane of the sheet but the c-axes are now inclined
at about 45° to the initial draw direction.

The dynamic mechanical behaviour of these sheets reflects
both the lamellar orientation and the c-axis orientation (10).
The so-called β relaxation ca 0°C shows the anisotropy expected
for interlamellar shear i.e. maximum loss when the tensile stress
is applied at 45° to the lamellar planes.   The α - relaxation,
on the other hand, corresponds to c-shear with maximum loss when

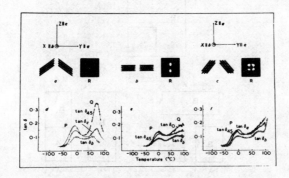

Figure 9.   Schematic structure diagrams and mechanical
             loss spectra a and d: for b-c sheet; b and e:
             for parallel lamellae sheet; c and f: for a-b
             sheet.
             P: interlamellar shear process;
             Q: c-shear process
             R: small angle X-ray pattern

the tensile stress is applied at 45° to the c-axis direction.
The magnitude of the losses can be shown to relate to $\overline{\cos^2\gamma\sin^2\gamma}$
where γ is the angle between the applied tensile stress and a
lamellar plane normal (28).   This assumed simple shear between
lamellae.   As we have seen, to interpret the data for the
parallel lamellae sheet,we must also invoke pure shear (29).   It
is then possible to show quantitative equivalence between these
different sheets at temperatures where interlamellar shear is the
predominant mechanism of deformation.   Further analysis of the
parallel lamellae sheet has included examination of the Poisson's
ratios (30) which confirms that pure shear occurs, but also
suggests the importance of the c-shear process in permitting
substantial interlamellar shear.

     Because the orientation of the lamellae has been the key
factor in the low density polyethylene sheets, there has been no
discussion of detailed morphology and in particular whether there
is not some degree of crystalline continuity or amorphous
continuity i.e. whether one or other of the phases may not be
continuous to some extent.   Takayanagi recognised this and
elaborated his simple model to permit continuity of one of the
phases, as shown in Figure 10, where the Series-Parallel and

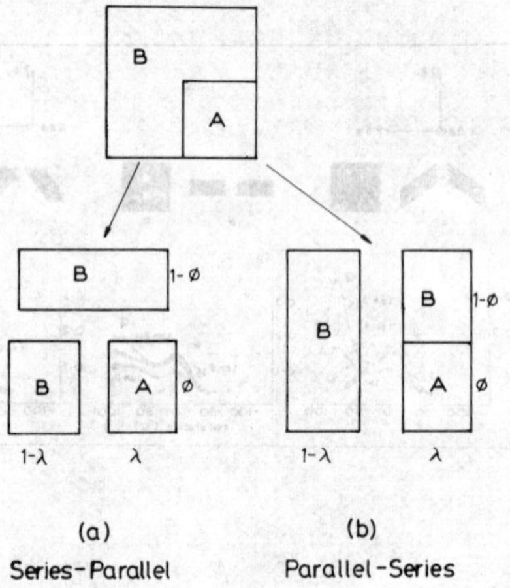

(a)
Series-Parallel

(b)
Parallel-Series

Figure 10.   The Series-Parallel and Parallel Series
             Takayanagi models.

Parallel-Series models are shown.   In stress-crystallised
polyethylenes (31) and in oriented nylons (32,33) application of
these models indicates that there must be some degree of amorphous
continuity i.e. that the lamellae are not infinite in lateral
extent.   The oriented nylons are interesting in that again
specially oriented sheets can be prepared with the hydrogen bonds
either in the plane of the sheet (α - nylon sheets) or inclined
at an angle (γ - nylon sheets).   The extensional moduli of these
sheets in the wet and dry states can be most simply analysed on
the basis of Takayanagi modelling.   The shear moduli suggests
chain slip (in this case b-shear) as a predominant process.
Differences between the anisotropy of two types of sheet can be
attributed partly to the different arrangement of hydrogen bonded
sheets and partly to the less regular chain in the γ nylon which
makes b-shear a rather more difficult process.

When we come to ultra-high modulus polyethylenes a Takayanagi
model can again provide the first moves in understanding, but in
this case we require some degree of continuity of the crystalline
phase.   Gibson, Davies and Ward (34) have proposed that the
crystal blocks become increasingly linked with intercrystalline
bridges as the draw ratio is increased.   On the assumption that

these bridges are randomly arranged their concentration is defined
by a single parameter p, the probability of linking two adjacent
crystal blocks.   The probability of linking blocks is then
$P_n = p^{n-1}(1-p)$ from which it can be shown that the volume fraction
of continuous phase (i.e. linking two or more adjacent blocks) is
given by $v_f = \chi p(2-p)$ and on the Series-Parallel assumption the
modulus is given by

$$E = \chi p(2-p)E_c + \frac{\{(1-\chi) + \chi(1-p)^2\}E_a}{(1-\chi) + \chi(1-p)^2 E_a/E_c}$$

On the Takayanagi model $E_a \to 0$ above the γ relaxation so
that at $-50^\circ C$, there is a plateau in modulus where

$$E_{-50} = \chi p(2-p)E_c$$

providing a simple test of the validity of this approach.
In Figure 11 results are shown for high modulus polyethylenes,

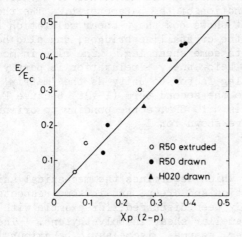

Figure 11. E/E_c as a function of $\chi p(2-p)$ for
             linear polyethylenes

the values of p being obtained from the integral breadth of the (002) reflection.   There is a good correlation, but the theory can be made more sophisticated by taking up the theory of short fibre composites.   It is assumed that the crystal sequences which link two or more adjacent lamellae act like reinforcing fibres in a matrix consisting of the remaining lamellar material and amorphous material.   The effectiveness of the crystalline bridge sequences is then limited by shear lag as described by the Cox model for a short fibre composite.   This makes possible a comprehensive approach which embraces both the extensional and shear deformation behaviour.   It is also more satisfying in physical terms than the Takayanagi model, because we recognise that polymers basically deform by shear mechanisms.   In broad outline we have

$$E = \chi p(2-p) \; \Phi' \; E_c + (1-\chi) \; E_m$$

where Φ' is the average shear lag factor for the assembly of crystalline bridge sequences and the second term $(1-\chi)E_m$ is just identical to the second term in the original Takayanagi model formulation.   Gibson, Jawad, Davies and Ward (35) have developed a theory along these lines, extending the Cox theory to deal with viscoelastic behaviour of the matrix.   The temperature dependence of the storage modulus and the magnitude of the loss modulus are predicted as a function of the parameter p.   The theory is instructive in that it shows how the c-shear relaxation reduces the effectiveness of the crystalline bridges, and also how the γ - relaxation leads to some shear lag.   The rise in modulus as the temperature falls below the γ - relaxation is partly due to increasing the shear lag factor to unity as the 'matrix' stiffens and partly due to the second term $(1-\chi)E_m$ because E_a is now significant, and values ~ 10 GPa, corresponding to oriented amorphous material, are shown for E_a.

CREEP BEHAVIOUR

The third factor which determines the mechanical anisotropy of polymers is relaxation and creep, and in some respects this has, of course, already been discussed, for example with regard to c-shear and interlamellar shear in polyethylene.   The discussion has, however, assumed viscoelastic behaviour i.e. essentially reversible deformation.   It is also possible that permanent deformation i.e. flow or plasticity occurs and such processes can be extremely important in practical applications. The area of non-linear viscoelasticity and plasticity in polymers is another complex topic and it is only possible here to illustrate the connection between viscoelasticity and plasticity in polymers by citing an important example, that of ultra high modulus polyethylene.

Although the dynamic modulus at $-50^{\circ}C$, the plateau modulus, depends only on draw ratio, the creep behaviour and hence creep moduli of oriented polyethylenes depend on other factors, especially molecular weight and branch content. This has been illustrated in several publications (36,37,38). In all cases at high stresses permanent flow is observed, which we believe corresponds to the c slip process. The measured activation volume of about $50\overset{\circ}{A}^{3}$ could well correspond to the movement of a Reneker defect through the crystal lattice. Coupled in parallel with this process is a network deformation process which is vital to providing a mechanism for recovery at low stresses. The process relates to molecular weight, irradiation cross-linking and branch content. Such considerations add further dimensions to the understanding of mechanical anisotropy in polymers, and show that there is a molecular as well as a morphological aspect to this area.

ACKNOWLEDGEMENTS

Grateful acknowledgement is made to the following publishers and Societies for permission to reproduce illustrations.

Book or Journal	Publisher	Figure
Polymer	IPC Business Press Ltd.	2,3
I.M.Ward, Mechanical Properties of Solid Polymers	Wiley Interscience	5,7,8
Physics Bulletin	Institute of Physics	9

REFERENCES

1. Nye, J.F.: 1957, 'Physical Properties of Crystals'
   Clarendon Press, Oxford.
2. Ward, I.M.: 1971, 'Mechanical Properties of Solid Polymers'
   Wiley, London.
3. Hadley, D.W., and Ward, I.M.: 1973, Rep. Prog. Physics
   38, 1143
4. Ward, I.M.: 'Structure and Properties of Oriented Polymers'
   1975, Applied Science Publishers, London
5. Darlington, M.W., and Saunders, D.W.: 1975, 'Structure and
   Properties of Oriented Polymers' Chapter 10 Ed. I.M.Ward,
   Applied Science Publishers, London
6. Ward, I.M.: 1977, Plastics and Rubber: Materials and
   Applications, p.141
7. Kwan, S.F., Chen, F.C., and Choy, C.L.: 1975, Polymer 16,
   481
8. Chan, O.K., Chen, F.C., Choy, C.L. and Ward, I.M.: 1978,
   J. Phys. D., Applied Physics 11, 617
9. Takayanagi, M.: 1963, Mem. Fac. Engng, Kyushu Univ. 23,41
10. Stachurski, Z.H., and Ward, I.M.: 1968, J. Poly. Sci., A2,
    6, 1817
11. Meyer, K.H., and Lotmar, W.: 1936, Helv. Chem.Acta, 19,68
12. Lyons, W.J.: 1958, J. Appl. Phys. 29, 1929;1958, 30, 796
13. Treloar, L.R.G.: 1960, Polymer, 1, 95/279/290
14. Tashiro, K., Kobayashi, M. and Tadokoro, H., 1977,
    Macromolecules, 10, 413
15. Odajima, A., and Maeda, M.: 1966, J. Polym.Sci., C15,55
16. Sakurada, I., Ito, T., and Nakamae, K.: 1966, J. Polym. Sci.,
    C15, 75
17. Dulmage, W.J., and Contois, L.E.: 1958, J.Polym.Sci., 28,275
18. Clements, J. Jakeways, R. and Ward,I.M.:1978,Polymer,19,629
19. Mizushima S., and Shimanouchi T.: 1967, J. Chem.Phys.
    47, 3605
20. White J.W.: 1976, Structural Studies of Macromolecules by
    Spectroscopic Methods ed. K.J. Ivin, Wiley, London.
21. Ward, I.M.: 1962, Proc. Phys. Soc. 80, 1176
22. Ward, I.M.: 1977, J. Polym.Sci., C, 58, 1
23. Raumann, G., and Saunders, D.W. 1961, Proc. Phys. Soc.,
    77, 1028
24. Northolt, M.G., and Van Aartsen, J.J.: 1977, J. Polymer Sci.
    C58, 283
25. Charch W.H., and Moseley W.W.: 1959, Text. R.J. 29, 525
26. Ward, I.M.: 1964, Text. Res. J. 34, 806
27. Samuels, R.J.: 1974, Structured Polymer Properties, Wiley,
    New York
28. Davies, G.R., Owen, A.J., Ward, I.M., and Gupta, V.B.: 1972
    J. Macromol Sci., B6, 215
29. Owen, A.J., and Ward, I.M.: 1971, J. Mater. Sci., 6, 485

30. Richardson, I.D., and Ward, I.M.: 1978, J. Polymer Sci., Polym. Phys. Edn., 16, 667
31. Kapuscincki, M., Ward, I.M., and Scanlan, J.: 1975, J. Macromol. Sci. Phys. B11, 475
32. Lewis, E.L.V., and Ward, I.M.: 1980, J. Macromol. Sci.Phys. B18, 1
33. Lewis, E.L.V., and Ward, I.M.: 1981, J. Macromol.Sci.Phys. B19, 75
34. Gibson, A.G., Davies, G.R., and Ward, I.M.: 1978, Polymer, 19, 683
35. Gibson, A.G., Jawad, S.M., Davies, G.R., and Ward, I.M.: Polymer (in press)
36. Wilding, M.A., and Ward, I.M.: 1978, Polymer, 19, 969
37. Wilding, M.A., and Ward, I.M.: 1981, Polymer, 22, 870
38. Wilding, M.A., and Ward, I.M.: 1981, Plastics and Rubber Processing and Applications, 1, 167

DYNAMIC MECHANICAL PROPERTIES OF AMORPHOUS POLYMERS

J. Heijboer

Plastics and Rubber Research Institute,
TNO, Delft, The Netherlands

ABSTRACT

This paper describes various experimental techniques to determine
dynamic mechanical properties together with some of their limitations. In the glassy region shifting procedures are shown to be
of limited applicability.
A short survey is given of the various types of molecular motions
giving rise to mechanical loss peaks. Local main chain motions are
particularly sensitive to antiplasticization. The effect of free
volume on side chain motions is small. Motions in small molecules
dissolved in a solid polymer matrix can also be studied by mechanical measurements. Some motions at low temperature do not give
rise to mechanical loss peaks.
Arrhenius plots of the temperature of secondary loss maxima
(T_m^{-1} vs log ν) extrapolate to log ν = 13 to 15 for T $\rightarrow \infty$, local
main chain motions giving the higher value. In their frequency
dependence the γ-maxima of polyethylene and polyoxymethylene behave like glass transitions.
Although the molecular relaxation strength is governed by the
energy difference between equilibruim positions, current theories cannot explain the dependence of relaxation strength on
temperature.

INTRODUCTION

Polymers are viscoelastic materials: when they are subjected to
a periodic stress their deformation shows a phase lag with respect to the stress. This phase difference between stress and
deformation affords information about the molecular processes
taking place in the polymers.

R. A. Pethrick and R. W. Richards (eds.), Static and Dynamic Properties of the Polymeric Solid State, 197–211.

The best mechanical method from a physical point of view would
be to measure the mechanical properties at a given temperature
over a very broad frequency range, covering all molecular processes at that temperature. In practice, however, the frequency
range of mechanical measurements is rather narrow. As a rule one
uses a single method and varies the temperature to adapt the
frequency range of the molecular processes to that of the mechanical experiment. To obtain data over a broad frequency range
one then applies shifting procedures (time-temperature superposition (1,2)). The accuracy of the modulus data obtained by these
procedures, particularly for the glassy region, has not yet been
established.

EXPERIMENTAL TECHNIQUES FOR MEASURING MODULUS AND DAMPING (3, 4)

A broad frequency range cannot be covered in a single experimental technique; a whole series of techniques is required.
The best known and most accurate instrument is the torsion pendulum (5), which affords data in the region of 0.1 to 10 Hz.
From 3.5 to 110 Hz the Rheovibron is a very useful instrument
to determine transitions, but the accuracy of the modulus data
it provides is often overrated (6). The same applies even more
so to the Dupont dynamic mechanical analyser, because the

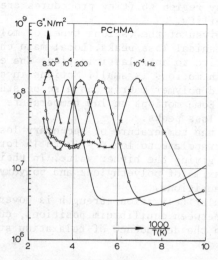

Fig. 1   Loss modulus in shear G" as a function of 1/T for
          poly(cyclohexyl methacrylate) at five frequencies (11).

deformation mode of the specimen in this instrument is not well defined (7). Accurate determinations are described by Read and Dean (3). In the MHz region the complete set of elastic contants can be obtained by means of wave propagation (8). At low frequencies, data can be obtained from creep measurements; to compare the creep compliance with torsion pendulum measurements, some assumptions have to be made (9). In the glass-rubber transition range these assumptions have been shown to be justified (10). Figure 1 shows the G" loss peak of poly(cyclohexyl methacrylate) (PCHMA) at five frequencies. The various curves are obtained from wave propagation, longitudinal vibration, flexural vibration, torsion pendulum and torsional creep measurements. From 10^4 Hz downwards the height of the peak remains constant, the greater height at 8×10^5 Hz being due to interference with the glass transition. If the inverse temperature of the maximum is plotted vs log ν a straight line results; from this line an activation energy is obtained of 47 kJ/mol, the same as the NMR value for the flip-flop motion of cyclohexane.

SHIFTING PROCEDURES IN THE GLASSY REGION (12)

Since G_m'' remains constant with frequency, it seems justified to shift G" data. The result for the measurement in torsion is shown in Fig. 2 and compared with measurements in creep (left) and flexural vibration (right) at -80 °C.

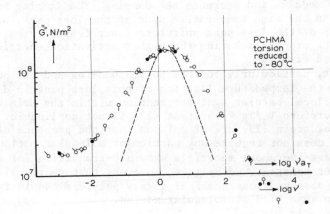

Fig. 2   Reduction of G" for PCHMA to -80 °C. Filled circles are derived from measurements at -80 °C. Dashed line: maximum for a single relaxation time.

The correspondence between shifted and direct measurements is
satisfactory. The shift of the high-temperature side of the peak
yields a higher activation energy than does that of the low-tem-
perature side. This means that even a narrow secondary peak has
a distribution of activation energies.
Shifting uncorrected G' values or G' values corrected for the
general temperature dependence outside the relaxation region
leads to erroneous values of the activation energy (11, 13).

MOLECULAR MOTIONS AND MECHANICAL LOSSES

Many examples of mechanical loss peaks and their suggested molecular mechanisms have been compiled by McCrum, Read and Williams
(12). A recent survey is given by Cowie (14).
There are various types of molecular motions that give rise to
mechanical losses in the glassy region (13):

1. Cooperative motions of large parts of the main chain

These motions are responsible for the glass-rubber transition.
They do not stop abruptly below the glass transition but cause
the low temperature tail of the main damping peak. These losses
are particularly sensitive to physical aging (15).

2. Local main chain motions

Characteristic examples are the β-maximum in PVC and the γ-transition in polycarbonates (16, 17). These transitions are particularly sensitive to antiplasticization (18, 19); the losses are
reduced by addition of so-called antiplasticizers, additives that
increase the modulus and decrease the damping. The damping begins
to decrease on the high temperature side of the loss peak, and
the remainder of the loss peak shifts to lower temperatures. This
means that the processes having the higher activation energies
are obstructed first.
It is evident, particularly for the polycarbonates and the poly-
esters, that the temperature location of this loss peak is determined by the local barrier against rotation within the main chain
(20, 21). Therefore T_β/Tg = constant (22) is not applicable.
Johari and Goldstein (23) are of opinion that the presence of a
β-transition does not require any particular molecular motion and
that all glassy amorphous materials show a β-transition for which
T_β/Tg is about constant. We believe, however, that nearly all experimentally observed loss peaks of glassy polymers can in fact
be related to some kind of molecular motion.

3. Motions of side groups about the bonds linking them to the main chain

The best studied example of such a motion is the β-peak of the poly(methacrylates), which is situated at about 290 K, 1 Hz, irrespective of the glass transition temperature of the poly(alkyl methacrylate) (12, 24). In this series T_β/Tg is again not constant. The peak is caused by a rotation, probably only partial, of the alkoxycarbonyl group. The methyl groups on the main chain constitute the barrier to the rotation. Bulky alkyl groups (cycloalkyl, tert.butyl) depress the peak. A comparison with dielectric measurements shows that the side group of atactic poly(methyl methacrylate) (PMMA) develops a considerable freedom of movement at the β-transition; this means that the main chain participates in the motion. The side group of poly(methyl acrylate), by contrast, is much more restricted in its motion, the corresponding maximum being much smaller than that of PMMA; in addition it also lies at a much lower temperature (about 150 K at 1 Hz).

4. Internal motion within the side group itself, without interaction with the main chain

There are a great many examples of this type of motion, e.g. that of the alkyl group in poly(alkyl methacrylates) (24, 25, 26) (See Fig. 3), in n-alkyl substituted poly(phenylene oxides) (27) and in poly (itaconic acid esters) (14), cycloalkyl groups in poly(cycloalkyl methacrylates) (11, 14, 28, 29), substituted alkyl groups in poly(methacrylic acid esters) (30, 31, 32).

Many of these loss peaks do not respond to the molecular environment; the location of the loss peak is determined by the local barrier within the group itself and is independent of intermolecular interaction. A typical example is the flip-flop motion of the cyclohexyl ring. The temperature of the corresponding loss peak remains constant in a series of copolymers, and does not depend on the glass transition of the polymer (11, 33), nor does annealing have any effect on this secondary maximum. On the other hand, in the series of copolymers of methyl methacrylate with n-butyl methacrylate, the temperature of the loss peak due to the motion of the n-butyl group increases with n-butyl concentration. This has been explained as a free volume effect (34). This explanation, however, is rather improbable, since for polymers of greatly varying free volume, the temperature of the n-butyl peak depends only on the concentration of the n-butyl group (26) (See Fig. 4).

In a series of 2-methyl 6-alkyl substituted poly(phenylene ethers) the temperature of the loss peak also increases with the n-alkyl content (27).

By adding small molecules, Janacek and coworkers (31, 35) observed interesting effects on this type of motion. The maximum disappeared and a new maximum appeared at a higher temperature. This can be explained by the small molecule associating with the side group, the motion within the side group is thereby replaced by a motion of a larger group.

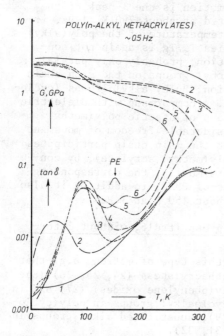

Fig. 3 Shear modulus G' and
damping tan δ of poly(n-alkyl
methacrylates) as functions
of temperature. PE indicates
the location of the loss peak
of polyethylene (26).

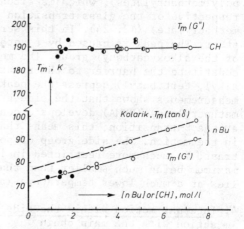

Fig. 4 Temperatures T_m of the
maxima of the loss modulus of
copolymers containing cyclohexyl (CH) and n-butyl (nBu)
groups as a function of the
concentration of the groups.
Filled points: moving group
within a small molecule.
For comparison maxima of
tan δ in a copolymer series
n-butyl methacrylate-MMA are
given (34).

5. Motions within a small molecule dissolved in a solid polymer matrix

Many motions of this type have been studied by mechanical measurements on solid solutions of small molecules in PMMA (11, 13),
e.g. dibutyl phthalate, cyclohexyl esters, bisdioxans, caprolac-
tam. The loss peaks of groups in small molecules lie at the same
temperature as those of the same groups in a polymer molecule.
This means that molecular motions within small molecules can be
studied by mechanical (or dielectric (36)) measurements on solutions of these molecules in a solid polymer matrix. For this
purpose, however, the small molecule must be anchored in the
polymer matrix. When the molecule as a whole moves at a lower
temperature than the group within the molecule, the motion of
the group cannot be observed. See Fig. 5 as an example: the
rotation of dioxane proceeds more readily than its flipping

motion. Obviously, the bisdioxane is not free to rotate and
is showing the maximum corresponding to the flipping motion.

MOLECULAR MOTIONS NOT GIVING RISE TO MECHANICAL LOSS PEAKS

The seven-membered cycloheptyl ring is very flexible; in its
lowest energy form it can pseudorotate over a large number of
conformations, the barrier to this motion being smaller than
6 kJ/mol (37, 11). This pseudorotation is accompanied by a
considerable mass transfer. One would expect a loss peak below
25 K at 1 Hz (see next section). Figure 6, however, shows that
the expected loss peak is absent in poly(cycloheptyl methacrylate).

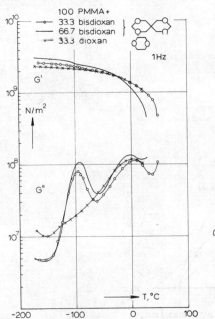

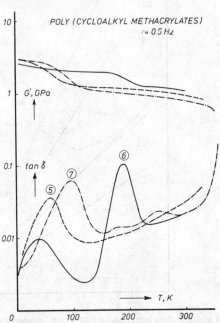

Fig. 5 Storage modulus G' and
loss modulus G" at 1 Hz as a
function of temperature for
PMMA containing 25 % bisdioxane, 40 % bisdioxane, and
25 % dioxane, respectively.

Fig. 6 Shear modulus G' and
damping tan δ for poly(cyclo-
pentyl methacrylate), poly(cycloheptyl methacrylate) and poly-
(cyclohexyl methacrylate) as
functions of temperature T (26).

DEPENDENCE OF LOSS PROCESSES ON FREQUENCY

Most secondary maxima obey an Arrhenius equation

$$\ln \nu = \ln \nu_o - E_a/RT_m ; \qquad\qquad\qquad\qquad (eq. 1)$$

where ν is the frequency of the measurement, E_a the activation energy, R the gas constant and T_m the temperature of the loss maximum obtained from a plot of loss modulus or tan δ vs T. Figure 7 (38) shows a few examples of maxima measured over a wide frequency range.

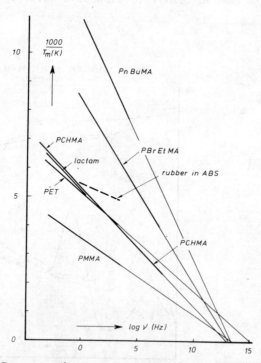

Fig. 7   Arrhenius plots of secondary maxima.
         PnBuMA : γ-maximum of poly(n-butyl methacrylate)
         PBrEtMA: γ-maximum of poly(2-bromoethyl methacrylate)
         PCHMA : cyclohexyl maximum of poly(cyclohexyl
                     methacrylate)
         Lactam : loss maximum of 20 % of caprolactam
                     dissolved in PMMA
         PET   : β-maximum of poly(ethylene terephthalate)
         PMMA   : β-maximum of poly(methyl methacrylate)
Thick parts of the lines indicate the region of the measurements; the thin parts are extrapolated.

The straight lines usually extrapolate to a value of $\log \nu_0 = 13.5 \pm 1$, for some secondary processes, however, $\log \nu_0$ is slightly higher, e.g. 15. This agrees with the expected entropy contributions in the relevant transitions (11). The behaviour of glass transitions is completely different, as is shown by the example of the rubber in ABS in Fig. 7. The slope of the line is very much less than those of corresponding secondary maxima, indicating a much higher activation energy. Assuming that $\log \nu_0 = 13$, the activation energy E_a (in kJ/mol) for the secondary maxima can be estimated from the maximum T_m at 1 Hz by equation 2:

$$E_a = 0.25 \, T_m \, (1 \text{ Hz}) \qquad\qquad (\text{eq. 2})$$

or at another frequency by equation 3:

$$E_a = (0.25 - 0.019 \log \nu_m) \, T_m \, . \qquad (\text{eq. 3})$$

The behaviour of secondary maxima can be described by a thermally activated process involving redistribution of groups across a potential barrier between two equilibrium positions. Application of stress disturbs the equilibrium. The rate of redistribution is determined by the temperature and the height of the barrier; the latter equals the activation energy E_a. This molecular picture implies that the location of the secondary loss peak is determined by the local barrier within the molecule. ν_0 is the frequency at which the groups vibrate in their equilibrium position.
The accuracy of the estimate of Eq. 2 is illustrated in Fig. 8 for data obtained from measurements at different temperatures over a frequency range of at least three decades.
Nearly all side-chain motions and motions within a solute obey equation 2 to within 10 %, whereas local main-chain motions tend to somewhat higher E_a values, which do not, however, deviate by more than 20 %. The glass transition of rubber in ABS falls completely outside this range; it has another molecular mechanism than the secondary maxima. The γ-maxima of polyethylene and polyoxymethylene (POM) behave like a glass transition.

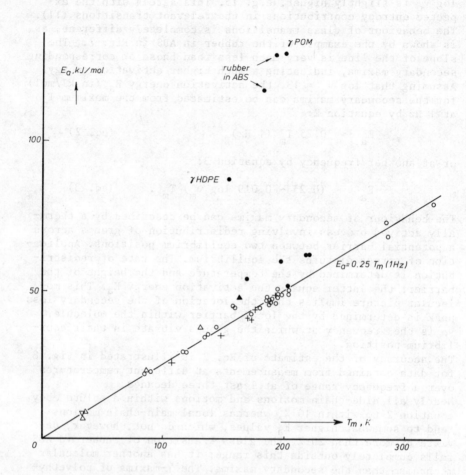

Fig. 8   Activation energy E_a as a function of temperature T_m of the secondary loss maxima at 1 Hz.

Filled symbols: main-chain motions
Open symbols : side group motions
Crosses   : motions within dissolved molecules
Triangles   : literature data.

DIFFERENCES BETWEEN SECONDARY MAXIMA AND THE GLASS TRANSITION

Secondary transitions and glass transitions differ not only in
their dependence on frequency, but also in the effect of plast-
ization and steric hindrance (13). The glass transition temperature Tg of a polymer is strongly decreased by addition of a
plasticizer, whereas secondary transitions often stay at the
same temperature (e.g. the β-transition of PMMA and the γ-transition of PCHMA). A more bulky polymer chain shifts Tg upwards
(compare polystyrene with polypropylene), but more bulky alkyl
side groups in poly(methacrylates) do not shift the β-transition upwards, nor does methyl substitution in the cyclohexyl
ring shift the γ-transition. In principle steric hindrance de-
presses secondary maxima, but does not shift them. This agrees
with the viewpoint that secondary transitions are governed by a
local intramolecular potential barrier. Any considerable intermolecular interaction blocks the motion; the motion is not gradually shifted to higher temperatures.

THE DEPENDENCE OF THE RELAXATION STRENGTH ON TEMPERATURE

The relaxation strength of a loss process is the decrease in modulus $\Delta G'$ due to that loss process. It follows from Fig. 9 that
the molecular relaxation strength $\Delta G'_m$ ($\Delta G'$ for 1 mol/l of the
moving group) is fairly constant for the cyclohexyl motion in
cyclohexyloxycarbonyl groups, irrespective of whether the moving
group is present in a plasticizer or in a polymer molecule (11).

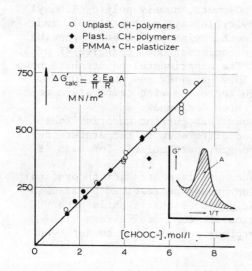

Fig. 9 Calculated relaxa-
tion strength $\Delta G'$ at 1 Hz
of the cyclohexyl motion as
a function of cyclohexyl-
oxycarbonyl group concen-
tration for polymeric com-
positions containing the
latter group in the polymer
or in the plasticizer.

Insert: definition of the
      area A.

ΔG_m depends strongly on the conformational energy of the group, i.e. the difference in free energy ΔF between the two chair conformers of the cyclohexyl ring.

For ΔF values higher than 5 kJ/mol, owing to the change with temperature in distribution of the groups over the two positions, ΔG_m should depend strongly on temperature. It can be derived (11) that ΔG_m is given by

$$\Delta G_m' = \frac{<b^2>}{2RT} \; \frac{1}{\cosh^2 \dfrac{\Delta F}{2RT}} \qquad\qquad (eq. 4)$$

The only assumption in the derivation, apart from a Boltzmann distribution, is that the relative change in depth of the potential wells is proportional to the deformation of the polymer, the proportionality constant being b. $<b^2>$ is the average of b^2 over all directions.

Other derivations (39, 40, 41) lead to essentially the same expressions for temperature dependence all of which governed by

$$\left[RT \cosh^2 \frac{\Delta F}{2RT} \right]^{-1} \qquad\qquad (eq. 5)$$

Cyclohexyl compounds are particularly suitable for testing this relationship. For small molecules the differences in free energy ΔF of the two chair conformers have been determined by NMR. In view of the independence of the relaxation process of the molecular environment, it is unlikely that in polymers containing the corresponding side groups, the differences in free energy would be affected by the environment.

Figure 10a shows the molecular relaxation strength $\Delta G_m'$ as a function of temperature for three polymers, namely poly(cyclohexyl methacrylate), poly(cis-4-methylcyclohexyl methacrylate), and poly(trans-4-methylcyclohexyl methacrylate). The corresponding small cyclohexyl compounds have conformational energies ΔF of 3, 4 and 10 kJ/mol, respectively. The experimental points have been obtained from measurements at 10^{-3} Hz, 1 Hz and about 10^4 Hz. Clearly the relaxation strength decreases with temperature to about the same extent for all three compounds.

On the other hand, Fig. 10b shows that current theories (see eq. 5) predict a very pronounced difference in temperature dependence of the relaxation strength, particularly for high ΔF values.

Thus the experimental data fundamentally contradict theoretical expectations. This contradiction cannot be resolved by assuming a distribution of ΔF values. We conclude that a satisfactory theory describing the interaction of mechanical stress with molecular motion in glassy amorphous polymers has so far been lacking.

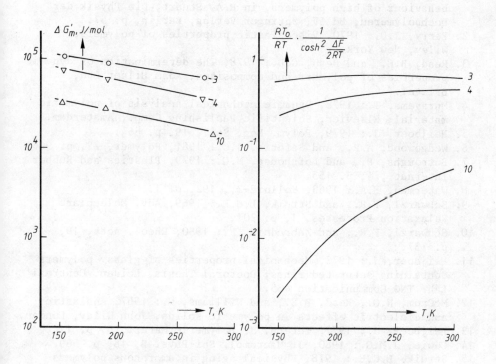

Fig. 10 Dependence of relaxation strength on temperature.

    a.   Experimental data of the molecular relaxation strength
$\Delta G_m'$ for poly(cyclohexyl methacrylate) (3), poly(cis-
4-methylcyclohexyl methacrylate) (4) and poly(trans-
4-methylcyclohexyl methacrylate) (10).

    b.   Theoretical course of the relative molecular relaxation
strength of three cyclohexyl compounds with conformational energies of 3, 4 and 10 kJ/mol.

REFERENCES

1. Staverman, A.J., and Schwarzl, F.R.: 1956, Linear deformation
   behaviour of high polymers, in H.A. Stuart, Die Physik der
   Hochpolymeren, Bd IV, Springer Verlag, Berlin, p. 57.
2. Ferry, J.D.: 1970, Viscoelastic properties of polymers,
   Wiley, New York.
3. Read, B.E., and Dean, G.D.: 1978, The determination of dynamic
   properties of polymers and composites, Adam Hilger Ltd,
   Bristol.
4. Murayama, T.: 1978, Dynamic mechanical analysis of polymeric
   material, Elsevier, Scientific Publishing Comp., Amsterdam.
5. Heijboer, J.: 1979, Polym. Eng. Sci., 19, p. 664.
6. Wedgewood, A.R., and Seferis, J.C.: 1981, Polymer, 22, p. 966.
7. Burroughs, P., and Lofthouse, M.G.: 1979, Plastics and Rubber
   Internat., 4, p. 155.
8. Waterman, H.A.: 1963, Kolloid-Z., 192, p. 1.
9. Schwarzl, F.R., and Struik, L.C.E.: 1969, Adv. Molecular
   Relaxation Processes, 1, p. 201.
10. Schwarzl, F.R., and Zahradnik, F.: 1980, Rheol. Acta, 19,
    p. 137.
11. Heijboer, J.: 1972, Mechanical properties of glassy polymers
    containing saturated rings, Doctoral Thesis, Leiden. Centraal
    Lab. TNO Communication 435.
12. McCrum, N.G., Read, B.E., and Williams, G.: 1967, Analastic
    and dielectric effects in polymeric solids, John Wiley, Londen.
13. Heijboer, J.: 1977, Intern. J. Polymeric Mater., 6, p. 11.
14. Cowie, J.M.G.: 1980, J. Macromol. Sci-Phys. B, 18, p. 569.
15. Struik, L.C.E.: 1978, Physical aging in amorphous polymers
    and other materials, Elsevier, Amsterdam.
16. Illers, K.H., and Breuer, H.: 1961, Kolloid-Z, 176, p. 110.
17. Yee, A.F., and Smith, S.A.: 1981, Macromolecules, 14, p. 54.
18. Pezzin, G., Ajroldi, G., and Garbuglio, C.: 1967, J. Appl.
    Polym. Sci., 11, p. 2553.
19. Petri, S.E.B., Moore, R.S., and Flick, J.R.: 1972, J. Appl.
    Phys., 43, p. 4318.
20. Bussink, J., and Heijboer, J.: 1965, Proc. Int. Conf. Physics
    Noncrystalline Solids, N. Holland Publ. Co., Amsterdam, p. 388.
21. Reding, F.P., Faucher, J.A., and Whitman, R.D.: 1961, J.
    Polym. Sci., 54, p. S 56.
22. Boyer, R.F.: 1975, J. Polym. Sci. C., 50, p. 189.
23. Johari, G.P., and Goldstein, M.: 1971, J. Chem. Phys. 55,
    p. 4245.
24. Heijboer, J.: 1965, Proc. Int. Conf. Physics Noncrystalline
    Solids, N. Holland Publ. Co., Amsterdam, p. 231.
25. Shimuzu, K., Yana, O., Wada, Y., and Kawamura, Y.: 1973,
    J. Polym. Sci-Phys. 11, p. 1641.

26. Heijboer, J., and Pineri, M.: 1981, Proc. Int. Cryogenic
    Mat. Conference, Geneva, August 1980, Plenum Press,
    New York, in print.
27. Cayrol, B., Eisenberg, A., Harrod, J.F., and Rocanieri, P.:
    1972, Macromolecules 5, p. 676.
28. Frosini, V., Magagnini, P., Butta, E., and Baccaredda, M.:
    1966, Kolloid-Z, 213, p. 115.
29. Heijboer, J.: 1968, J. Polym. Sci. C., 16, p. 3413.
30. Janacek, J., and Kolarik, J.: 1967, J. Polym. Sci. C., 16,
    p. 279.
31. Janacek, J., and Kolarik, J.: 1976, Intern. J. Polymeric
    Mater., 5, p. 71.
32. Waterman, H.A., Struik, L.C.E., Heijboer, J., and
    Van Duijkeren, M.P.: 1972, Amorphous Materials, Third Int.
    Conf. Phys. Noncrystalline Solids, Wiley, p. 29.
33. Heijboer, J.: 1978, in Molecular basis of transitions and
    relaxations, ed. by Meier, D.J., Midland Macromol. Mono-
    graphs, Gordon and Breach, Londen, 4, p. 297.
34. Kolarik, J.: 1971, J. Macromol. Sci. Phys. B, 5, p. 355.
35. Kolarik, J., and Janacek, J.: 1972, J. Polym. Sci. A-2, 10,
    p. 11.
36. Davies, M., and Swain, J.: 1971, Trans. Faraday Soc., 67,
    p. 1637.
37. Hendrickson, J.B.: 1967, J. Am. Chem. Soc., 89, p. 7043.
38. Heijboer, J.: 1977, Fourth Int. Conf. Physics of Non-Crys-
    talline Solids, G.H. Frischat, Ed., Trans. Tech. Publica-
    tions, Aedermannsdorf, Switserland, p. 517.
39. Hoffman, J.D., Williams, G., and Passaglia, E.: 1966,
    J. Polym. Sci. C., 14, p. 173.
40. Matsui, M., Masui, M., and Wada, Y.: 1971, Polymer J., 2,
    p. 134.
41. Gisolf, J.H.: 1975, Colloid & Polymer Sci., 253, p. 185.

DIELECTRIC RELAXATION OF SOLID POLYMERS

Graham Williams

Edward Davies Chemical Laboratories, University College
of Wales, Aberystwyth SY23 1NE, Dyfed, U.K.

The essential features of the dielectric relaxation behaviour
selected amorphous and crystalline polymers are considered. For
amorphous polymers it is shown that three dipole relaxation
processes are generally observed whose magnitudes and frequency
locations depend upon temperature and applied pressure. The
phenomenological aspects of the observed behaviour are considered
and the similarities to the multiple dielectric relaxations for
small-molecule glass-forming systems are emphasized. The molecular
theory for dipole relaxation in amorphous polymers is briefly
considered. It is shown that the β-process is due to the partial
relaxation of dipoles in a range of temporary local environments
whilst the α and (αβ) processes are due to gross micro-brownian
motions of chains, or, for small-molecule glass-forming systems,
are due to overall motion of whole molecules, activated by local
thermal fluctuations in the system. The dielectric behaviour of
certain partially-crystalline polymers, especially polyethylene
terephthalate, poly(ethylene), poly(vinylidene fluoride) and
oxide polymers, is briefly discussed.

I.   EXPERIMENTAL METHODS

The complex dielectric permittivity $\varepsilon(\nu)$ is related to the
normalized transient-response current function $\Phi(t)$ following a
step-applied voltage by the Fourier transform relation (1)

$$\frac{\varepsilon(\nu) - \varepsilon_\infty}{\varepsilon_0 - \varepsilon_\infty} = \int_0^\infty dt \; \Phi(t)\exp(-i2\pi\nu t) \tag{1}$$

213

R. A. Pethrick and R. W. Richards (eds.), Static and Dynamic Properties of the Polymeric Solid State, 213–239.
Copyright © 1982 by D. Reidel Publishing Company.

ε_o and ε_∞ are the limiting low and high frequency permittivities.
Measurements of $\varepsilon(\nu)$ are conveniently made in the range
10^{-1}Hz $< \nu < 10^{10}$ Hz using a variety of bridge, resonance circuit,
transmission line and cavity-resonator techniques (1,2). For the
low-frequency range 10^{-5}Hz $< \nu < 10^{-1}$Hz, transient charging (or
discharging) currents give $\Phi(t)$ and hence $\varepsilon(\nu)$ via eqn. (1),
usually achieved using modifications of the Hamon approximation to
the Fourier transform (1), which allow point-by-point conversion
of (Φ,t) to (ε,ν). Most dielectric studies are made in the range
10^{-2}Hz $< \nu < 10^6$Hz.

II.   AMORPHOUS POLYMERS

II.A. Phenomenological Aspects

Two groups of polymers may be recognized: those with dipoles
rigidly attached to the main chain (e.g. polyethylene terephthalate,
poly(vinyl halides), poly(vinylidene halides)), and those with
flexible dipolar side-groups (e.g. poly(alkyl methacrylates),
poly(vinyl esters)). Extensive dielectric data for such systems
are described in the texts by McCrum, Read and Williams (1),
Karasz (3) and Hedvig (4) and in the reviews by McCall (5),
Ishida (6), Saito (7), Sasabe (8), Wada (9) and Williams (10). For
all dipolar amorphous polymers an α-process is observed above the
operationally-defined glass-transition temperature T_g and a β-
process is observed below T_g. In the vicinity of T_g the α and β
processes may coalesce so at elevated temperatures a new process,
called the $(\alpha\beta)$ processes emerges (10).

II.A.1. Polymers with dipoles rigidly attached to the main chain. As one example we may quote the data for amorphous poly-
ethylene terephthalate (1) which exhibits a well-defined α-
relaxation in the ranges 350 - 400 K, 10^{-1} - 10^5 Hz and a small
and broad β-relaxation in the ranges 240 - 300 K, 10^{-1} - 10^6 Hz.
For each process the plots of permittivity ($\varepsilon'(\nu)$) and loss-factor
($\varepsilon''(\nu)$) against $\log \nu$ give three items of information at each
temperature:
(i) the magnitude $\Delta\varepsilon_i$ for the process (i is α or β) where

$$\Delta\varepsilon_i = (\varepsilon_o - \varepsilon_\infty)_i = (2/\pi)\int \varepsilon_i''(\nu)d\nu/\nu \qquad (2)$$

(ii) the frequency of maximum loss, ν_{mi} which defines an average
relaxation time $<\tau>_i = (2\pi\nu_{mi})^{-1}$ for the process.

(iii) the overall shape of the process - as assessed bt its half-
width $(\Delta \log \nu)_i$ and its asymmetry in the $\varepsilon''(\nu)$ - vs - $\log \nu$ plot.

Measurements as a function of temperature give $\Delta\varepsilon_i(T)$, $<\tau(T)>_i$ and

$(\Delta \log \nu)_i$. The quantity $\{\partial \ln <\tau(T)>_i /\partial T\}_p = -Q_{pi}(T)/RT^2$, where Q_{pi} is the constant pressure activation energy for process i. In the case of polyethylene terephthalate (1) $\varepsilon''_{m\alpha} \simeq 0.4$, $\varepsilon''_{m\beta} \simeq 0.07$ which leads to $(\Delta\varepsilon_\beta /\Delta\varepsilon_\alpha) \simeq 0.44$, the exact value depending upon the reference temperatures taken. $(\Delta \log\nu)_\alpha \simeq 2-3$, $(\Delta \log \nu)_\beta \simeq 6$ for this polymer (1) and such values of $(\Delta\varepsilon_\beta /\Delta\varepsilon_\alpha)$ and the half-widths are typical of this class of amorphous polymer (1). With the exception of amorphous polychlorotrifluoroethylene (1, 11) all amorphous polymers having dipoles rigidly attached to the main-chain give $\Delta\varepsilon_\beta << \Delta\varepsilon_\alpha$. Also $Q_\alpha >> Q_\beta$, with Q_α being strongly dependent on temperature (decreasing with increasing T in accord with the well-known Williams-Landel-Ferry equation (see ref.1,6-8)) and Q_β being approximately independent of temperature. Inspection of the transition maps of $\log \nu_{mi}$ - vs - K^{-1} for such polymers(1,5) suggests that α and β processes coalesce at temperatures a little above T_g.

Studies have been made of the effect of an applied hydrostatic pressure for the α-process of poly(propylene oxide) (12), the α and β processes of poly(vinyl chloride) (8,13-15) the α and β processes of polyethylene terephthalate (8,15,16) and acrylonitrile-butadiene copolymers (17). Such studies have been reviewed by Williams and Watts (see ref. 3, p.17). It is found that pressure has a large effect on $\log \nu_{m\alpha}$ but only a small effect on $\log \nu_{m\beta}$. The quantity $(\partial \log\nu_{m\alpha}/\partial P)_T$ lies in the range-1 x 10^{-3} to-4 x 10^{-3} atm $^{-1}$, being dependent upon chemical structure, pressure and, especially, temperature. For the β-relaxation $(\partial \log\nu_{m\beta}/\partial P)_T$ lies in the range-0.1 x 10^{-3} to-0.5 x 10^{-3} atm^{-1} being dependent upon chemical structure but not especially upon temperature or pressure. Of special interest is the case of poly(vinyl chloride). The α-process in this polymer is unusually broad (1) and lowering of temperature suggests that a β-process resolves at higher temperatures with respect to the primary maximum. Sasabe and coworkers (8,15) show for a sample at 376 K, that increase in pressure from 1 atm to 1440 atm moves the large peak ($\varepsilon''_m \simeq 1.5$) rapidly to lower frequencies at a rate $(\partial \log\nu_m/\partial P)_T \simeq -4$ x 10^{-3} atm^{-1},and as this occurs, the loss peak narrows and leaves behind, as a clearly observed high frequency tail, a small process whose location is only slightly dependent upon pressure. Such behaviour is consistent with an $(\alpha\beta)$ process converting into α and β processes with increased pressure, and this is a very significant result. It means that α and β processes may coexist for T > Tg for polymers having dipoles rigidly attached to the main chain (see ref. 10 for further discussion).

With regard to the shape and breadth of α and β processes, it is found that the α-process is fairly narrow $((\Delta\log\nu)_\alpha$ in the range 1.8 to 2.5) and asymmetrical whilst the β-process is extremely broad $((\Delta\log\nu)_\beta$ in the range 4 to 6) in plots of $\varepsilon''(\nu)$ against $\log \nu$ at fixed temperatures. The empirical relaxation function (18-20)

$$\Phi(t) = \exp(-t/\tau)^{\beta}; \quad 0 < \beta \leqslant 1 \tag{3}$$

used in eqn. 1 yields asymmetrical loss curves of the type found
in practice for the α process and reasonable fits to experimental
data are obtained with β in the range 0.38 to 0.65 (18-20, see
also Williams in ref. 3). No significance should be attached to
the ability of eqn. (3) to fit such data: it simply means that
we may approximately represent $\varepsilon(\nu)$ (= $\varepsilon'(\nu) - i\varepsilon''(\nu)$) for an α-
process at a given temperature and pressure by a minimal basis set
($\Delta\varepsilon_\alpha$, $\nu_{m\alpha}$ and β_α). Clearly, fitting data with a more extended
basis set would give a more accurate representation of data.

   II.A.2   Polymers with dipolar groups flexibly attached to
the main chain.   Among the well-known examples are 'conventional'
poly(methyl methacrylate) and poly(vinyl acetate), whose dielectric
relaxations are well-documented (1). Interestingly these polymers
show sharply contrasting behaviour. Poly(vinyl acetate) exhibits
a large well-defined α-relaxation, whose frequency-temperature
variation and dispersion contour are similar to those observed for
polyethylene terephthalate, and a small, very broad β-relaxation
(1). 'Conventional' poly(methyl methacrylate) is atactic, but with
a tendency to prefer the syndiotactic sequences along the chain
(see e.g. refs. 1,8,21,22). Shindo, Murakami and Yamamura (21)
made an extensive dielectric study of this material and poly(methyl
methacrylates) of very different tacticities. They showed that
the 'conventional' material exhibited α and β relaxations where
$\Delta\varepsilon_\beta > \Delta\varepsilon_\alpha$, and that the two processes tend to coalesce a little
above T_g. They also showed that variation of tacticity from
predominantly syndiotactic polymer through to predominantly
isotactic polymer results in ($\Delta\varepsilon_\beta/\Delta\varepsilon_\alpha$) $\approx$ R, say, going from R > 1
to R < 1: i.e. the predominantly isotactic polymer behaves in a
manner similar to poly(vinyl acetate) (1), many other polymers
with rotatable dipolar side-groups, and to polymers with dipoles
rigidly attached to the main-chain (A.1 above) as far as ($\Delta\varepsilon_\beta/\Delta\varepsilon_\alpha$)
is concerned. The studies of Shindo and coworkers (21) confirm
and extend the earlier studies of poly(methyl methacrylate) of
different tacticities (see ref. 1) and are crucial to our
understanding of multiple relaxations in amorphous polymers, as
we shall see below in A.3. Other 'atactic' poly(alkyl methacryl-
ates) show α and β processes for which ($\Delta\varepsilon_\beta/\Delta\varepsilon_\alpha$) > 1. Poly(ethyl
methacrylate) and poly(n-butyl methacrylate) both show such
behaviour (see ref. 1) and, again, the two processes coalesce to
form an ($\alpha\beta$) process at temperatures close to T_g (1,8,22,23,24).
Whilst the frequency-temperature location of the α-process is
greatly affected by going from methyl to ethyl to n-propyl, since
T_g varies rapidly for this sequence, the corresponding location of
the β-process is essentially unchanged for the 'atactic' polymers.
These factors led Williams and Edwards (22) and Williams (23) to
carry out a study of the effect of pressure on the relaxations of
poly(n-butyl methacrylate) PnBMA, and poly(ethyl methacrylate) PEMA,
respectively. Whilst the data for PnBMA could be analyzed in terms
of the α, β and ($\alpha\beta$) relaxations, the frequency-range covered was

very limited. The later studies of PEMA were carried out over
wide ranges of frequency, temperature and pressure both above and
below the apparent $T_g(23)$ and the following was demonstrated
for PEMA.

(i) α and β relaxations occur: the β process was observed
both above and below T_g and α and β processes coexist for a limited
range above T_g before they coalesce to form the $(\alpha\beta)$ process.

(ii) As pressure increases resolution of α and β processes is
achieved, $(\partial \log v_{m\alpha}/\partial P)_T$ being extremely large and comparable
with values obtained for other α-relaxations (see A.1 above) whilst
$(\partial \log v_{m\beta}/\partial P)_T$ is extremely small.

(iii) As pressure increases $\Delta\varepsilon_\beta$ decreases rapidly e.g. at 369.6 K
$\varepsilon''_{m\beta}$ decreases from 2.0 at 1200 atm to 1.1 at 2700 atm, the
location, $\log v_{m\beta}$, being essentially unchanged. Concurrent with
the decrease in $\Delta\varepsilon_\beta$ is an increase in $\Delta\varepsilon_\alpha$ indicating that a
conservation rule applies:-

$$\Delta\varepsilon_\alpha + \Delta\varepsilon_\beta = \Delta\varepsilon_{total} \simeq constant \tag{4}$$

Thus α and β processes are coupled via their relaxation strengths.
If pressure acts to decrease $\Delta\varepsilon_\beta$ then there is a corresponding
increase in $\Delta\varepsilon_\alpha$. This leads us back to the behaviour of poly
(methyl methacrylates) of different tacticities as described by
Shindo and coworkers (21). The root-mean-square dipole moment
per repeat unit of isotactic polymer (μ_{app} = 1.4D) and syndiotactic
polymer (μ_{app} = 1.27D) are not very different (25) so $\Delta\varepsilon_{total}$ for
both polymers will be similar in magnitude. It is a striking
feature of the data (21) that as $\Delta\varepsilon_\beta$ increases on going from
predominantly isotactic to predominantly syndiotactic polymer,
$\Delta\varepsilon_\alpha$ shows a corresponding decrease in order to conserve $\Delta\varepsilon_{total}$ $\simeq$
constant. Thus the data for poly(ethyl methacrylate) under
pressure and for poly(methyl methacrylates) of varying tacticity
suggest that the β-process partially relaxes $\Delta\varepsilon_{total}$, the
remainder relaxing with the α-process, and that if variables
(e.g. pressure, temperature, tacticity) act in such a way as to
decrease/increase $\Delta\varepsilon_\beta$ then there will be a corresponding increase/
decrease in $\Delta\varepsilon_\alpha$ so as to conserve $\Delta\varepsilon_{total}$. We note that for the
higher poly(n-alkyl methacrylates) the process normally thought to
be the α-process may be the $(\alpha\beta)$ process. Thus Williams and
Watts (26) showed for poly(n-nonyl methacrylate) that the $(\alpha\beta)$
process gave α and β processes at high applied pressure where
$\Delta\varepsilon_\beta$ decreased with increasing pressure.

The effects of varying temperature and pressure for the
methyl, ethyl, n-butyl, n-octyl and n-lauryl methacrylate polymers
were investigated by Sasabe and Saito (8,24) who confirmed and
extended the earlier studies (23,24). They clearly demonstrated

for poly(n-butyl methacrylate) that as $\Delta\varepsilon_\beta$ decreased, $\Delta\varepsilon_\alpha$ increased, with increasing pressure.

II.A.3 <u>Small-molecule glass-forming systems</u>. The observations by Johari, Goldstein and Smyth (27-31) and subsequently by Williams and coworkers (see e.g. ref. 32 and refs. therein) that small molecule glass-forming systems exhibit α, β and (αβ) relaxations which are entirely similar to those for polymers having $\Delta\varepsilon_\alpha > \Delta\varepsilon_\beta$ is of great significance for the molecular interpretation of the processes for polymers, as we shall see below. Johari showed that α and β processes coexist for $T > T_g$ for such systems as pyridine/toluene and chlorobenzene/decalin mixtures. The essential features of the shape of the processes in the ε″ - vs - logν plots, and the apparent activation energies Q_α and Q_β and their dependencies upon temperature for the small-molecule systems are very similar to those observed for such polymers as amorphous polyethylene terephthalate and poly(vinyl acetate). Since the dipolar molecules studied by Johari and coworkers (27 - 31) and Williams and coworkers (32) (e.g. chlorobenzene, nitrobenzene, pyridine) were rigid, it is evident that the β-process for these systems could not be assigned to internal motions, so the question arises - may the α, β, (αβ) pattern of relaxations for both polymer and small-molecule glass-forming systems be interpreted in a general way that transcends the details of molecular structure and also in a way that does not require specific models for motion?

II.A.4 <u>Molecular aspects of dielectric relaxation</u>. The theory of the static permittivity ε_0 and the complex permittivity $(\varepsilon(\nu))$ for systems composed of dipolar polarizable molecules is well-documented (1,2,10,33,34). We may write (34)

$$\Delta\varepsilon = (\varepsilon_0 - \varepsilon_\infty) = \frac{4\pi}{3kT} \left\{ \frac{3\varepsilon_0(2\varepsilon_0 + \varepsilon_\infty)}{(2\varepsilon_0 + 1)^2} \right\} \frac{\langle\vec{M}(0).\vec{M}(0)\rangle}{V} \quad (5)$$

ε_0 and ε_∞ are defined as for eqn. (1) above, $\vec{M}(0)$ is the instantaneous dipole moment of a sphere of volume V and $\langle\vec{M}(0).\vec{M}(0)\rangle$ is the mean-square dipole moment, where the average is taken over all configurations of the ensemble of molecules. For N equivalent molecules in the volume V we may write (1,33)

$$\frac{\langle\vec{M}(0).\vec{M}(0)\rangle}{V} = c_p \langle P^2\rangle g_1 \quad (6)$$

where c_p is the number of molecules per unit volume, $\langle P^2\rangle$ is the mean-square dipole moment of any representative molecule and g_1 is the Kirkwood-Fröhlich correlation factor, which is given by

$$g_1 = 1 + \frac{1}{N} \sum_i \sum_{\substack{i' \\ i' \neq i}} <\vec{P}_i(0).\vec{P}_{i'}(0)> / <P^2>$$

$$= 1 + \sum_{\substack{i \\ i' \neq 1}} <\vec{P}_1(0).\vec{P}_i(0)> / <P^2> \tag{7}$$

i and i refer to different molecules, 1 is any given reference
molecule. Thus g_1 expresses the vector correlations between the
dipolar molecules. For small-molecule glass-forming systems in
which the dipolar solute is at low concentration (32), the magnitude of the cross-correlation terms will be small so that $g_1 \simeq 1$.
For flexible polymers e.g. PVC, PET, PMMA, the angular correlation
factors between dipolar groups along a representative chain decrease
in magnitude with increasing separation of the groups, as model
calculations confirm (10,34). It is thus useful to express
$<\vec{M}(0).\vec{M}(0)>$ in terms of the individual dipole moments of these
component groups. Ignoring cross-correlation terms between
chains (although they may be included if we desired) we write (10,
34)

$$<\vec{M}(0).\vec{M}(0)> = c_r \mu^2 \{1 + \sum_{\substack{k' \\ k' \neq k}} <\vec{\mu}_k(0).\vec{\mu}_{k'}(0)>/\mu^2$$

$$\equiv c_r \mu^2 g_1'(0) \tag{8}$$

where c_r is the number of dipole units per unit volume, $\vec{\mu}_k(0)$ is
the instantaneous dipole moment of a given reference unit, k, and
$\vec{\mu}_{k'}(0)$ is a similar quantity for a different unit k' along the
same chain as the unit k. Thus eqns. (5) and (8) show that the
total magnitude of the dipole relaxation, $\Delta\varepsilon$, is proportional to
the concentration of repeat dipoles, c_r, the square of the dipole
moment of the repeat dipole, μ^2, and to g_1 where the latter
factor involves the angular correlations along any representative
chain. Eqn. (8) allows us to consider bulk polymer by focussing
our attention on any reference dipole-unit k and looking out to
the angular correlations it has with its neighbouring dipoles along
the chain. This gives a simple physical approach to the polymer
systems which is not possible from eqn. (7). Cross-correlation
terms between chains may be included (1) in order to generalize
eqn. (8) but it seems likely that in bulk amorphous polymers the
terms included in g_1' are the most important. The above can
be generalized to include a distribution of molecular weight, but
for sufficiently high molecular weights we may readily show that
eqn. 8 is again obtained. The essential features for a comparison
of small-molecule glass-forming systems with polymer systems is
now apparent. For the former systems (e.g. chlorobenzene/decalin
mixtures) the lack of direct chemical bonding between the dipoles
ensures that g_1 (eqn. 7) will approximate to unity in the region

$T > T_g$. So as far as $\Delta\varepsilon$ is concerned the essential difference
between the small molecule and polymer systems is that the cross-
correlation terms $<\vec{\mu}_k(0).\vec{\mu}_{k'}(0)>$ are of appreciable magnitude for
the polymer systems. Estimates of their magnitude and sign may
be obtained with the help of models for the statistical conformation
of chains, as developed by Volkenstein (35) and Flory (36).
Williams and Cook (34) have adapted the theory of Read (37) for
the case of model linear polyethers and have shown how $<\vec{\mu}_k(0).\vec{\mu}_{k'}(0)>$
varies with molecular structure and the separation of the dipoles
along the chain. It is shown that $<\vec{\mu}_k(0).\vec{\mu}_{k'}(0)>$ may be positive
or negative as k' is varied for a given structure of chain. Thus
$g_1'(0)$ is to be thought as being made up of an addition of terms,
some of which may be negative, and although the terms decrease in
magnitude as $|k - k'|$ is increased nearest neighbours and next-
nearest neighbours may make substantial contributions to $g_1'(0)$
(and hence $\Delta\varepsilon$) for flexible-chain polymers. Whilst there has been
some interest in analyzing the static permittivity data for
amorphous solid polymers in terms of the location of dipole moments
along the chain and chain conformation (1,3,6,9,38-41, see also
ref. 42 and refs. therein), it is fair to say that most dielectric
studies have been primarily concerned with (i) the documentation
of behaviour (ii) the analysis of the dynamic behaviour in terms
of assumed models for motion.

The complex dielectric permittivity for a dipolar system is
most conveniently expressed in relation to molecular behaviour
using the time-frequency Fourier transform relation involving
the generalized time-correlation function $\Lambda(t)$ for fluctuations in
$M(t)$ (2,10,34,42)

$$\left[\frac{\varepsilon(\nu) - \varepsilon_\infty}{\varepsilon_0 - \varepsilon_\infty}\right].p(\omega) = \int_0^\infty \{\frac{-d\Lambda(t)}{dt}\}\exp(-i\omega t)dt$$

$$= 1 - i\omega\int_0^\infty \Lambda(t)\exp(-i\omega t)dt \qquad (9)$$

$$\Lambda(t) = \frac{<\vec{M}(0).\vec{M}(t)>}{<\vec{M}(0).\vec{M}(0)>} \qquad (10)$$

and $p(\omega)$ is an internal field factor (2,10,34,42). $\omega = 2\pi\nu$. By
the same reasoning as that above for the equilibrium case, for
flexible-chain polymers excluding angular correlations between
different polymer chains we may write

$$\Lambda(t) = \frac{<\vec{\mu}_k(0).\vec{\mu}_k(t)> + \sum_{k'\neq k} <\vec{\mu}_k(0).\vec{\mu}_{k'}(t)>}{\mu^2 + \sum_{k'\neq k} <\vec{\mu}_k(0).\vec{\mu}_{k'}(0)>}$$

$$\equiv g'(t)/g'(0) \qquad (11)$$

Eqns. (9-11) indicate that measurement of $\varepsilon(\nu)$ gives information on the motions of a reference dipole unit (via the auto-correlation function $\langle \vec{\mu}_k(0).\vec{\mu}_k(t) \rangle /\mu^2$) and the relative motions of dipole units (via the cross correlation functions $\langle \vec{\mu}_k(0).\vec{\mu}_{k'}(t) \rangle$). The auto-correlation function (ACF) has a positive sign but the cross-correlation functions (CCF) may be positive or negative. For non-associated small-molecule systems $\Lambda(t) \simeq \langle \vec{\mu}_k(0).\vec{\mu}_k(t) \rangle /\mu^2$ i.e. only the auto-correlation function is involved.

It is evident from the above equations for the equilibrium property (eqns. 5-8) and the dynamic property (eqn. 9-11) that the dielectric properties of amorphous solid polymers involve auto- and cross-correlation terms for the individual dipole groups along a chain. Given the complexities of chain structure and chain dynamics, it might appear to be an entirely daunting task to attempt to unravel the observed behaviour and arrive at its rationalization in simple physical terms. However, certain crucial observations suggest that rationalization is possible at a general level, independent of the details of structure and dynamics. These are the observations detailed in Sections A.1 and A.2 above which summarized are:

(i) α, β, ($\alpha\beta$) processes generally occur for polymers and small molecule systems, α and β processes coexisting for a small range of temperature above T_g.

(ii)   The conservation rule, eqn. 4 applies to polymer relaxation data.

This means that α and β processes are coupled so that $\Lambda(t)$ is <u>partially relaxed</u> by the β-process, the remainder relaxing with the α-process. Below T_g this means that only the β-process will be observed. Any decrease in $\Delta\varepsilon_\beta$, brought about by increase in pressure, will leave the remainder of $\langle \vec{M}(0).\vec{M}(0) \rangle$ to be relaxed by the α-process. Physically it means that $\langle \vec{\mu}_k(0).\vec{\mu}_k(t) \rangle$ and the terms $\langle \vec{\mu}_k(0).\vec{\mu}_{k'}(t) \rangle$ partially relax (β-process) from their initial values by limited spatial motion of the dipole vectors and that the remainder relaxes via gross-micro-brownian motions giving the α-process. For polymers such as poly(vinylchloride), polyethylene terephthalate and isotactic poly(methylmethacrylate) $\Delta\varepsilon_\beta \ll \Delta\varepsilon_\alpha$ so the limited motions of the dipole vector span, on average, only a limited solid angle, whilst for syndiotactic poly(methylmethacrylate) and for the 'atactic' the dipole vector spans a much larger solid angle giving rise to $\Delta\varepsilon_\beta > \Delta\varepsilon_\alpha$. Two cautionary points should be made. First, the motion of a group and the motion of the dipole vector of that group should be carefully distinguished. Thus in PVC, motion of the CH-Cl group and the dipole vector are synonymous but for PMMA motion of the C_aH-($C_b OOC_c H_3$) group can occur in several ways. Rotation about the C_a - C_b bond only moves a component of the dipole vector. Secondly, for 'conventional' PMMA, the remarkable decrease in $\Delta\varepsilon_\beta$, which is accompanied by the

increase in $\Delta\varepsilon_\alpha$, with pressure, may be interpreted in terms of a
'blocking' effect on the limited motions of the reference side
group – which requires that gross-microbrownian motions would
still relax the remainder of $<\mu^2>$ of that group. However, the
cross-correlation terms $<\vec{\mu}_k(0).\vec{\mu}_{k'}(t)>$ may be very important for
PMMA. Whilst the equilibrium terms $<\vec{\mu}_k(0).\vec{\mu}_{k'}(0)>$ may not depend
much upon applied pressure, the time-dependence of $<\vec{\mu}_k(0).\vec{\mu}_{k'}(t)>$
may be significantly affected by pressure, in some as yet
undetermined way. Thus whilst the 'one-body' interpretation would
rationalize the behaviour of PBMA, the complete explanation must
take into account the cross-correlation terms.

Clearly then, interpretations of dielectric relaxation
behaviour are made complicated by the presence of cross-
correlation terms in addition to the auto-correlation terms. These
additional terms contribute a strength-factor to $\Delta\varepsilon$ and time-
functions which may differ from that of the auto-correlation time-
function. In view of these factors it might be thought that the
results of dielectric measurements may be more difficult to
interpret in molecular terms than the results of, say, fluorescence,
NMR and ESR studies. However no individual technique is without
its problems in experimentation or interpretation at the molecular
level. The fluorescence-depolarization studies using the steady-
state or time-resolved methods are limited by the life-time
of the fluorophores and there are complications in interpretation
involving the reference coordinates for the initial excitation
and subsequent fluorescence processes. Internal energy-transfer
processes and anisotropic rotation may also complicate the
analysis of data. The ESR and NMR studies are made at selected
frequencies covering only a small range in practice – although
there are some studies where a wide range has been covered (43,44).
Also for proton NMR studies, questions may arise for the mechanism
of nuclear relaxation (e.g. spin-diffusion may obscure the study
of the reorientational motions of groups). Rather than say that
any one technique is superior to another, it is best to regard all
results as providing evidence for the mechanisms of the observed
motional processes. We will show that it is not possible, on the
basis of data obtained from a single experimental technique, to
deduce the mechanism for relaxation. Instead all available data
have to be used together to obtain the required information. We
note that a feature of the dielectric technique is that measurements
may be readily carried out in the range 10^{-4} to 10^6Hz, and, with
greater difficulty, in the range $10^6 - 10^{12}$ Hz and beyond. Wide
temperature and applied pressure ranges can be covered with relative ease. As a result there now exists a large body of experimental data for amorphous solid polymers (1,3-10) which may be
used in support of interpretations of data using fluorescence,
NMR and ESR techniques.

One further point should be noted. Leffingwell and Bueche (45)

found that the characteristic asymmetric shape of the ε''-vs- log ν
plot for the α-relaxation in styrene-p-chlorostyrene random
copolymers was essentially unchanged as the dipolar content of
the chain was raised (by increase in p-chlorostyrene component).
Clearly, for small dipole concentration in the random copolymer
the dielectric behaviour is dominated by the auto-correlation
terms in eqn. 11, whilst at larger dipolar concentrations the
cross-correlation terms will make an important contribution to
$\Lambda(t)$. The lack of variation in the shape of the loss-plots may be
taken to indicate that the time-dependencies of the auto- and cross-
correlation functions in eqn. 11 are approximately the same,
thus collapsing $\Lambda(t)$ to $<\vec{\mu}_k(0).\vec{\mu}_k(t)>/\mu^2$. This has been
discussed by Williams and coworkers (see in ref. 3 p.17-44 and also
ref 46) and they have shown that if the motion of an individual
dipolar group occurs in an isotropic manner with respect to its
original direction, irrespective of the initial conformation, then
auto- and cross-correlation functions will have the same time
dependence. They further showed that the α-relaxations for poly
(vinylacetate), poly(vinyl octanoate), styrene-acrylonitrile
copolymers, acrylonitrile (40%)-butadiene copolymer and styrene-
p-chlorostyrene copolymers all gave quite similar (asymmetric)
ε''- vs - log ν plots, which were, in turn, quite similar to those
found for small-molecule glass-forming systems (e.g. anthrone
in o-terphenyl (32). All data could be approximately fitted
using eqn. (3) with β in the range 0.40 - 0.70, with several
being close to $\beta = 0.50$. These results suggest also that the
description of the essential mechanism which gives rise to the
α-relaxation in these systems does not require information on
the detailed chemical structure - a surprising result. A related
observation is that for the low-frequency (α) dielectric relaxation
of certain alkyl halides a little above their glass transition
temperatures the process is well-described by a Davidson-Cole
or Williams-Watts (eqn. 3) function. Mixtures of two alkyl
halides having quite different T_g values give a single well-
dfined α-process of the same shape as the components but give no
indication of a double peak or broadening of the process. The
mixtures have a single T_g value intermediate between the two
extremes of the components (for a review see ref. 32). Thus the
spatial extent of the fluctuations which give rise to the α-
relaxation must, for this case, greatly exceed the dimensions of
the individual molecules, resulting in both species relaxing at
the same average rate. It would appear that the same situation
prevails for the α-process in poly(vinyl acetate), poly ethylene
terephthalate, poly(vinyl chloride), styrene-p-chlorostyrene
copolymers and acrylonitrile (40%) - butadiene copolymer. The
same conclusions may apply to isotactic poly(methyl methacrylate)
since it has $\Delta\varepsilon_\alpha >> \Delta\varepsilon_\beta$ and an α-process which is similar in
form to that for the above polymers. Physically one may imagine a
given dipole group along the chain moving through a limited solid
angle which partially relaxes $<\mu^2>$ for the group (β-process) but

the group must wait for extremely long times for the micro-
brownian process which will completely relax the remaining dipole
vector $<\vec{\mu}>$ as $\{<\vec{\mu}>\}^2$. The processes for syndiotactic poly(
methyl methacrylate) and the other poly(alkyl methacrylates),
for which $\Delta\varepsilon_\beta > \Delta\varepsilon_\alpha$ are visualized as follows. The local
reorientational motions relax a large part of $<\mu^2>$ for the group
leaving only a small net vector (a 'blurred' dipole vector) to be
relaxed as $\{<\vec{\mu}>\}^2$ in the α-process. These are for 'one-body'
motions, i.e. auto-correlation functions. Since the repeat units
are chemically equivalent we might suppose that all $<\vec{\mu}_k(0).\vec{\mu}_k(t)>$
are equivalent. However this is not the case for polymers where
tacticity is a factor i.e. for many vinyl polymers. Groups
belonging to different triad sequences would be expected to relax
at different rates - and such information may possibly be gleaned
from ^{13}C - ^1H NMR relaxations. Clearly much further interest
should be attached to the complicated dielectric relaxation
behaviour of the lower poly(n-alkyl methacrylates).

The model of partial relaxation (β process) followed by the
α-process, where the former process may occur in a variety of
temporary local environments, leads to the following expression
for the auto-correlation function (10)

$$<\vec{\mu}_k(0).\vec{\mu}_k(t)> = \mu^2.\Phi_{\alpha kk}(t).\left[A_{kk} + B_{kk}.\Psi_{\beta(kk)}(t)\right] \quad (12)$$

$$A_{kk} = \sum_{r(k)} {}^0P_{r(k)}.q_{\alpha r(k)} \quad (13)$$

$$B_{kk}.\Psi_{\beta(kk)}(t) = \sum_{r(k)} {}^0P_{r(k)}.q_{\beta r(k)}.\Phi_{\beta r(k)}(t) \quad (14)$$

${}^0P_{r(k)}$ is the probability that dipole k is found in the environ-
ment r. $q_{\alpha r(k)} + q_{\beta r(k)} = 1$ and $q_{\alpha r(k)} = \{<\vec{\mu}_{r(k)}>\}^2/<\mu^2>$ is the
fraction of $<\mu^2>$ not relaxed by dipole k in environment   by the
β process (where the latter is characterized by $\Phi_{\beta r(k)}(t)$. A_{kk} is
the strength of the α process for dipole k. $\Psi_{\beta(kk)}(t)$ is the
effective relaxation function for the overall β-process. The
weighted sum of all elementary β-processes for the dipole k is
expressed in eqn. 14 and gives the magnitude as B_{kk}. $\Phi_{\alpha kk}(t)$ is
the effective relaxation function for the α-process - i.e. for
the relaxation of the residual quantity A_{kk} after the β process
has occurred. Eqn. 12 is simply an expression of partial and total
relaxation of the dipole k. It is readily generalized to the
case of cross-correlation functions (10). Eqn. 12 rationalizes
the occurrence of α, β and $(\alpha\beta)$ relaxations and also explains
the conservation rule eqn. 4. If $\Psi_{\beta kk}(t)$ is a faster function of
time than $\Phi_{\alpha(kk)}(t)$, eqn. 12 yields α and β processes. If B_{kk} is
decreased by increase in applied pressure, then A_{kk} is increased
in complementary manner since $A_{kk} + B_{kk} = 1$. If $\Phi_{\alpha kk}(t)$ is

increased in rate to the point that it becomes a far faster
function of time than $\psi_{\beta(kk)}(t)$, all of μ^2 is relaxed by the α
process. Thus the coalescence of α and β processes yields, in
this case, an $(\alpha\beta)$ process which is essentially an extrapolation
of the α process to higher temperatures.

This phenomenological approach to the multiple relaxations
rationalizes the observed behaviour in amorphous solid polymers
and in small-molecule glass-forming systems. For cases where
$\Delta\varepsilon_\alpha \gg \Delta\varepsilon_\beta$ it seems likely that the one-body approach to motion
is adequate and that the inclusion of the dipolar cross-
correlation terms would not substantially change the interpretation
of α, β and $(\alpha\beta)$ processes (10). For those poly(alkyl methacrylates)
having $\Delta\varepsilon_\beta > \Delta\varepsilon_\alpha$, whilst the one body approach rationalizes the
observed behaviour as a function of temperature and pressure, it
is likely that the inclusion of cross-correlation terms may
affect the physical interpretation. The data for these polymers
present us with a puzzle. Since $\Delta\varepsilon_\beta$ decreases rapidly with
increasing pressure does this imply that the conformation of
the chain is being significantly changed with pressure? Now
$(\Delta\varepsilon_\alpha + \Delta\varepsilon_\beta)$ is approximately conserved as pressure is raised and
since it is this total quantity which is related to the mean-
square dipole moment (see eqn. 8), it seems likely that $\mu^2 g_1'(0)$
is not greatly affected by pressure. The remarkable decrease in
$\Delta\varepsilon_\beta$ is associated with inter-molecular interactions which block
the ester motion on a time scale short compared with that for
backbone motion but $\mu^2 g_1'(0)$ will be the equilibrium average
taken over a time-scale far longer than that for side-group and
backbone motion in the bulk polymer. Such remarks are qualitative
and further experimental work on chain conformation and its
changes with pressure is required for those poly(alkyl methacrylates)
having $\Delta\varepsilon_\beta > \Delta\varepsilon_\alpha$.

To this point specific mechanisms for motion in the α and β
processes have not been given. This is an advantage since the
mechanisms may differ from one system to another. Implicit in
eqn. 12 is the idea that the β process would be a weighted sum
of elementary processes and would therefore be broad in the time or
frequency domains. Since the experimental β processes have a
half width $\Delta\log\nu \sim 4$ to 6 units of $\log\nu$, there seems to be little
more one can say about its specific mechanism in any given system.
The α-process, however, has a well-defined asymmetric shape with
β in the range $0.4 - 0.6$. In the next section we shall see how
far it is possible to go in deriving its mechanism given comprehensive dielectric data. Before doing so, it is instructive to work
through a very simple model that leads to partial and total
relaxation of a dipole vector.

Consider a two site model in which a dipole vector μ moves
from $\leftarrow$ to $\rightarrow$ over a barrier of height E say. If the energy of

sites 1 and 2 are equal then $\langle\vec{\mu}\rangle$ = 0, where the average is taken over the occupation of the two sites. So all of $\langle\mu^2\rangle$ is relaxed in the barrier system and the correlation function is (exp $-2kt$) where k is the transition probability for motion between the sites. Consider the same two site model except now we introduce an energy difference ΔE between the sites with site 1 having the lower energy. $\langle\vec{\mu}\rangle \neq$ 0 since the dipole spends more time in site 1 than site 2. It is readily shown (20) that $|\langle\vec{\mu}\rangle| = \{(1 - x)/(1 + x)\}$ where $x = \exp-(\Delta E/kT)$, and is the Boltzmann factor for the relative occupation of the sites. The dipole moment correlation function is now $\{\exp-(k_{12} + k_{21})t\}$, where k_{12} and k_{21} are the transition probabilities for motion between the sites, and only a part of $\langle\mu^2\rangle$, $\{4x/(1 + x)^2\}$, is relaxed by motion between the sites (20). This is a very simple example of partial relaxation and it may be generalized to multi-site models or to rotational diffusion models for motions within a restricted solid angle (47). The unrelaxed part, $\{\langle\vec{\mu}\rangle\}^2$, may be relaxed by a separate process e.g. rotation of the reference coordinate frame or 'collapse' of the local environment which restricted the motion. In this way a variety of models may be generated, all of which conforming to the pheno-menological relation for partial and total relaxation, eqn. 12. We now consider specific models for motion.

II.A.5 <u>Models for molecular motion</u>. The experimental dielectric data for α, β and $(\alpha\beta)$ relaxations in amorphous polymers and small-molecule glass-forming systems may be fitted using a variety of models. For the moment consider one body motions e.g. of a given repeat unit in a polymer chain or a whole molecule in a small-molecule glass-forming system. As emphasized earlier (10) such one body motions may be modelled in terms of

(1) Barrier Theories - rotation in a local barrier system (Hoffman and coworkers (48,49,20)

(2) Diffusion Theories - small-step rotation in a viscous continuum (Debye (50)), or in a viscoelastic medium (Di Marzio and Bishop (51)) or large-step rotation in a continuum (Ivanov (52), Anderson (53)).

(3) Defect Diffusion Theories - rotation resulting from 'defects' moving through a liquid (Glarum, (54), Phillips and coworkers (55)).

Such models may always be fitted to a given set of data since they contain adjustable parameters. If the loss-curves are broad, a 'distribution of relaxation times' may be arbitrarily introduced to help fit the data. Given the fact that relaxation curves are at least as broad as an exponential decay in time, it is clear that discrimination between the different models is not possible without further information about a given system. For polymer chains the basic concepts involved in (1) - (3) above may be

incorporated into master equations for the chain dynamics. Thus
Yamafuji and Ishida (56) considered rotational diffusion for a
flexible dipolar chain, Jernigan (57) and later Beevers and
Williams (58) considered the conformational transitions, using
the Flory rotational isomer model, for a flexible dipolar chain.
Monnerie and coworkers (59,60) considered local conformational
transitions for a chain on a tetrahedral lattice. There is little
doubt that several of the models for one-body motion and for chain
motion have considerable justification in their application to
real systems. However, no matter how much we may improve the
precision of the data and its range in the time or frequency
domains, it is not possible, a priori, to distinguish between
different models especially if a distribution of relaxation times
is allowed. Such a dilemma does not arise, for example, for
high-resolution molecular vibration-rotation spectra since the
spectra are rich in detail and may only require a small number of
parameters in order to fit a large number of observed frequencies.
How may one state the basic problem in the elucidation of the
mechanism of motion in a formal manner? The answer lies in the
orientational distribution functions which express the angular
motion of a body in space. In general, the motion of a rigid
body may be expressed in terms of the Wigner rotation matrices
$D_{K,M}^{J}(\Omega)$ which describe the spatial distribution, in Euler space,
and time correlation functions $<D_{K,M}^{J}(\Omega(t))D_{K,M}^{J*}(\Omega(0))>$. A simpler
case is the anisotropic motion of a vector, which may be expressed
in terms of the spherical harmonics $Y_{m}^{l}(\Theta,\emptyset)$, which describes the
spatial distribution in $(\Theta,\emptyset)$ space, and time-correlation functions
$<Y_{m}^{l}(\Theta(t),\emptyset(t))Y_{m}^{l*}(\Theta(0),\emptyset(0))>$. The simplest case of all is the
axially symmetric rotation of a vector, which may be expressed in
terms of Legendre polynomials, $P_{n}(\cos \Theta)$, which describe the
spatial distribution, and time-correlation functions
$<P_{n}(\cos\Theta(t) P_{n}(\cos\Theta(0))>$. The essential problem of elucidating a
relaxation mechanism from dielectric data may be explained in
terms of either of these three cases but is most simply done for
the case of axially-symmetric motion. For this case we write the
conditional probability density of obtaining the vector in the
range Ω to $\Omega + d\Omega$ at time t given the vector was in the range
Ω_0 to $\Omega_0 + d\Omega_0$ at time t = 0 as

$$f(\Omega,t|\Omega_0,0) = \frac{1}{4\pi} \sum_{n=0}^{\infty}(2n + 1)P_{n}(\cos\Theta)\Psi_{n}(t) \qquad (15)$$

where Θ is the angle between the vectors at t = t and t = 0. The
$\Psi_{n}(t)$ are time-correlation functions of the motion and are given by

$$\Psi_{n}(t) = \int f(\Omega,t|\Omega_0,0)P_{n}(\cos\Theta)d\Omega d\Omega_0 \qquad (16)$$

Now dielectric measurements give $\Psi_1(t)$ thus only give a part of
the expansion in eqn. 15. In order to establish the mechanism for
motion, $f(\Omega,t|\Omega_0,0)$ is required but we only have a single term in

its expansion. Therefore we <u>cannot</u> establish the mechanism from
dielectric measurements alone, no matter how precise and extensive
such measurements might be. This is the formal statement of the
basic problem in relaxation phenomena, and if we so wished we
could state it in more general terms if $f(\Omega,t|\Omega_0,0)$ was expanded
in terms of the Wigner rotation matrices and their associated
correlation functions.

The only way to establish the relaxation mechanism is to
obtain as many terms $\Psi_n(t)$ in the expansion, eqn. 15, and, since
the $P_n(\cos \Theta)$ decrease rapidly with increasing n, it is possible
that $f(\Omega,t|\Omega_0,0)$ could be determined with reasonable accuracy.
Experimentally $\Psi_2(t)$ is accessible from fluorescence depolarization
experiments in the time-domain and dynamic Kerr-electro-optical
decay transients in the time domain. It is not clear how $\Psi_n(t)$
for $n \geq 3$ could be obtained experimentally. Thus at the present
time it seems that the best one can do is to measure $\Psi_1(t)$ and
$\Psi_2(t)$ for the same vector using dielectric and fluorescence or
Kerr-effect techniques and then compare them in terms of the
various models for motion. Now both dielectric and Kerr-effect
measurements for polymers have problems in interpretation due to
(i) cross-correlation terms contributing equilibrium and dynamic
factors to the correlation function (ii) complicating effects of
anisotropic rotation. Thus perhaps the most significant tests of
the mechanisms for motion in glass-forming systems were those
made by Williams and coworkers (61) for (a) a rigid dipolar
solute, fluorenone, in supercooled o-terphenyl and (b) the ion-
pair, tri-n-butyl ammonium picrate, in supercooled o-terphenyl.

For case (a) the α-process was studied by dielectric and Kerr-
effect techniques. It was found that the rise and decay transients
were entirely equivalent at each temperature and that the
derived relaxation functions, $\zeta(t)$ say, had the same non-exponential
form and average relaxation time as that obtained from dielectric
studies made concurrently. Now for small-step rotational
diffusion

$$\Psi_n(t) = \exp(-n(n + 1)D_R t) \tag{17}$$

where D_R is the rotational diffusion coefficient. This would
lead to a Kerr-effect rise-function slower than the decay function
and to the condition $\tau(\text{dielectric}) = 3\tau(\text{Kerr-effect})$. The fact
that the measurements for (a) do not agree with this model allows
one to rule-out small-step diffusion as the mechanism for the
α-process in the fluorenone/o-terphenyl system. A mechanism
which is entirely consistent with the data is the 'fluctuation-
relaxation' mechanism (62). Here it is considered that a
reference molecule is subjected to local 'fluctuations', being
the random forces applied to the molecule by its neighbours, and
it is assumed that when the molecule moves it does so instantly

and randomizes its orientation completely (i.e. it jumps through
angles of arbitrary size). The orientation distribution function
for this 'strong-collision' model is given by

$$f(\Omega,t|\Omega_0,0) = \delta(\Omega - \Omega_0)\zeta(t) + \frac{1}{4\pi} (1 - \zeta(t)) \qquad (18)$$

and the correlation functions $\Psi_n(t) = \zeta(t)$; i.e. they are all
equivalent. In particular $\Psi_1(t)$ & $\Psi_2(t)$ are equivalent, as found
for the fluorenone/o-terphenyl system. It is important to emphasize
that this model for motion is only one of several which gives
$\Psi_1(t) = \Psi_2(t)$. For example the motion of a dipole vector on a
tetrahedral lattice, as described by Monnerie and coworkers (60)
also gives $\Psi_1(t) = \Psi_2(t)$ and prescribes the form of $\Psi_1(t)$ as

$$\Psi_1(t) = \exp(t/\tau)\mathrm{erfc}(t/\tau)^{\frac{1}{2}}$$

Note that $\zeta(t)$ remains to be prescribed in eq. 18. The
experimental data for the α-process in fluorenone/o-terphenyl may
be fitted by the empirical relaxation function $\exp(-t/\tau)^\beta$ with
$\beta \approx 0.5$. It remains to be seen if models for fluctuations in
viscous organic liquids can be devised which would give a function
of this form. For the α-relaxation in system (b) it was found
that the orientation mechanism for the ion-pair changed with
temperature. At the lowest temperatures studied, i.e. a little
above T_g, the dielectric and Kerr-effect data conform to the
'fluctuation-relaxation' model in all respects but on increasing
the temperature the rise transient of the Kerr-effect became
slower than the decay transient and the behaviour tended to that
of a single relaxation time process with τ(Kerr-effect) $\approx (^1/3)$
τ(dielectric). This is exactly that expected for small-step
rotational diffusion of a dipolar molecule so it appears that
for this solute the α-process at high temperatures involves
small-step diffusion but as temperature is decreased the solvent
molecules 'clamp-in' on the solute leading to cooperative motions
like those of enmeshing cogs in which molecules jump through
angles of arbitrary size.

Some comparative studies of the dielectric and Kerr-effect
relaxations of bulk amorphous polymers have been made. The
α-relaxation of a poly(phenyl methyl siloxane) (63) and of a
series of low molecular weight poly(propylene glycols) were
studied (64) but the behaviour is complex, showing evidence for
marked anisotropy for motion of the chains and making interpretations difficult. The author suggests that interpretations
of motions for polyether chains or poly siloxane chains in the
bulk (or in solution) should be quite different from those of
PET, PVAc, and poly(alkyl methacrylates). The motions of oxide
polymers on tetrahedral lattices have been modelled by Monnerie
and Geny (60) but these calculations may be most appropriate to
the solution phase. We shall further discuss the motions of

fairly 'smooth' chains such as polyethylene, poly(trichloro-
fluoroethylene), poly(vinylidene fluoride) and poly(vinyl chloride)
when describing the behaviour of partially crystalline polymers.

In summary we note that the mechanisms for α, β and $(\alpha\beta)$
relaxations in certain amorphous solid polymers appear to be
quite similar to those in small-molecule glass-forming systems,
and that cross-correlation functions are a complicating factor,
but not an essential one, for the interpretation of relaxations of
polymers having $\Delta\varepsilon_\alpha \gg \Delta\varepsilon_\beta$. For the poly(n-alkyl methacrylates)
having $\Delta\varepsilon_\beta > \Delta\varepsilon_\alpha$ although the dielectric behaviour is easily
rationalized within the one-body representation it is possible
that the cross-correlation terms play a determining note. It is
emphasized that the relaxation mechanism may only be understood
if the results of complementary experiments such as dielectric
and Kerr-effect relaxation are compared. A start has been
made in this direction for the α-process in small-molecule glass-
forming systems but much remains to be done for polymer systems.
For certain small-molecule glass-forming systems the observation
that $\Psi_1(t) \simeq \Psi_2(t)$ rules out small-step diffusion and favours a
'fluctuation-relaxation' mechanism for dipole reorientation.

Finally we note that the free-volume theory of the dependence
of relaxation time upon temperature does not explicitly involve
time in its derivation. The WLF equations (65) may be derived
from the empirical relation $\tau = \tau_0 \exp\{B/(T - T_\infty)\}$ or by first
relating τ to a free-volume v_f and then writing v_f as a linear
function of temperature. It is found that the ratio of constant
volume to constant pressure activation energy, $\{Q_V(T,V)/Q_P(T,P)\}$
is in the range 0.7 to 0.8 for the α-process in amorphous polymers
(66). It is difficult to reconcile the usual free-volume
theories with this result. If the volume V is the sum of a
'bound' volume, v_b, and a 'free'-volume, v_f, and v_b is a constant,
then increase in temperature for constant V would give no change
in v_f and hence $Q_V(T,V)$ would be zero. If v_b increases with
increasing temperature (which is physically reasonable) then
increasing temperature for constant V might lead to a <u>decrease</u>
in v_f and hence $Q_V(T,V)$ would be <u>negative</u>. The difficulty with
the free volume theories is that they attempt to relate a
relaxation time (a dynamic quantity) to a time-averaged quantity
'free-volume' (an equilibrium quantity) and they do not explicitly
take into account the temporal fluctuations in volume that
actually lead to relaxation. Keeping the macroscopic volume
constant while temperature is raised does not keep the local volume
fixed. That might be possible in a crystal but is not possible
in a glass-forming liquid. The constant-macroscopic-volume
constraint as temperature is raised slows down the increase in
relaxation rate over that for constant pressure, giving
$Q_V(T,V) < Q_P(T,P)$, but $Q_V(T,V)/Q_P(T,P)$ is so large that it
emphasizes that the relaxation occurs by <u>thermally</u> activated

fluctuations and is not describable in terms of time-averaged
free volume models.

III   PARTIALLY CRYSTALLINE POLYMERS

II.A   General Remarks

     Broadly speaking, melt-crystallized or quench-annealed-
crystallized polymers fall into two general classes:- those of
degree of crystallinity $\chi \lesssim .55$ and those having $\chi \gtrsim .80$. It is
appropriate to consider their dielectric behaviours separately.
Owing to the large literature which has accumulated, our account
is highly selective and is limited, as far as possible, to
recent studies which build on the earlier works.

III B   Polymers having medium degrees of crystallinity.

     In this category we have melt-crystallized poly ethylene
terephthalate, poly(vinylidene fluoride), poly(chlorotifluoro-
ethylene), polycarbonates, and the nylons. The essential features
of the dielectric behaviours are summarized in the book by
McCrum, Read and Williams (1). For the present account we consider
the behaviour of partially crystalline polyethylene terephthalate
and the nylons. We note that Williams (10) recently summarized
the data for polyethylene terephthalate and for poly(vinylidene
fluoride). The latter polymer has attracted great interest in
recent years due to its remarkable properties as a piezoelectric
material.

     First we note that the term 'amorphous' is commonly used in
connexion with the structure and properties of partially
crystalline polymers. The well-known morphological studies
of partially crystalline polymers of medium and high degrees of
crystallinity suggest that the crystalline and non-crystalline
regions of the material within spherulites are in intimate
contact. It is therefore possible that the non-crystalline regions
are in a quite different state from that obtaining in the melt or
in any wholly amorphous polymer. Rather than use the term
'amorphous' we shall refer to the non-crystalline regions as
being 'disordered'.

     For polyethylene terephthalate it is well established that
the amorphous polymer gives a sharp, well-defined dielectric
α-process and a small, broad β process (see ref. 1 and Section
II.A.1 above). For melt crystallized PET with $\chi = 0.55$ the α
peak is broadened and smaller and is shifted to lower frequencies
(1,10). This might be expected since dipole reorientation in the
crystalline regions is very restricted, giving a small $\Delta\varepsilon$, and

since the disordered regions are ill-defined in the spherulites.
Tidy and Williams (67) studied the dielectric relaxation of
amorphous (quenched) polymer as the material was crystallized by
heating the glass rapidly from just below T_g to 106.7°C. At this
temperature the crystallization proceeds by growth of spherulites
in the supercooled liquid polymer until impingement occurs, and
the time scale (0 - 200 min) of crystallization allows one to
follow the loss in the range 10^2 - 10^5Hz. Tidy and Williams
showed that as crystallization proceeded the normal 'amorphous'
α relaxation decreased in magnitude without a shift in $\log\nu_m$ and
concurrently there grew a new process, the α' relaxation say,
which occurred at $\log\nu_m$ about 2 below that of the normal 'amorphous' α relaxation. Thus as crystallization proceeds the normal
amorphous phase between spherulites disappears leading to a
reduction of the usual α process and the disordered regions
within the growing spherulites give the small, broad and
relatively slow α' relaxation. A detailed analysis shows that
$\Delta\varepsilon_\alpha$, at long times is less than that expected on the basis of the
known degree of crystallinity ($\chi \simeq 0.5$) which indicates that a
part of the disordered regions within the spherulites is
immobilized, presumably due to contact with the crystalline
regions. The remainder of the disordered regions within the
spherulites are dynamically disordered and give rise to a very
broad α' process. This breadth arises due to the constraints
imposed by the crystalline regions on the disordered regions,
leading to a much lower average mobility for the chain segments
(thus a higher T_g) and a whole range of relaxation times. Thus
the disordered regions within the spherulites may not be
regarded as a normal amorphous phase such as one has in the bulk
amorphous polyethylene terephthalate.

Another polymer group having only medium degrees of
crystallinity is the nylons. The complicated behaviour of
several nylons (nylon 6,6, nylon 6, nylon 6,10) is summarized and
analyzed in McCrum, Read and Williams (1). More recently Boyd
and Porter (68) and Yemni and Boyd (69) have studied nylon 6-10
and the nylons 7-7 and 11 respectively. Boyd and Porter (68)
showed for a 50% crystalline sample of nylon 6-10 that the
α-process increased in its magnitude, $\Delta\varepsilon_\alpha$, up to the melting
point (225°C). The process was continuous into the melt with
respect to location ($\log\nu_m$) and shape of the loss peak but the
magnitude $\Delta\varepsilon_\alpha$ was discontinuous on melting with $\Delta\varepsilon(\text{melt}) > \Delta\varepsilon_\alpha$.
They conclude that the process occurs in the amorphous regions
(we would say 'disordered' regions) and, near T_{melting}, the
melting process creates more of this material but does not
greatly alter its nature. A feature of the results is that the
half-width $\Delta\log\nu$ of the ε_α'' - vs - $\log\nu$ plots decreases rapidly
with increasing temperature. The strength of the α-process
increases by a factor of approximately two in the melting range
200 - 225°C being indicative of the creation of more disordered

phase in the material at the expense of the crystalline phase. This α-process is, however, unlike that for the α-process in wholly amorphous polymers such as polyethylene terephthalate and polyvinyl acetate in two respects: (a) the distribution parameter characterizing the width of the peak increases markedly with increasing temperature (loss peak narrowing) and (b) the apparent activation energy, Q_{app}, is approximately constant over a wide range of temperature (thus does not follow a WLF-type relation) and is small (Q_{app} - 80 kJmol^{-1}). Such differences suggest that the mechanism for motion of the - CONH - dipole groups in the disordered regions is quite different from the 'fluctuation-relaxation' α process - however strongly perturbed by the crystalline regions - in partially-crystalline polyethylene terephthalate. Indeed, it suggests that the motions in the nylon 6-10 are to do with conformational changes in disordered regions which are influenced to some degree by the crystalline regions. We shall return to this question again when considering polyethylene relaxations. Yemni and Boyd (69) made a dielectric study of the odd-numbered polyamides nylon 7-7 and nylon 11. These have dipoles with components that point in the same direction, i.e. perpendicular to the long-axis of the planar zig-zag chain, but parallel to each other, rather than alternating parallel and anti-parallel as for the even-numbered polyamides. They found that the α-process gave a normalized $\Delta\varepsilon_\alpha$ which was markedly higher than that in the even-numbered nylons. They also studied the oriented materials and suggest that the interpretation of the anisotropy in the dielectric α-relaxation does not require the disordered regions to be orientated (and thus be anisotropic) but rather that the anisotropy is due to composite nature of the crystalline and disordered material in an oriented sample.

We note that the 'α_a' process in poly(vinylidene fluoride) (for a review see ref. 10), is characterized by ε_α'' - vs - logν plots which narrow rapidly with increase in temperature (Δlogν decreases from about 6 to 2 on going from 238 to 272K), $\Delta\varepsilon_\alpha$ increasing rapidly in the range just above T_g and Q_{app} which has rather small values (80 - 120 kJ mol^{-1} in the range 273 to 253K). Similar behaviour maintains for the α processes in poly(chlorotrifluoroethylene) (11) and oxidized polyethylene (see below). Such similarities are intriguing since in all cases one is dealing with 'smooth' or 'thread-like' chains, which are very different from polyethylene terephthalate, poly(vinyl acetate) and the alkyl methacrylate polymers.

III.C Polymers having high degrees of crystallinity.

Poly(oxymethylene), poly(ethylene oxide), poly(trimethylene oxide) and poly(tetramethylene oxide) crystallized from the melt or from dilute solution or by solid-state polymerization all give

highly crystalline samples in which dielectric absorptions are
observed. Their properties have been reviewed by McCrum, Read
and Williams (1), Ishida (6) and by Williams (10). The most
important feature is that solution crystallized or solid state
polymerized samples, which have higher order than the melt-
crystallized samples, exhibit a small broad relaxation e.g. for
a single crystal melt sample of poly(ethylene oxide) T_{max} is at
188K for 12.8 kHz (6). Melting a single crystal mat or
solid-state-polymerized sample yields two processes, one as in
the original sample and another which occurs at higher temperatures:
e.g. for poly(ethylene oxide) T_{max} is 240K at 12.8 kHz for this
second process. The higher temperature process (γ' say) appears
to be due to motion in regions of considerable disorder between
the lamella crystals. It can be shifted (plasticized) by
swelling with dioxane for poly(oxymethylene) (70) again suggesting
that its origin is in regions between the crystals. It is
important to note that the breadth of the γ' process narrows
considerably as temperature is increased, just as is observed
for the glass-transition process for PVF_2 the nylons and for the
γ process in poly chlorotrifluoroethylene) mentioned above.

Polyethylene is essentially non-dipolar, but the studies of
lightly oxidized or chlorinated polymer are amongst the most
extensive in the polymer literature (1,4-6). In 1966, Hoffman,
Williams and Passaglia (71) presented a number of models for the
α, β and γ processes in partially crystalline polymers, with
special emphasis being given to the behaviour of poly(ethylene)
and poly(chlorotrifluoroethylene). A comprehensive and detailed
study of the dielectric behaviour of lightly oxidized and lightly
chlorinated samples of low and high density poly(ethylenes)
of varying thermal histories has been given by Ashcraft and
Boyd (72). In addition, Sayre, Swanson and Boyd (73) have
studied the effects of hydrostatic pressure on the α and γ
relaxations, Mansfield and Boud (74) have carried out conformational energy calculations for motions in paraffin and polyethylene
crystals and, very recently, Boyd and Sayre (75) have examined
the dielectric evidence for the thermal creation of defects in
poly(ethylene) crystals. The work contained in this series of
papers is of the greatest importance for our understanding of the
α, β and γ relaxations in polyethylene. The main conclusions
relate to the nature of the α and γ relaxations. Firstly, it
appears that the C = O or C - Cl dipoles are preferentially
rejected from the crystals. Also the strength of the γ process
is well below that for full participation by all the dipoles in
the disordered regions where that process occurs (72). With
regard to the activation energy of the α-process its dependence
on chain length (for long chain esters and ketones) and its
dependence upon lamella thickness (for polyethylenes) we suggest
that the earlier models (71) for reorientation within the
crystal are at least true in a semi-quantitative sense. These

earlier models, due to Fröhlich and subsequently to Hoffman and
coworkers (71,76), proposed that for short chains, the reorientation in the crystal occured as a rigid rod process but for long
chains the chains twisted as they rotated and translated by C/2
along the chain direction. The calculations were made quantitative
by Mansfield and Boyd (74) using a realistic intermolecular potential for a model of a $C_{22}H_{46}$ crystal. They showed that the
dependence of activation energy on chain length resulted not from
a rotational lattice mismatch of the twisted region but from
a translational lattice mismatch induced by the π-rotation of one
planar zig-zag stem relative to the other. The twisted region
propagates smoothly without local barriers to its motion thus it
may be thought of as a transition state (saddle-point state) rather
than a hopping defect. Using the pressure dependence of the
relaxation time for the α process in order to convert from
constant pressure to constant volume apparent activation energies,
it was shown that the theoretical relation for apparent activation
energy as a function of chain length for the α-process, gave
a good representation of the experimental data (see ref. 74,p.1243).
Very recently Boyd and Sayre (75) have further examined the
possibility that defects (or 'kinks') created thermally may
contribute to the overall α-process by providing an alternative
relaxation path involving a defect or by interfering with the
normal process described above. This latter study was prompted by
the observation of Sayre and coworkers (73) that $\Delta\varepsilon_\alpha$ x T decreased
markedly with increasing temperature. Such behaviour indicated
that a portion of the relaxation strength was being taken by a
much faster process in the crystal which robs the normal α-process
of its relaxation strength. Such a process could arise from
thermally generated defects, and Boyd and Sayre (75) conclude that
defects of energy 12 - 25 kJ mol^{-1} and having a high degeneracy,
if allowed, would satisfy the observed behaviour (73). These
defects could be kinks in the normal trans-chain e.g. ...TTGTG'TT...
where T and G indicate trans and gauche sequences. They suggest
that the thermally created defect is unlikely to be a Reneker
defect owing to the higher energy of such defect (∿50 kJ mol^{-1}).
Thus the studies over many years, culminating in the recent works
of Boyd and coworkers, appear to have led to a good understanding
of the α process for dielectric relaxation in poly(ethylenes).

The γ process was also studied and analyzed by Boyd and
coworkers (72,73). Ashcraft and Boyd (72) clearly demonstrated
that the calculated fraction, f_γ, of dipoles participating in the
γ process was in the range 0.1 - 0.2, and f_γ decreased with
increasing degree of crystallinity. Their plot of f_γ against
crystalline fraction does not extrapolate to give $f_\gamma \approx 0$ for
χ = 100% so it is still unclear if the γ process contains
contributions from within the lamella crystals (γ_c process as
considered by Hoffman and coworkers (71). However Sayre and
coworkers (73) found that the shape of the γ peak was essentially

unchanged with increased pressure giving no indication of a
resolution of component processes and thus inferring that the
γ process is a single very broad process. Ashcraft and Boyd
reason that the γ process is due to local conformational
reorganizations of the chains, being dependent upon internal
chain energetics and partially upon environmental packing factors.
At low temperatures the variety of environments leads to broad
loss curves but on increasing the temperature the influence of
local packing diminishes and the loss process becomes narrower,
tending at the highest temperatures to reflect the intramolecular
conformational changes. Clearly the γ process takes place in
disordered regions of the polymer, but it is still unclear as to
their exact location (surface of lamellae, between lamellae?).
The similarity of the activation energies and the temperature
dependence of the shape of the ε_γ'' - vs - $\log\nu$ curves for polyethylene, for poly(chlorotrifluoro ethylene (γ process) and for
the β process in poly(vinylidene fluoride) and the α process in
the nylons, as mentioned above, is striking. In all cases one
is dealing with chains which may undergo local conformational
changes, requiring only small internal energy changes for the
chain, in which the dipole vectors are partially relaxed. It is
interesting to speculate that the low values of f_γ for polyethylene
(oxidized) simply reflect <u>partial</u> reorientation of C = O dipole
vectors between energetically equivalent states involving a wide
range of local barriers due to inter and intra-molecular factors.
However such a statement is too general. It is necessary to
devise a theory which will allow a collection of isolated dipole
vectors (no cross-correlation dipole factors) attached to chains
to give a broad spectrum of relaxation times. As far as the
author is aware, this has not been done. It seems likely that
at the highest temperatures a dipole vector partially relaxes due
to fluctuations in energy that are spatially localized near the
vector, but at lower temperatures the energy changes become
progressively delocalized along and between chains so that the
partial motions are distributed across a sequence of collective
events. How this may be formulated into a model theory for dynamics
remains an important and unsolved problem.

IV. CONCLUDING REMARKS

It should be clear from the above account that, following the
extensive documentation of the dielectric behaviour of solid
polymers, there are now emerging a number of situations where
the mechanism for relaxation is well-understood. Much remains
uncertain, especially the understanding of those extremely broad
relaxations which are observed for a molecularly simple system
e.g. C = O motions which give the γ process in oxidized bulk poly-
(ethylene). Given the variety of chemical composition and
structure of amorphous polymers it is surprising how similar the

observed behaviour is for α, β, (αβ) relaxations so general phenomenological relations such as discussed here (eqn. 12) are a pre-requisite to our understanding of the detailed mechanisms. However, it should be clear from eqn. 15, and its generalization to the anisotropic motion of whole units, that insufficient information is contained in the results of one experiment (e.g. dielectric, NMR, fluorescence, ESR) to deduce the mechanism for motion. It is essential that all available data be brought together and rationalized as a whole in order to understand the origin and nature of given relaxations. When doing so it is well-worth remembering the principle of Joel H. Hildebrand stated in a paper on the use of models in Physical Science to the American Philosophical Sciety in 1964 (77) 'A model should be regarded as suspect if it yields inferences in serious conflict with any of the pertinent properties of a system, regardless of how closely it can be made to agree with some, especially if there are adjustable parameters. A model that is consistent with all properties, even if only approximately, can probably be made more precise; but if it is in irreconcilable conflict with any part of the evidence, it is destined to be discarded, and in the meantime, predictions and extrapolations based upon it should be regarded as unreliable'.

REFERENCES

1. McCrum, N.G., Read, B.E., and Williams, G., 1967, *Anelastic and Dielectric Effects in Polymeric Solids,* Wiley, London and New York.
2. Hill, N., Vaughan, W.E., Price, A.H., and Davies, M., 1969, *Dielectric Properties and Molecular Behaviour,* Van Nostrand, New York.
3. Karasz, F.E., ed., 1972, *Dielectric Properties of Polymers,* Plenum, New York.
4. Hedvig, P., 1977, *Dielectric Spectroscopy of Polymers,* Adam Hilger, Bristol.
5. McCall, D.W., 1969, in *Molecular Dynamics and Structure of Solids,* ed. Carter, J., and Rush, J.J., Nat. Bur. Stand. Special Publ. No. 301, Washington, D.C.
6. Ishida, Y., 1969, J. Polymer Sci., A2, 7, 1835.
7. Saito, S., 1964, Res. Electrotecha Lab., Tokyo, Japan, No. 648.
8. Sasabe, S., 1971, Res. Electrotech. Lab., Tokyo, Japan, No. 721.
9. Wada, Y., 1977, in *Dielectric and Related Molecular Processes* ed. M. Davies, Spec. Period. Report, The Chem. Soc., London, 3, 143.
10. Williams, G., 1979, Adv. Polymer Sci., 33, 59.
11. Scott, A.H., Scheiber, D.J., Curtis, A.J., Lauritzen, J.I., and Hoffman, J.D., 1962, J. Res. Nat. Bur. Stand., 66A, 269.
12. Williams, G., 1965, Trans. Faraday Soc., 61, 1564.

13. Koppelmann, J., and Gielessenn, J., 1961, Z. Elek., 65, 689.
14. Williams, G. and Watts, D.C., 1971, Trans. Faraday Soc.,
    67, 1971.
15. Saito, S., Sasabe, H, Nakajima, T., and Yada, K., 1968,
    J. Polym. Sci., A-2, 6, 1297.
16. Williams, G., 1966, Trans. Faraday Soc., 62, 1321.
17. Williams, G., Watts, D.C., and Nottin, J.P., 1972, J. Chem.
    Soc., Faraday Trans. II, 68, 16.
18. Williams, G., and Watts, D.C., 1970, Trans. Faraday Soc.,
    66, 80.
19. Williams, G., Watts, D.C., Dev, S.B., and North, A.M., 1971,
    Trans. Faraday Soc., 67, 1323.
20. Williams, G., and Cook, M., 1971, Trans. Faraday Soc., 67, 990.
21. Shindo, H., Murakami, I., and Yamamura, H., 1969, J. Polym.
    Sci., A-1, 7, 297.
22. Williams, G., and Edwards, D.A., 1966, Trans. Faraday Soc.,
    62, 1329.
23. Williams, G., 1966, Trans. Faraday Soc., 62, 2091.
24. Sasabe, H., and Saito, S., 1968, J. Polymer Sci., A-2, 6, 1401.
25. Pohl, H.A., Bacskai, R., and Purcell, W.P., 1960, J. Phys.
    Chem., 74, 1701.
26. Williams, G., and Watts, D.C., 1971, Trans. Faraday Soc.,
    67, 2793.
27. Johari, G.P., and Smyth, C.P., 1969, J. Amer. Chem. Soc.,
    91, 5168.
28. Johari, G.P., and Goldstein, M., 1970, J.Phys. Chem., 74, 2034.
29. Johari, G.P., and Goldstein, M., 1970, J. Chem. Phys., 53,2372.
30. Johari, G.P., 1973, J. Chem. Phys., 58, 1766.
31. Johari, G.P., 1976, Ann. New York Acad. Sci., 279, 117.
32. Williams, G., 1975, in Dielectric and Related Molecular
    Processes, ed. M. Davies, Spec. Period, Rep., The Chem. Soc.,
    London, 2, 151.
33. Böttcher, C.J.F., Theory of Electric Polarization, Second Ed.,
    (a) Vol. I, 1973, (b) Vol. II, 1978, Elsevier, Amsterdam.
34. Cook, M., Watts, D.C., and Williams, G., 1970, Trans. Faraday
    Soc., 66, 2503.
35. Volkenstein, M.V., 1963, Configurational Statistics of
    Polymeric Chains, Wiley - Interscience, New York.
36. Flory, P.J., 1969, Statistical Mechanics of Chain Molecules,
    Wiley - Interscience, New York.
37. Read, B.E., 1965, Trans. Faraday Soc., 61, 2140.
38. Williams, G., 1963, Trans. Farady Soc., 59, 1397.
39. Williams, G., 1964, Trans. Faraday Soc., 60, 1556.
40. Williams, G., 1965, Trans. Faraday Soc., 61, 1564.
41. Work, R.N., 1974, J. Chem. Phys., 60, 3078.
42. Williams, G., 1972, Chem. Rev., 72, 55.
43. Kimmich, R., 1977, Polymer, 18, 233.
44. Kimmich, R., and Schmauder, K.L., 1977, Polymer, 18, 239.
45. Leffingwell, J., and Bueche, F., 1968, J. Appl. Phys., 39,
    5910.

46. Williams, G., Cook, M., and Hains, P.J., 1972, J. Chem. Soc.,
     Faraday Trans. II, 68, 1045.
47. Wang, C.C., and Pecora, R., 1980, J. Chem. Phys., 72, 5333.
48. Hoffman, J.D., and Pfeiffer, H.G., 1954, J. Chem. Phys., 22,
     132.
49. Hoffman, J.D., 1955, J. Chem. Phys., 23, 1331.
50. Debye, P., 1929, *Polar Molecules*, Chemical Catalog Co. New York
51. Di Marzio, E.A., and Bishop, H., 1974, J.Chem. Phys., 60, 3802.
52. Ivanov, E.N., 1964, Sov. Phys., JETP, 18, 1041.
53. Anderson, J.E., 1972, cited in J. Chem. Soc., Faraday Symp.
     6, 90.
54. Glarum, S.H., 1960, J. Chem. Phys., 33, 639.
55. Phillips, M.C., Barlow, A.J., and Lamb, J., 1972, Proc. Roy.
     Soc., A 329, 193.
56. Yamafuji, K., and Ishida, Y., 1962, Koll. Z., 183, 15.
57. Jernigan, R.L., 1972, in ref. 3., p.99.
58. Beevers, M.S., and Williams, G., 1975, Adv. Molec. Relax.
     Processes, 7, 237.
59. Valeur, B., and Monnerie, L., 1976, J. Polymer Sci. (Polymer
     Phys. Edn.), 17, 11, 29.
60. See e.g. Geny, G., and Monnerie, L., 1979, J. Polymer Sci.,
     (Polymer Phys. Edn.,) 17, 131, 147, 173 and refs therein.
61   Beevers, M.S., Crossley, J., Garrington, D.C., and Williams,
     G., 1976, J. Chem. Soc., Faraday Symp., 11, 38.
62. Beevers, M.S., Crossley, J., Garrington, D.C., and Williams,
     G., 1976, J. Chem. Soc., Faraday Trans. II, 72, 1482.
63. Beevers, M.S., Elliott, D.A., and Williams, G., 1980, Polymer,
     21, 279.
64. Beevers, M.S., Elliott, D.A., and Williams, G., 1980, Polymer,
     21, 13.
65. Ferry, J.D., 1961, *Viscoelastic Properties of polymers,*
     J. Wiley, New York.
66. Williams, G., 1964, Trans. Faraday Soc., 60, 1548, 1556.
67. Tidy, D.T., and Williams, G., manuscript in preparation.
68. Boyd, R.H., and Porter, C.H., 1972, J. Polymer Sci., A-2,
     10, 647.
69. Yemni, T., and Boyd, R.H., 1979, J. Polymer Sci., Polymer
     Phys. Edn., 17, 741.
70. Read, B.E., and Williams, G., 1961, Polymer, 2, 239.
71. Hoffman, J.D., Williams, G., and Passaglia, E., 1966, J.Polymer
     Sci., C, 14, 173.
72. Ashcraft, C.R., and Boyd, R.H., 1976, J.Polymer Sci., Polymer
     Phys. Edn., 14, 2153.
73. Sayre, J.A. Swanson, S.R., and Boyd, R.H., 1978, J. Polymer
     Sci., Polymer Phys. Edn., 16, 1739.
74. Mansfield, M., and Boyd, R.H., 1978, J. Polymer Sci., Polymer
     Phys. Edn., 16, 1227.
75. Boyd, R.H., and Sayre, J.A., 1979, J. Polymer Sci., Polymer
     Phys. Edn., 17, 1627.
76. Williams, G., Lauritzen, J.I., and Hoffman, J.D., 1967,
     J. Appl. Phys., 38, 4203.
77. Hildebrand, J.H., 1964, Amer. Phil. Soc., 108, 411.

ACOUSTIC STUDIES OF POLYMERS

R.A. Pethrick

Department of Pure and Applied Chemistry, University of
Strathclyde, Glasgow G1 1XL, Scotland, U.K.

ABSTRACT

Acoustic propagation studies of dilute, semi-dilute and
concentrated solutions of polystyrene in toluene are used to
illustrate how this technique can be used to characterize high
frequency normal mode and segmental relaxation of the polymer
backbone.   The amplitude of the acoustic relaxation is observed
to vary in a non linear manner with increasing concentration of
polymer and is indicative of the onset of polymer-polymer interactions.   Investigations of the effect of variation of the
molecular weight of the acoustic relaxation in polydimethylsiloxane
are interpreted in terms of the effects of polymer entanglement.
In the case of solid polymers, relaxation features observed in
the megahertz frequency range correlate closely with similar
phenomena detected using dielectric and dynamic mechanical relaxation methods and are characteristic of local and side chain
motions of the polymer.   Marked increases in the attenuation are
observed when the wavelength of the sound wave becomes comparable
with the domain size and scattering occurs.   Scattering in the
case of isotropic polyethylene can be attributed to the effects of
domains, whereas in the case of drawn polyethylene it can be
associated with the presence of air voids.   Investigations of
polar diagrams allows identification of crystal twinning in low
draw ratio polyethylene.   The use of high frequency sound waves
in the detection of phase separation in solutions of polystyrene-
butadiene-styrene triblock copolymer is also described.

R. A. Pethrick and R. W. Richards (eds.), Static and Dynamic Properties of the Polymeric Solid State, 241–249.
Copyright © 1982 by D. Reidel Publishing Company.

INTRODUCTION

Measurements of acoustic attenuation in polymer solutions and solids have been reported for over thirty years.   However, it is only relatively recently that the molecular origins of these processes have begun to be understood(1,2).   A sound wave propagates through a liquid adiabatically and produces a periodic variation in the translational temperature.   Molecules are continuously undergoing collisions, elastic or inelastic.   The former allow the maintenance of a normal distribution of energy. Alternatively, inelastic collisions provide a means of converting part of the translational energy into internal vibrational energy of the molecule.   For most organic molecules, the lowest vibrational mode involves torsional motions about an internal carbon-carbon bond.   The synchronous excitation and deactivation of this mode can lead to the observation of conformational change between rotational isomeric states.   The energy lost from the translational energy is associated with the observed acoustic attenuation.   If the rate of perturbation of the translational energy becomes very much faster than the rate of exchange between them.   Details of the theoretical treatment of ultrasonic relaxation and its application to the investigation of conformational change have been published elsewhere (1,2).

DILUTE POLYMER SOLUTIONS

An example of the type of behaviour observed in polymer solutions is presented in figure 1.   Increasing the temperature in the case of solutions of polystyrene in toluene leads to a decrease in the amplitude of the relaxation and a shift of the dispersion to higher frequency.   Examination of the effect of change in the molecular weight at constant composition indicates that the amplitude increases up to a value of M_n equal to 10,000 whereupon it becomes approximately independent of molecular weight (3).   Calculation of the contribution due to normal mode relaxation to the dispersion in the megahertz frequency range indicates that co-operative modes cannot explain the observed effects.   Recently ultrasonic studies of the polystyrene oligomers (4,5) have indicated that the amplitude of the relaxation is expected to be very low and the frequency high at room temperature.   Incorporation of these data with those obtained previously, figure 2, indicates that the origin of the molecular weight effect in the case of polystyrene is associated with changes in the rotational energy profile with increasing chain length. In the case of short chains the barrier hindering internal rotation is lower and the energy difference is smaller than in the case of the longer chains.   At a molecular weight of approximately 10,000 the contribution of the more mobile end groups is negligible compared with that from the hindered elements of the main chain(6).

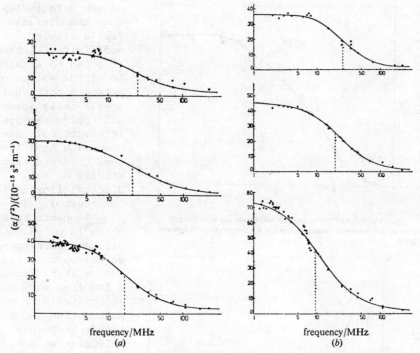

Fig. 1 —(α/f^2) plotted against frequency for (a) 4000 and (b) 97 000 M_n polystyrene at three temperatures. Top 40°C; middle 30°C; bottom 20°C; solid lines are computer fits to the experimental data, dashed lines indicate f_c.

An investigation of the acoustic relaxation in dilute solutions at high temperatures (6) has indicated the existence of a change in the number of monomer units involved in the relaxation. Similar and related phenomena have been reported from ^{13}C nmr relaxation and light scattering studies of solutions of isotactic polystyrene(7,8). It is believed that at approximately 333K or just above the isotactic sequences become mobile and release the helical structure which is frozen at lower temperatures(6).

SEMI DILUTE AND CONCENTRATED SOLUTIONS

Investigations of the effects of increasing the concentration of polymer in solution have been reported by Dunbar et al(9). It is found that up to a concentration of 10 w % the observed curves can be generated theoretically by combining a contribution associated with segmental relaxation based on dilute solution data and a normal mode contribution based on shear viscosity measurements. Above 10 w % the compressibility of the solution is observed to decrease and the acoustic attenuation to increase markedly with increasing concentration. This can be associated with the occurrence of polymer-polymer interactions. Analysis

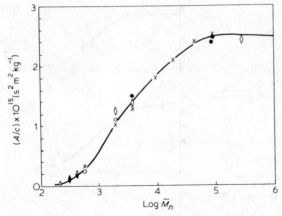

Variation of the relaxation amplitude per unit concentration as a function of molecular weight for polystyrene in toluene at 293K.

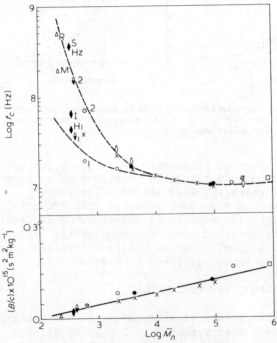

Variation of relaxation frequency and residual absorption per unit concentration as a function of molecular weight for polystyrene in toluene at 293K.

Fig. 2

of the data indicates that the deviations from a simple additivity relationship can be attributed to either volume or entropic contributions specifically associated with polymer-polymer interactions. Similar phenomena have also been observed from studies of poly-dimethylsiloxane in toluene(10).

Acoustic relaxation in polydimethyl-siloxane corresponds to the combination of two separate processes; a low frequency motion associated with normal mode motion and a higher frequency molecular weight independent process involving segmental relaxation(11). The lower frequency process can be directly computed from shear relaxation data and for low molecular weights a good correspondence is observed, figure 3. Increase in molecular weight leads to a difference between prediction and experiment which can be attributed to the relaxation of incipient polymer-polymer entanglements. The high frequency relaxation of short segments of the polymer chain is observed to obey 'starvation' kinetics and demonstrates the principle of a system being controlled by the rate of energy being fed into a reaction centre

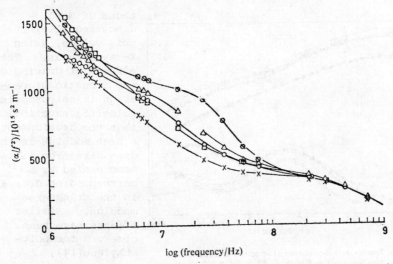

Variation of ultrasonic attenuation with molecular weight for poly(dimethyl siloxane) in the frequency range 1-1000 MHz at 303 K.   Values of $\overline{M}_n$:   ×, 1.54×10^4;   ○, 2.10×10^4;   △, 3.77×10^4;   ⊘, 6.33×10^4 and □, 7.94×10^4.

rather than by the probability of its acquiring sufficient energy to achieve a particular activated state(11).

SOLID STATE

Acoustics has traditionally been used as a method of obtaining high frequency dynamic mechanical test data.   In the case of polyethylene it allows examination at around room temperature of processes which are normally frozen out at liquid nitrogen temperatures, figure 4.   The velocity of sound in an isotropic polyethylene was found to be dependent upon the crystallite size and directly correlates with the degree of crystallinity, figure 5(12).   Examination of the effect of drawing on the acoustic attenuation in the case of polypropylene(13) indicates that the generation of air voids leads to a marked increase in the acoustic attenuation, figure 6.   The observed temperature dependence of the acoustic attenuation is consistent with the presence in these samples of voids of average size one micron.   This prediction was confirmed using scanning electron microscopy on fractured samples of the drawn polymer.   Rotation of the orientated sample in the plane of the propagation of the sound beam allows exploration of the direction of molecular orientational dependence of the sample.   These 'pole' diagrams indicate that in low draw ratio (4-7) samples of polyethylene crystal twinning can occur.   Thermal treatment of the samples can completely destroy this morphology.   However, certain pressure treatments can induce reorientation of the crystallites in the sample(14).   These observations are in agreement with

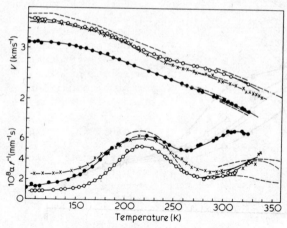

those of Ward discussed elsewhere in this volume, using X-ray methods. The effects of drawing on the acoustic propagation in polypropylene films(15) have indicated that the development of a high modulus in the draw direction is accompanied by a corresponding decrease in the transverse modulus. Similar effects have been observed for polyethylene(14).

Temperature dependence of velocity v and attenuation divided by frequency, α/f, at 5 MHz. O, P1; X, P4; ●, P8. Broken lines show the literature values for velocity and α/f data; high density refs. 3, 4, 5, 6, 8, 9 and low density refs. 5, 8, 11 polyethylene

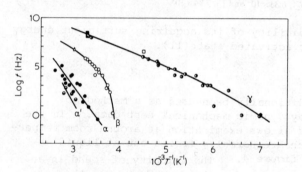

Arrhenius correlation plots for the α, α', β and γ processes (references 13–20, 27–31). Present data: □, and △, from temperature dependence of α/f. ■, and □, from frequency dependence of α/f

Fig. 4

ACOUSTIC STUDIES OF PHASE SEPARATED SYSTEMS

Investigation of the anisotropy of sound propagation in extruded polystyrene-butadiene-styrene triblock copolymer indicates that propagation along the alignment of the styrene cylinders occurs more easily than transverse to it(16). This reflects the relative magnitudes of the modulii of the styrene and butadiene phases. Examination of the temperature dependence of the attenuation in a series of samples indicates that the position of the butadiene relaxation is sensitive to the size of the domain and also to the presence of cross links in the case of

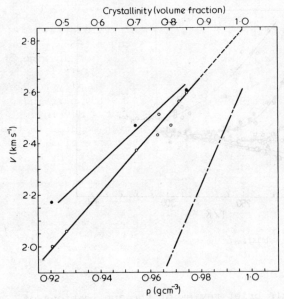

Correlation of the velocity with density ○, 4.96 MHz;
●, 15.3 MHz. Broken line, data by Davidse *et al.* at ~5 MHz

Fig. 5

a star block copolymer. Examination of the temperature dependence of the high frequency (750 MHz) attenuation in solutions of the triblock copolymer allows identification of the onset of formation of a heterophase, figure 7 (18). The formation of a micelle structure in solution leads to a high degree of scattering of the acoustic wave and the generation of a marked temperature dependence of the loss. This method may, in the future, provide a sensitive method for the investigation of heterophase structures, a topic which is currently of considerable theoretical interest.

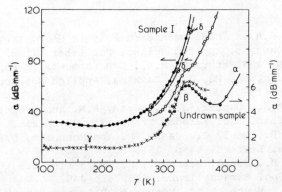

Temperature dependence of attenuation at 5 MHz for a
drawn opaque sample (Sample I of *Table 1*) and an undrawn sample.
●, Sample I; ○, Sample I annealed at 373K for 5800 min; X, undrawn sample; ○, undrawn sample annealed at 373K for 4000 min.
The broken line is calculated from equation (3) using the inter-
polated velocity data on drawn polypropylene

Fig. 6

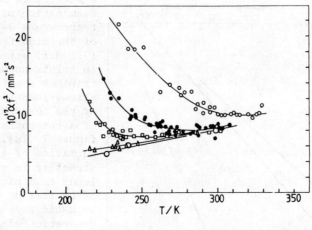

Fig. 7

CONCLUSIONS

It is hoped that this brief review of the applications of ultrasonics for the investigation of polymer systems may have indicated to the reader the potential and limitations of this technique. No single method can provide all the answers. Ultrasonics used with neutron scattering, nmr, esr and dielectrics may allow in the future a more unified approach to polymer dynamics to be developed.

REFERENCES

1. Pethrick, R.A., *Developments in Polymer Characterization*, Ed. J.V. Dawkins, Applied Science Publishers (1982) Vol. IV.
2. North, A.M. and Pethrick, R.A., *Developments in Polymer Characterization*, Ed. J.V. Dawkins, Applied Science Publishers (1980) Vol. II.
3. Cochran, M.A., Dunbar, J.H., North, A.M. and Pethrick, R.A.: 1972, J. Chem. Soc. Faraday Trans. II, 70, p 215.
4. Froelich, B., Noel, C., Jasse, B. and Monnerie, L.: 1976, Chem. Phys. Letters, 44 p. 159.
5. Froelich, B., Jasse, B., Noel, C., Monnerie, L.: 1978, J. Chem. Soc. Faraday Trans. II, 74 p 445.
6. Dunbar, J., North, A.M., Pethrick, R.A. and Poh, B.T.: 1980, Polymer, 21, p 764.
7. Ono, K., Shintani, H., Yano, O. and Wada, Y.: 1973, Polymer J., 5, p 164.
8. Reiss, C. and Benoit, C.R.: 1961
9. Dunbar, J.H., Steinhauer, D.B., North, A.M. and Pethrick, R.A.: 1977, J. Polymer Sci. Polymer Phys., 15, p. 263.
10. Daly, J., Bell, W., North, A.M., Pethrick, R.A. and Poh, B.T.:

1979, J. Chem. Soc. Faraday Trans II, 75, p 1115.

11. Bell, W., North, A.M., Pethrick, R.A. and Poh, B.T.: 1979, J. Chem. Soc. Faraday Trans. II, 75, p 1452.

12. Adachi, K., Lamb, J., Harrison, G., North, A.M. and Pethrick, R.A.: 1981, Polymer, 22, p 1032.

13. Adachi, K., Lamb, J., Harrison, G., North, A.M. and Pethrick, R.A.: 1981, Polymer, 22, p 1026.

14. Pethrick, R.A., unpublished data.

15. Ward, I. and Gupta, V.B.: 1968, J. Macromol. Sci. Phys. B2(1), p. 89.

16. Datta, P.K. and Pethrick, R.A.: 1978, Polymer, 19, p 145.

17. Datta, P.K. and Pethrick, R.A.: 1977, Polymer, 18, p 324

18. Pethrick, R.A., unpublished data.

NUCLEAR MAGNETIC RESONANCE OF SOLID POLYMERS : AN INTRODUCTION

F. Heatley

Department of Chemistry,
University of Manchester,
Manchester, M13 9PL,
U.K.

A simple introduction to nuclear magnetic resonance in solid polymers is given. The basic theory of the n.m.r. phenomena of importance (dipole-dipole coupling, chemical shift anisotropy and magnetic relaxation) is described, and the application of these effects to polymers is illustrated by selected examples. The review covers both well-established broadline n.m.r. and recently developed solid-state high-resolution n.m.r.

1. INTRODUCTION

This article has been written with two objectives, firstly to give an elementary introduction to the phenomena controlling nuclear magnetic resonance (n.m.r.) in solids, and secondly to illustrate these phenomena by describing some applications of n.m.r. to polymers. It is hoped that this approach will commend itself to post-graduate students beginning research and to more experienced scientists wishing to broaden their knowledge.

In general, the information provided by n.m.r. can be divided into two types, structural (phase structure, orientation, conformation) and dynamic (relaxation mechanisms, correlation frequencies). Accordingly, the article is divided into two main sections under those headings, though the areas are by no means mutually exclusive. The majority of existing n.m.r. studies of solid polymers have employed well-established broadline techniques which do not resolve chemically distinct nuclei. In recent years, techniques for distinguishing individual nuclei in solids have been developed ('high-resolution' n.m.r.) which promise significant advances. Both broadline and high-resolution methods are described here. For

251

R. A. Pethrick and R. W. Richards (eds.), Static and Dynamic Properties of the Polymeric Solid State, 251–270.
Copyright © 1982 by D. Reidel Publishing Company.

further study, the interested reader is referred to recent excellent
advanced reviews (1-3). Those by McBrierty and Douglass(1) cover
all n.m.r. techniques, while those by Lyerla(2) and Schaefer and
Stejskal(3) deal only with high-resolution ^{13}C n.m.r.   McCall(4)
has provided an elementary account of n.m.r. relaxation in solid
polymers. A comprehensive annual survey of n.m.r. in polymers is
published by the Royal Society of Chemistry in the Nuclear Magnetic
Resonance series of their Specialist Periodical Reports.

It is assumed that the reader has some familiarity with the
basic principles of n.m.r. such as may be obtained from one of a
number of basic texts (5-8).

2.   STRUCTURE DETERMINATION

The application of n.m.r. to structural problems depends on
two phenomena, dipole-dipole coupling and chemical shift anisotropy.
(Dipole-dipole coupling is also the principal source of dynamic
information and is further discussed in section 3).

2.1 Dipole-dipole Coupling

This effect may be simply explained in terms of the 'local
field' concept whereby a nucleus 'sees' not only the principal
polarising magnetic field B_0 but also local magnetic fields from
neighbouring magnetic nuclei. For nuclei with a spin quantum
number of 1/2, (notably ^1H, ^{13}C and ^{19}F), there are only two
allowed orientations of the nuclear magnetic moment relative to
B_0, termed α and β spin states. Hence the local field generated
by one such nucleus (A) at the site of another (B) also has two
possible values, giving two resonance conditions for B i.e. the B
signal is split into a doublet. Conversely, the A resonance is
identically split by the local field from B. The doublet splitting
is termed the dipole-dipole coupling constant, D_{AB}, and according
to this simple model is given in frequency units by

$$D_{AB} = (\frac{\mu_0}{4\pi}) \frac{\gamma_A \gamma_B \hbar}{2\pi r^3} (3\cos^2 \theta - 1) \qquad \cdots \cdot (1)$$

where γ_A and γ_B are nuclear magnetogyric ratios, r is the inter-
nuclear distance and θ is the angle between r and B_0.

Properly, dipole-dipole coupling is dealt with using a quantum-
mechanical formalism. The energy levels and transitions are
calculated using a spin Hamiltonian

$$\mathcal{H} = \mathcal{H}_0 + \mathcal{H}_D$$

where $\mathcal{H}_0$ represents the interaction with B_0 and $\mathcal{H}_D$ represents the dipole-dipole interaction. $\mathcal{H}_D$ is given by

$$\mathcal{H}_D = (\frac{\mu_o}{4\pi})\gamma_A\gamma_B\hbar^2\{\frac{3\cos^2\theta-1}{r^3}\}(\hat{I}_A\cdot\hat{I}_B-3\hat{I}_{ZA}\hat{I}_{ZB}) \qquad \ldots\ldots(2)$$

where $\hat{I}_A$ and $\hat{I}_B$ represent A and B spin operators. This coupling term produces a splitting of the resonance, identical to equation (1) if A and B are different nuclear species, but greater than equation (1) by a factor of 3/2 if A and B are the same species.

In general, a nucleus experiences numerous inter and intra-molecular couplings and also most samples are powders with inter-nuclear vectors oriented isotropically. The vast number of different coupling constants overlap in a single broad resonance, typically 10 to 30 kHz wide. Experimentally, the lineshape may be characterised by several parameters:

(i)   the full width at half-height, $\Delta\nu_{\frac{1}{2}}$ (frequency) or $\Delta B_{\frac{1}{2}}$ (field) depending on the spectrum sweep mode;

(ii)   the spin-spin relaxation time, T_2, related to the linewidth by $T_2 = 1/(\pi\Delta\nu_{\frac{1}{2}})$. (See Section 3 for the formal definition of T_2

(iii)   the second moment, M_2, defined by

$$M_2 = \int_0^\infty I(B)(B-\bar{B})^2 dB / \int_0^\infty I(B)dB$$

where $I(B)$ is the spectrum intensity at field B and $\bar{B}$ is the peak centre.

The lineshape is determined by the distribution of dipolar coupling constants i.e. the geometrical parameters r and θ, but this structural information is of very coarse quality. For a known molecular and crystal structure, the second moment can be calculated(9). For a rigid N-proton system, the general expression is

$$M_2 = \frac{9}{8}(\frac{\mu_o}{4\pi})^2\frac{\gamma_H^2\hbar^2}{N}\sum_{j>k} r_{jk}^{-6} \qquad \ldots\ldots(3)$$

For an isotropic polymer, averaging of equation (3) over all orientations produces

$$M_2 = \frac{9}{10}(\frac{\mu_o}{4\pi})^2\frac{\gamma_H^2\hbar^2}{N}\sum_{j>k} r_{jk}^{-6} \qquad \ldots\ldots(4)$$

If molecular motions are present, the dipolar coupling constant becomes time-dependent through θ and r. If the motion is

sufficiently rapid, averaged coupling constants are observed, the
spread of which is less than in the rigid system. Thus $\Delta v_{\frac{1}{2}}$ and
M_2 are decreased by an amount depending on the amplitude of the
motion. For 3-dimensional motion, the coupling constant averages
to zero, but spatially restricted motions in solids lead only to
a partial reduction. The reduction in linewidth does not proceed
continuously as the rate of motion (i.e. temperature) increases,
but occurs as a transition, generally over 20-30°, as shown
schematically by the T_2 and M_2 curves in Figure 1.

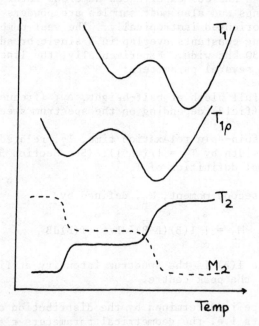

Figure 1. Schematic diagram of the temperature dependence of T_1,
$T_{1\rho}$, T_2 and M_2 for a polymer undergoing two types of motion with
widely different frequencies. The vertical disposition of the T_1,
$T_{1\rho}$ and T_2 curves represent their relative magnitudes.

The transition occurs when the correlation frequency v_c reaches
the critical value given by $v_c = (\gamma/2\pi)\sqrt{M_2^R}$, where M_2^R is the rigid
lattice second moment before the transition. Multiple motions
produce successive transitions if their frequencies are widely
spaced (Figure 1). Correlation frequencies in the range 5-30 kHz
are accessible by this technique.

2.2 Chemical Shift Anisotropy

The chemical shift effect stems from the fact that nuclei are
shielded from the external B_o field by surrounding electrons, to

an extent which depends on the electronic structure. If the
nuclear site symmetry is not spherical, the shielding will depend
on the molecular orientation in B_0, leading to the shielding
powder patterns for rigid systems shown in Figure 2. The frequency

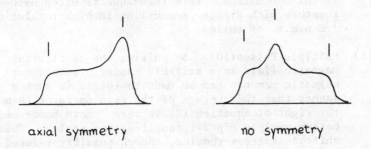

axial symmetry   no symmetry

Figure 2. Schematic chemical shift anisotropy powder patterns in
solids. The vertical bars indicate the locations of the principal
elements of the chemical shielding tensor.

spread depends on the nucleus, the field B_0 and the molecular
structure, but is usually less than the dipolar linewidth. It is
negligible for ^1H nuclei but may be up to 250 p.p.m. for ^{13}C.
Anisotropic shielding is represented (for one nucleus) by the
Hamiltonian

$$\mathcal{H}_o = -\gamma \,\hat{I}.(\underline{1} - \underline{\sigma}).\underline{B}_o$$

where σ is the shielding tensor. Molecular motion averages $\underline{\sigma}$ in
a similar manner to dipolar coupling. Three-dimensional rotation
reduces the powder pattern to a single peak with average shielding
constant $<\sigma> = (1/3)\mathrm{Tr}\underline{\sigma}$. Rotation about an axis reduces the no-
symmetry pattern to axial symmetry.

2.3 High-Resolution N.M.R. in Solids

In order to resolve individual nuclei in solids, dipole-
dipole coupling must be removed. This can be achieved in three
ways:

(i) Dipolar Decoupling. This applies to heteronuclear
interactions only. One of the coupled nuclei is
irradiated with a magnetic field B_2 oscillating at its
resonance frequency such that $\gamma B_2/2\pi > \Delta\nu_{\frac{1}{2}}$. A common
use is to remove ^1H dipolar broadening from ^{13}C spectra.
Chemical shift anisotropy effects are unaffected.

(ii)   Magic-Angle Rotation. The sample is rotated at a
       frequency ν_r about an axis inclined at $54°44'$ to B_o
       such that $\nu_r > \Delta\nu_{\frac{1}{2}}$. This procedure gives a zero time
       average for the $(3\cos^2\theta-1)$ term in equations (1) and
       (2). Both hetero- and homo-nuclear dipolar coupling
       vanish. Chemical shift anisotropy is also averaged
       as for 3-D motion. This technique is often used
       together with dipolar decoupling in high-resolution
       ^{13}C n.m.r. of solids.

(iii) Multiple Pulses(10). By pulsing the oscillating
       magnetic field in a suitable manner, the nuclear
       magnetic moments can be made to rotate in such a
       manner that the average of the spin operator term on
       the right of equation (2) is zero. Both homo- and
       hetero-nuclear dipolar coupling thus vanish. Chemical
       shift anisotropy remains, though possibly reduced.
       This technique has not been used for polymers as
       extensively as (i) and (ii).

For high resolution ^{13}C n.m.r. in solids, the technique of
cross-polarisation (2,3) is often used to enhance sensitivity.
This method relies on the transfer of polarisation from abundant
^1H nuclei to rare ^{13}C nuclei which can occur very rapidly if
certain conditions are met while both nuclei are polarised along
a magnetic field precessing about B_o at the appropriate resonance
frequency. In this state, the nuclear magnetisation is said to be
polarised in the rotating frame, and the method of achieving it
is outlined in Section 3.2.

2.4 Structural Applications

(i)   <u>Phase Structure of Partially Crystalline Polymers</u>. This
important topic is well illustrated by the much-studied example of
polyethylene (PE). Figure 3 shows the ^1H spectrum (recorded as
the derivative) of a linear PE at 25°C (11). Two components,
broad and narrow, are clearly visible, and are assigned to crystalline and non-crystalline regions respectively. Apparently at this
temperature, the non-crystalline region has sufficient mobility
to average dipolar coupling partially. At some lower temperature,
this motion 'freezes out' and the non-crystalline component broadens
into the crystalline. At a higher temperature, the crystalline
region melts and its spectrum merges with the non-crystalline. A
simple two-component analysis of the spectrum gives crystallinities
in reasonable agreement with X-ray data, but greater consistency
is obtained by the three-component decomposition (12) illustrated
in Figure 4. Obviously the poor resolution forbids an unconstrained
decomposition and therefore in this analysis it was assumed that
the broad component lineshape is represented by the low temperature
(rigid) spectrum, the narrow component by a Lorentzian lineshape

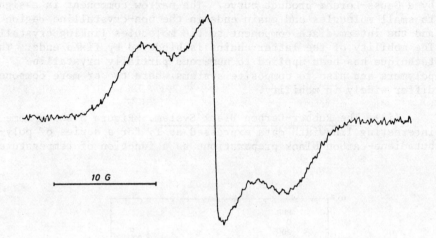

Figure 3. The ^{1}H spectrum of partially crystalline polyethylene
at 27°C. The spectrum is recorded as the derivative. (Reproduced
with permission from reference 11).

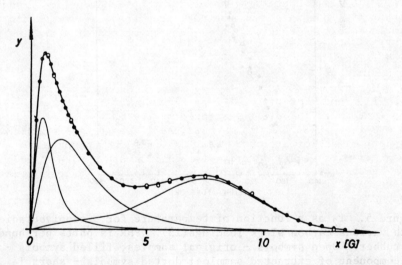

Figure 4. Three component analysis of the spectrum in Figure 3.
Symbols: o measured points; ● calculated points; ⊙ measured =
calculated points. (Reproduced with permission from reference 12).

expected for a highly mobile species, and the intermediate component
by a Gauss-Lorenz product curve. The narrow component is assigned
to small molecules and chain ends in the non-crystalline region
and the intermediate component to tie molecules linking crystallites.
The mobility of the latter chains is hindered by fixed ends. This
technique has been applied to numerous partially crystalline
polymers and also to composite systems where two or more components
differ widely in mobility.

(ii) The Rubber-Carbon Black System. Figure 5 shows some
interesting linewidth data expressed as T_2 for a series of poly-
butadiene-carbon black preparations as a function of temperature(13).

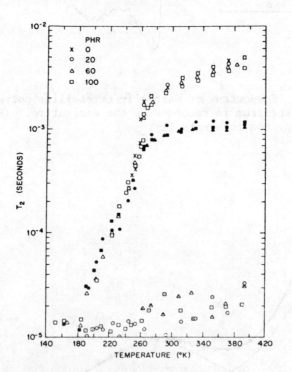

Figure 5. T_2 as a function of temperature for cis-polybutadiene
with various carbon black loadings(13). PHR is parts per hundred
of rubber. Open symbols - original samples; filled symbols - long
T_2 component of extracted samples; dotted symbols - short T_2
component of extracted samples. (Reprinted with permission from
Macromolecules 9, 653 (1976). Copyright 1976 American Chemical
Society).

Above 200 K, short and long T_2 components are observed corresponding
to 'rigid' and 'mobile' species. The proportion of the rigid

component increases with increasing carbon black loading, but the
location of the T_2 transition is unaffected. Following extraction
with boiling toluene, similar T_2 data are observed except that the
mobile component shows a lower high temperature T_2 plateau and
diminishes in intensity. The molecular interpretation of these
data is as follows. The 'rigid' component arises from rubber
segments bound to the carbon black and from highly restricted
segments adjacent to the bound segments. The 'mobile' component
in the extracted material arises from segments in loose loops
and long cilia attached to the filler at one end (cilia) or both
ends (loops). In the unextracted material, an unbound fraction
is present, with a longer high temperature T_2 than loops and cilia
but not sufficiently distinct to appear as a separate component.
The component intensities give the fraction of bound segments.

(iii) <u>Molecular Orientation in Drawn Polymers</u>. Equation (3)
shows that the second moment M_2 depends on the distribution of
orientation angles θ_{jk}. For an anisotropic system such as a drawn
polymer, M_2 will therefore depend on the sample orientation relative
to B_0. Figure 6 shows M_2 for polyethylene as a function of draw
ratio and the angle γ between the draw direction and B_0 (14).

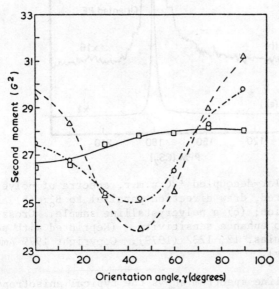

Figure 6. The second moment for drawn polyethylene as a function
of the angle, γ, between the draw direction and the magnetic
field(14). Draw ratios: □, 1.3; o, 2.3; Δ, 3.7. (Reprinted with
permission from reference 14. Copyright 1968 The Institute of
Physics).

As the draw ratio increases, the anisotropy in M_2 increases, indicating greater degrees of molecular orientation. The data of Figure 6, together with the spectrum fourth moment, has allowed the derivation of distribution functions for the PE chain axis using an extension of equation (3) to cover partially oriented systems(15).

Orientation in PE has also been investigated recently using high-resolution ^{13}C n.m.r.(16). Figure 7 shows ^{13}C spectra of polycrystalline and fibre-oriented PE samples obtained using dipolar decoupling but not magic-angle rotation.

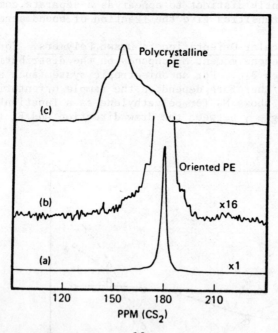

Figure 7. Dipolar-decoupled ^{13}C n.m.r. spectra of polyethylene(16). (a) fibre-oriented, draw direction parallel to B_o; (b) as (a) but x 16 amplification; (c) a polycrystalline sample. Cross-polarisation was used to enhance sensitivity. (Reprinted with permission from Macromolecules, 12, 1232 (1979). Copyright 1979 American Chemical Society).

The polycrystalline spectrum shows the typical anisotropy expected for a nuclear site with no symmetry. In the fibre, a single peak is seen at the high field extremity which corresponds to the resonance position when the axis of the PE chain is oriented parallel to B_o. In this particular sample shown, the degree of orientation is high - the amplified fibre spectrum shows only scant evidence of molecules at other orientations, though such molecules could

easily be detected. This example demonstrates the high potential
of high resolution ^{13}C n.m.r. in tackling such problems.

    (iv) <u>Chemical and conformational structure by high resolution</u>
^{13}C n.m.r. Two examples illustrate the chemical information
obtainable from high resolution ^{13}C n.m.r. using dipolar decoupling
and magic-angle rotation for maximum resolution. Figure 8 shows
the ^{13}C spectrum of poly(<u>p</u>-hydroxybenzoic acid)(2).

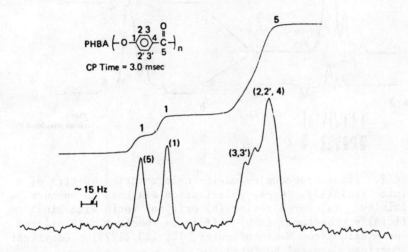

Figure 8. Dipolar-decoupled/magic-angle rotation ^{13}C n.m.r.
spectrum of poly(p-hydroxybenzoic acid) at 27°C. Line assignments
are given, together with the spectral integration and relative
intensities. The CP time referred to is the contact time in the
cross-polarisation technique used to enhance sensitivity.
(Reproduced with permission from reference 2).

Several ^{13}C nuclei are clearly resolved. An interesting feature
is the observation of two resonances for C_3 and $C_{3'}$, showing that
the ring is held in a conformation in which these two positions
are non-equivalent. The rate of ring rotation must be less than
the frequency separation of the two peaks. Figure 9 compares the
spectra of an acetylene-terminated polyimide resin before and after
curing(17). The difference spectrum shows the disappearance of the
acetylene peak at δ_C 84.0 and the growth of several olefinic signals
at lower fields. Comparison of the shifts of these new peaks with
those of model compounds showed that in the curing process, ca. 30%
of the acetylene groups trimerised to an aromatic ring while the
remainder underwent a substitution reaction to give ethylenic
structures.

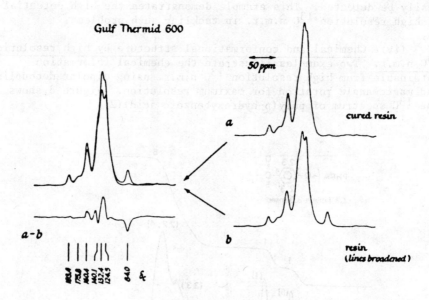

Figure 9. Dipolar-decoupled/magic-angle rotation spectra of a
polyimide resin(17). Cross-polarisation was used to enhance
sensitivity. (a) cured resin; (b) original resin with lines
artificially broadened; (a-b) difference spectrum. (Reprinted
with permission from Macromolecules, 12, 423 (1979). Copyright
1979 American Chemical Society).

3. MOLECULAR DYNAMICS

As described in Section 2, some limited data on molecular
dynamics can be obtained from linewidths, but more detailed
information can be obtained from magnetic relaxation experiments.

3.1 Nuclear Magnetic Relaxation - Basic Principles.

Figure 10 illustrates the phenomenen of nuclear magnetic
relaxation. At thermal equilibrium, the net magnetisation of a
collection of nuclei, M_O, lies parallel to B_O. If the magnetisation
is displaced from equilibrium, it will return by two distinct
relaxation processes:

    (i) Longitudinal or spin-lattice relaxation. The magneti-
          sation component parallel to B_O returns to its
          equilibrium value M_O with a time constant denoted T_1.

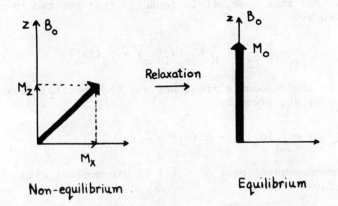

Figure 10.   Phenomenological definition of longitudinal and transverse relaxation (Section 3.1).

   (ii)   Transverse or spin-spin relaxation. The magnetisation component perpendicular to B_0 decays to its equilibrium value zero, usually by an exponential decay with time constant denoted T_2.

Longitudinal and transverse relaxation are different because only the former leads to changes in the total spin energy. In solid polymers, the most important source of relaxation is dipole-dipole coupling. Longitudinal relaxation proceeds by transitions between the nuclear spin states induced by time-dependent local fields arising from molecular motion. Transverse relaxation proceeds by the dephasing of individual nuclear magnetic moments due to differences in precession frequencies arising from a spread of local fields. T_2 is thus inversely related to the spread of local fields as measured by the linewidth $\Delta\nu_{\frac{1}{2}}$ (Section 2.1), and gives similar information.

   T_1 and T_2 are conveniently measured by pulse techniques (4-8). T_2 is obtained from the decay time of the free induction signal following a $\pi/2$ pulse. T_1 is obtained by applying a perturbing pulse (π or $\pi/2$), waiting a short time for partial relaxation, then monitoring the partially relaxed longitudinal magnetisation with a second $\pi/2$ pulse.

   T_1 depends on the rate of transitions between spin states induced by the fluctuating part of the dipolar interaction. Calculation of the transition rate by time-dependent perturbation theory requires the specification of the autocorrelation function, $G_n(\tau)$, of second-order spherical harmonic functions, $Y_{2,n}(\theta,\phi)$, where θ and ϕ describe the orientation of the internuclear vector.

For spatially restricted motions in a solid, $G_n(\tau)$ does not vanish as $\tau \to \infty$. For this case, it is found(1) that for two ^{1}H nuclei, T_1 is given by

$$\frac{1}{T_1} = \frac{9}{8} \left(\frac{\mu_o}{4\pi}\right)^2 \gamma_H^4 \hbar^2 \{J_1(\omega_o) + J_2(2\omega_o)\} \qquad \ldots \ldots (5)$$

where ω_o is the resonance frequency and the spectral density terms $J_{1,2}(\omega)$ are given by

$$J_n(\omega) = \int_{-\infty}^{+\infty} [G_n(\tau) - G_n(\infty)] \, e^{i\omega\tau} d\tau$$

If the time-dependent part of $G_n(\tau)$ is exponential with correlation time τ_c, $J_n(\omega)$ becomes

$$J_n(\omega) = \frac{2\tau_c}{1+\omega^2\tau_c^2} \, <|F_n|^2> \qquad \ldots \ldots (6)$$

with

$$<|F_n|^2> = C_n \{<|Y_{2,n}(\theta,\phi)/r^3|^2_{LT}>$$

$$-<|Y_{2,n}(\theta,\phi)/r^3|^2_{HT}>\} \qquad \ldots \ldots (7)$$

where the suffixes LT and HT signify low temperature and high temperature averages i.e. with no motion and with rapid motion respectively. For multinuclear systems with multiple motions equation (6) is summed over all nuclear pairs and types of motion.

The dependence of T_1 on τ_c (i.e. temperature) predicted by equations (5) and (6) is illustrated schematically in Figure 1, for a system with two types of motion. T_1 passes through a minimum whenever a correlation time meets the condition $\omega_o\tau_c = 0.616$. Thus T_1 measurements allow the detection of processes occurring at frequencies of the order of ω_o (10 to 100 MHz). Changing ω_o moves the minimum along the temperature axis, giving several points on the motional frequency - temperature map. The minimum value of T_1 also depends on ω_o, decreasing as ω_o decreases.

3.2 Spin-lattice Relaxation in the Rotating Frame

In this technique(1,4,18), relaxation of the nuclear magnetisation is followed while aligned along a magnetic field B_1 which precesses about B_o at the nuclear resonance frequency. The initial alignment is achieved by a $\pi/2$ pulse followed by a $\pi/2$ phase shift of B_1. Analysis of the relaxation shows that it is a spin-lattice process with a relaxation time designated $T_{1\rho}$. The effective frequency of the transitions is $\omega_1 = \gamma B_1$ which usually lies in the range 10-100 kHz. For two protons, $T_{1\rho}$ is given to a very good approximation by the expression (1)

$$\frac{1}{T_{1\rho}} = \frac{9}{32} \left(\frac{\mu_o}{4\pi}\right)^2 \gamma_H^4 \, \hbar^2 \, J_o(\omega_1) \qquad \ldots\ldots (8)$$

with $J_o(\omega_1)$ defined by equations (6) and (7).

As illustrated in Figure 1, $T_{1\rho}$ varies with temperature in a similar manner to T_1. Because of the lower effective frequency, the $T_{1\rho}$ minima are deeper and at a lower temperature than the corresponding T_1 minima.

3.3 Spin Diffusion (1)

Spin-lattice relaxation is the process by which the spin system exchanges energy with the molecular thermal energy bath (the 'lattice'). However, the same fluctuating field which promotes this energy exchange can also promote energy exchange between spins. This exchange is known as spin diffusion, and it occurs in a time of the order of the contribution to T_2 from the averaged dipole-dipole coupling between those spins. If the different spin populations are in the same molecule e.g. the aromatic and aliphatic protons in polystyrene, the dipolar coupling is large, and the spin diffusion rate is much greater than the spin-lattice relaxation rate of either set of spins. Thus a single averaged T_1 value is observed for the relaxation of the total spin magnetisation because spin diffusion maintains the spin system in internal equilibrium. If the different spin populations are in different molecules e.g. crystalline and non-crystalline regions or constituents of a polymer composite, the behaviour depends on the intimacy of mixing. If the molecules are in close contact, rapid spin diffusion will give a single T_1 value, but if the molecules are well separated, spin diffusion will be relatively inefficient and each constituent will relax with a T_1 value determined by its own motional and structural parameters. For polymer systems with two or more phases, as a rough order of magnitude, a single T_1 will be observed for protons if the domain size is less than 10 nm(1). Spin diffusion also operates in $T_{1\rho}$ relaxation to the same extent as in T_1 relaxation, but because $T_{1\rho}$ values are much less than T_1, the critical domain size for $T_{1\rho}$ averaging is ca. 1 nm (1).

3.4 Applications of Relaxation

(i) Natural rubber. Figure 11 compares [1]H T_2, $T_{1\rho}$ (42.6 kHz) and T_1 (30 MHz) data as a function of temperature for natural rubber(19). Two T_1 minima are observed at -170 and +10°C, the

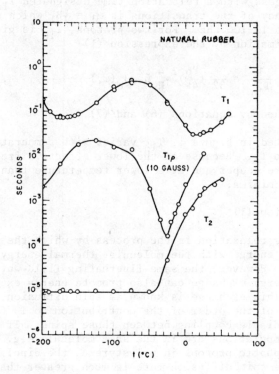

Figure 11. T_1, $T_{1\rho}$ and T_2 data for natural rubber as a function
of temperature. T_1 measured at 30 MHz and $T_{1\rho}$ at 42.6 kHz.
(Reprinted with permission from reference 19).

former arising from methyl rotation and the latter from backbone
segmental motion. Within the temperature range investigated, the
$T_{1\rho}$ and T_2 data show a minimum and transition respectively at
-30°C, arising from the backbone motion. The methyl rotation has
a low activation energy, and reaches the critical frequency for a
$T_{1\rho}$ minimum or T_2 transition at a temperature beyond the lower
limit of this data. Correlation frequencies derived from n.m.r.
data such as this are consistent with those from other techniques
such as dielectric or viscoelastic relaxation(4), and may be used
to obtain activation energies for the molecular processes.

   (ii) Structural dependence of motion in rubbers. Figure 12
presents ^1H T_1 data for four rubbers(20). Polybutadiene shows a
single minimum arising from segmental motion. The other three
polymers all contain methyl groups, and show two minima, that at
lower temperature arising from methyl rotation and that at higher
temperature from segmental motion. The similarity in location of
the methyl minima for polypropylene and hevea rubber (poly-cis-

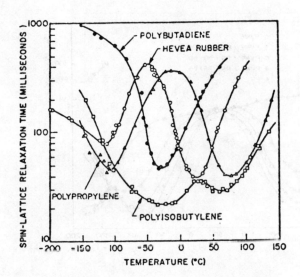

Figure 12. T_1 data for four rubbers as a function of temperature. The resonance frequency was 30 MHz. (Reprinted with permission from reference 20).

isoprene) is consistent with relatively unhindered methyl rotation of widely spaced methyl groups. In polyisobutylene, steric hindrance of geminal methyl groups displaces the minimum to a temperature 80°C higher. The frequencies of the backbone segmental motion decrease in the order polybutadiene > hevea rubber > polyisobutylene > polypropylene, and presumably reflect increasing barriers to rotation about backbone bonds. This order is the same as the glass transition temperature, indicating that the segmental process detected in n.m.r. is related to that occurring at T_g.

(iii) <u>Phase structure in blends of polystyrene (PS) and poly(vinyl methyl ether) (PVME)</u>. Figures 13 and 14 show T_1 and T_2 data respectively for a series of PS-PVME blends, together with data for the homopolymers for comparison(21). The mixtures were prepared by solvent evaporation from a toluene solution, and the measurements were made starting at low temperature. As indicated in Figure 13, up to 140°C, the blends showed only a single T_1 value intermediate between the values for the homopolymers. Above 140°C however, two T_1 components were observed corresponding closely to the homopolymers. At 170°C, when the component T_1 values differed sufficiently to allow a meaningful determination of component intensities, it was found that the component intensities corresponded to the blend composition. This data indicates

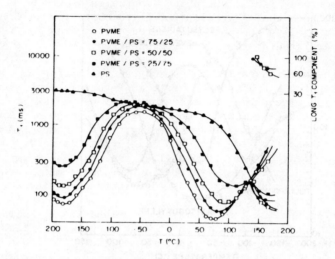

Figure 13. Temperature dependence of T_1 and fraction of long T_1 component for polystyrene, poly(vinyl methyl ether) and three PS/PVME blends. (Reprinted with permission from Macromolecules, 7, 667 (1974). Copyright 1974 American Chemical Society).

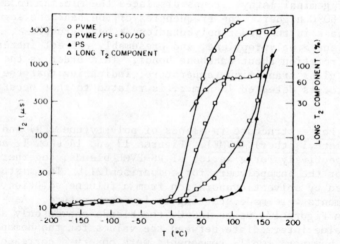

Figure 14. Temperature dependence of T_2 (o,□,▲) and fraction of long T_2 component (△) for polystyrene, poly(vinyl methyl ether) and a 50/50 PS/PVME blend. (Reprinted with permission from Macromolecules, 7, 667 (1974). Copyright 1974 American Chemical Society).

that there is considerable mixing of the two components in the
solvent evaporated blends, but phase separation occurs on heating
to 140°C when both chains are sufficiently mobile for segregation
to occur. On cooling the phase-separated blends, the two T_1
components persisted at the lower temperatures.

Although the T_1 data indicates substantial mixing in the
initially prepared blends, the T_2 data in Figure 14 show that
short-range inhomogeneities of the order of 1 nm may exist. For
a 50:50 blend, there are two T_2 components between 50 and 150°C,
which correspond closely, but not exactly, to the homopolymers.
T_2 values are sensitive to short-range interactions only, so this
observation indicates that some small degree of phase separation
may have occurred on solvent evaporation but the domains are not
large enough to prevent spin diffusion effects averaging T_1.

Many other applications of magnetic relaxation to solid
polymers have been reported, including such interesting features as
the effect of pressure on chain motions(22), the effect of cross-
linking(23) and the effect of radiation(24). The measurement of
^{13}C rotating-frame relaxation times(3) promises significant
developments in the molecular description of motions in the 10-
100 kHz frequency range which play an important role in the impact
strength of polymers.

REFERENCES

1. V.J. McBrierty and D.C. Douglass, Macromolecular Reviews,
    to be published; Phys. Rep., 63, 61 (1980).
2. J.R. Lyerla, Contemp. Topics Polym. Sci., 3, 143 (1979).
3. J. Schaefer and E.O. Stejskal, Topics C-13 N.M.R. Spect., 3,
    283 (1979).
4. D.W. McCall, Acc. Chem. Research, 4, 223 (1971).
5. F.A. Rushworth and D.P. Tunstall, 'Nuclear Magnetic Resonance',
    Gordon and Breach, London, 1973.
6. A. Carrington and A.D. McLachlan, 'Introduction to Magnetic
    Resonance', Harper and Row, New York, 1967.
7. R.J. Abraham and P. Loftus, 'Proton and Carbon-13 NMR Spectro-
    scopy', Heyden and Son Ltd., London, 1978.
8. D. Shaw, 'Fourier Transform N.M.R. Spectroscopy', Elsevier,
    Amsterdam, 1976.
9. J.H. Van Vleck, Phys. Rev., 74, 1168 (1948).
10. U. Haeberlen, 'High Resolution NMR in Solids', Academic Press,
    New York, 1976.
11. K. Bergmann, J. Polym. Sci., Polym. Phys., 16, 1611 (1978).
12. K. Unterforsthuber and K. Bergmann, J. Mag. Res., 33, 483 (1979).
13. J. O'Brien, E. Cashell, G.E. Wardell and V.J. McBrierty,
    Macromolecules, 9, 653 (1976).
14. V.J. McBrierty and I.M. Ward, J. Phys. (D), 1, 1529 (1968).

15. V.J. McBrierty, I.R. McDonald and I.M. Ward, J. Phys. (D),
    4, 88 (1971).
16. D.L. VanderHart, Macromolecules, 12, 1232 (1979).
17. M.D. Sefcik, E.O. Stejskal, R.A. McKay and J. Schaefer,
    Macromolecules, 12, 423 (1979).
18. T.M. Connor, NMR Basic Princ. Prog., 4, 247 (1972).
19. D.W. McCall and D.R. Falcone, Trans. Far. Soc., 66, 262 (1970).
20. W.P. Slichter, J. Polym. Sci., Part C, 14, 33 (1966).
21. T.K. Kwei, T. Nishi and R.F. Roberts, Macromolecules, 7,
    667 (1974).
22. J.E. Anderson, D.D. Davis and W.P. Slichter, Macromolecules,
    2, 166 (1969).
23. T.J. Rowland and L.C. Labun, Macromolecules, 11, 466 (1978).
24. A. Charlesby and R. Folland, Radiat. Phys. Chem., 15, 393
    (1980).

LOCAL MOLECULAR MOTIONS IN BULK POLYMERS AS STUDIED BY ELECTRON
SPIN RESONANCE (E.S.R.)

Lucien MONNERIE

Ecole Supérieure de Physique et de Chimie Industrielles
de Paris
10, rue Vauquelin
75231 Paris - France

ABSTRACT - The basic principles of E.S.R. are described in the
case of a free electron and the required instrumentation is schematically shown. Then, the g-tensor and the hyperfine A-tensor
are introduced for atoms and molecules. Typical E.S.R. spectra
obtained in solution and in oriented or nonoriented samples are
given in the case of nitroxide radicals. The effects of molecular
motions on line shape are given for fast and slow motions as well
as the expressions leading to the relaxation times. Some examples
of E.S.R. studies on bulk polymers are shown and discussed.

I - INTRODUCTION

The first observation of an E.S.R. spectrum was reported by
a Russian, E. Zavoisky, in 1945. In 1952, the first resonances
from organic free radicals were obtained and since this time che-
mistshave been interested in this technique. Paralleling the
growth of E.S.R. was that of nuclear magnetic resonance (N.M.R.).
Indeed, much of the fundamental physics is essentially the same
in the two techniques, in particular the Bloch and the Bloember-
gen, Purcell, Pound equations concern both N.M.R. and E.S.R.. In
the field of polymer, E.S.R. has been first used for studying polymerization and degradation (1). During the last ten years, E.S.R.
has been applied to local motions in polymer solutions or bulk
polymers, using mainly nitroxide radicals either covalently bound
to the polymer chain (spin label) or dispersed in the polymer
matrix (spin probe).

The purpose of this paper is to introduce the reader into
the field of E.S.R. spectroscopy and to show him how this technique can give information on molecular motions of radicals. The most

271

R. A. Pethrick and R. W. Richards (eds.), Static and Dynamic Properties of the Polymeric Solid State, 271–304.
Copyright © 1982 by D. Reidel Publishing Company.

important features will be pointed out but long theoretical calculations have been avoided, they can be found in ref. (2, 3, 4) .
Some typical examples of E.S.R. studies on bulk polymers will be
presented and discussed. An extensive review on this subject has
been recently published (5).

II - ELECTRON SPIN RESONANCE PHENOMENA

E.S.R. is a spectroscopic technique which deals with the transitions between the Zeeman levels of a paramagnetic system situated
in a static magnetic field.

II.1 - Case of a free electron

This is the simplest situation. Associated with the spin of an
electron, there is a magnetic moment, $\vec{\mu}$ which is a vector colinear
with the spin angular momentum $\vec{S}$, but antiparallel due to the
negative charge :

$$\vec{\mu} = - \gamma \vec{S} = - 2 \beta_e \vec{S}$$

$$\beta_e = e \, \hbar/2 \, m_e \, c$$

γ is the magnetogyric ratio, usually expressed in Bohr magneton,
β_e and m_e is the electron mass.

Actually theoretical treatment and experiments show that the
numerical factor is 2.00232, and it is called the free electron
g-factor, g_e. Thus, we have

$$(II.1) \quad \vec{\mu} = - g_e \, \beta_e \vec{S}$$

When a magnetic field, $\vec{H}$, is applied, its interaction with
$\vec{\mu}$ determines the energy. The corresponding quantum mechanical
Hamiltonian is

$$(II.2) \quad H = g_e \, \beta_e \vec{S} \cdot \vec{H}$$

this equation contains no operator acting on the spatial part of
the wave function and is called "spin Hamiltonian". This one is
the simplest possible example.

If $\vec{H}$ lies along z, this equation becomes

$$H = g_e \, \beta_e S_z \cdot H_z$$

and the associated energies are given by the 2 values of the spin
quantum number, m_s :

$$E = (^{\pm} 1/2) \, g_e \, \beta_e \, H_z$$

Thus, when a magnetic field is applied, the Zeeman interaction leads to two states (Fig. 1). The energy difference between the

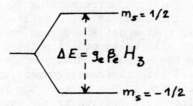

Fig. 1 - Zeeman splitting of an S = 1/2 state.

two states is

$$(II.3) \quad \Delta E = g_e \beta_e H_z$$

and it is possible to induce transitions between these two levels by irradiating at the frequency ν_o given by :

$$(II.4) \quad h\nu_o = g_e \beta_e H_z$$

As for nuclear magnetic moment in N.M.R., the spin magnetic moment $\vec{\mu}$ has a precession about the z axis of the magnetic field at the Larmor frequency, ω_o, given by :

$$h\omega_o/2\pi = h\nu_o = g_e \beta_e H_z$$

ω_o is just the frequency of radiation whose energy corresponds to the energy difference between the spin states of an electron in a magnetic fied. So, if we add a small magnetic field $H_1 (\ll H_o)$ rotating with angular frequency ω in the x-y plane perpendicular to $\vec{H}_o$, when $\omega = \omega_o$, there is a resonance phenomenon leading to an energy exchange between H_1 and the spin.

II.2 - Relaxation phenomena

We consider now a set of non-interacting spins in a static magnetic field, H_z. At thermal equilibrium, we expect the relative populations of the two levels (Fig. 1) to be given by the Boltzmann distribution :

$$n_u^o/n_\ell^o = \exp(-\Delta E/kT) = \exp(-g_e \beta_e H_z/kT)$$

where n_u^o and n_ℓ^o are the number of spins in the upper and lower levels, respectively. The magnetic energy is $0.3 - 1$ cm^{-1}, some hundred times smaller than kT, so that the difference in population of the magnetic levels is very small, but it is responsible for the detection of a E.S.R. signal. Indeed, when a rotating field H_1, at frequency ω_o, is applied, there is an energy absorption rising electron spins from the lower to the upper state;

the absorption rate is proportional to n_ℓ^o and to the intensity
of the radiation at ω_o. In the same time and with the same probability there is an induced emission corresponding to the inverse
transitions, but the rate of induced emission is lower by a
factor n_u^o/n_ℓ^o, in such a way that a net absorption occurs. Thus,
the effect of the radiation will be to equalize the populations
of the two levels, causing the signal to disappear (saturation).
Actually, the spins interact with their environment and relaxation
mechanisms occur which bring the system back to the Boltzmann
equilibrium populations after perturbation by the absorption of a
radiation. As for N.M.R., the relaxation processes lead to two
characteristic times, derived from the phenomenological Bloch
equations :

1/ spin-lattice or longitudinal relaxation time T_1, given by :

$$dM_z/dt = - (M_z - M_{eq})/T_1$$

where M_z is the macroscopic magnetization of the sample. T_1 is the
parameter characterizing the rate of decay of the bulk magnetization, M_z, when the magnetic field H_z is switched off. The spin-
lattice interaction comes from the fact that a spin in a radical
is subjected to a constantly fluctuating magnetic field arising
from its environment. Such a randomly fluctuating field at a particular electron spin will, in general, contain a component at the
Larmor frequency which will induce transitions between the levels.

2/ spin-spin or transverse relaxation time T_2, given by :

$$dM_x/dt = - M_x/T_2$$

T_2 is the parameter characterizing the rate of decay of the x
component of the bulk magnetization. The spin-spin interaction
does not involve the exchange of energy between the spin system and
the environment. It corresponds to the fact that the ensemble of
spins precessing in the applied field cannot precess in phase inde-
finitely. Due to magnetic interactions, the different spins experience slightly different local fields in the z-direction, thus
this results in a spread in the Larmor precession frequencies in
such a way that, even if at some times the spins all precess in
phase, there will be a gradual dephasing. These spin-spin relaxation phenomena are mainly responsible of the observed linewidth and
they produce an absorption curve that is described by a Lorentzian
function. On a angular frequency scale, the normalized shape function for a resonance line centered at ω_o is :

$$(II.5) \quad f(\omega) = (T_2/\pi)/(1 + T_2^2(\omega - \omega_o)^2)$$

II.3 - E.S.R. measurements

There are several commercial E.S.R. spectrometers available to-day and we will only recall the basic principle of measurements. More information can be found in (1, 3).

Substitution of the values of the physical constants in (II.4) shows that the frequency required for getting resonance is 2.8 MHz per gauss of magnetic field, so that for fields available in the laboratory, i.e. a few thousand gauss, the required frequency lies in the microwave region. Due to the breadth of the lines obtained with free radicals in solution, it is necessary to get a frequency stable to better than 2.10^{-5} in order to resolve these lines. As it is difficult to reach such a stability with a microwave source whose frequency can be varied smoothly and reproducibly in order to scan a spectrum, the technique used is to hold the microwave frequency constant and vary the magnetic field. Thus, the magnetic field is varied until it matches the resonance condition (II.4) and the spectrum is recorded as a function of the magnetic field. The two most common frequencies are 9.5 GHz (X-band) and 35 GHz (Q-band) corresponding to resonant field of 3,400 gauss and 12,500 gauss respectively. Concerning the magnetic field at the sample, it is modulated at 100 KHz in order to obtain a better signal amplification, thus from the spectrometer output it is convenient to trace the first derivative of the absorption line, rather than the absorption itself. An example is shown in Fig. 2 with the breadth parameters defined by :

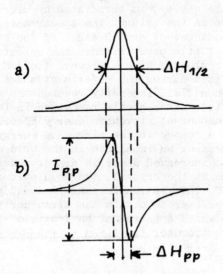

Fig. 2 - Presentation of spectral and breadth-parameters :
a/ absorption line, b/ first derivative. The lineshape function is Lorentzian.

- half-width at half-height of the absorption line : $\Delta H_{1/2} = 1/T_2$

- peak-to-peak width between the extreme of the first-derivative curve : $\Delta H_{pp} = 2/\sqrt{3}\ T_2$

- peak-to-peak amplitude of the first-derivative curve :
  $I_{pp} = 3\sqrt{3}\ T_2^2/4\ \pi$

    A block diagram of an E.S.R. spectrometer is shown in Fig. 3 with the essential components. The magnetic field at the sample

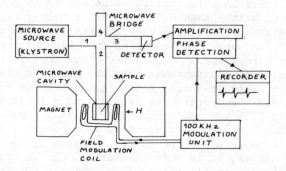

Fig. 3 - Block diagram of a simple E.S.R. spectrometer

is modulated by the 100 KHz signal through the cavity coils. The sample, positioned in the cavity, is irradiated by microware energy from the Klystron in the bridge. The spectrometer is cons- tructed so that the impedances of arms 2 and 4 of the bridge are nearly matched and very little power reaches the detector on arm 3. The cavity is a region of high microwave field H_1. The Klystron frequency is fixed during the experiment and the magne- tic field is scanned. When field intensity reaches the value required to induce electron paramagnetic resonance in the sample, a change occurs in the amount of microwave energy absorbed by the sample which causes a change in the microwave energy reflec- ted from the cavity, creating an imbalance in the bridge. This imbalance is detected as increased power in arm 3, modulated at 100 KHz. After detection at the crystal, the modulated signal which contains the E.S.R. information is amplified and phase- detected in a 100 KHz receiver to obtain the first-derivative of the absorption curve which appears on the Y-axis of the recor- der. The field scan potentiometer is linked to the recorder X-axis to calibrate the X-axis in gauss.

    Samples can be put into cylindrical tube or rectangular cells depending on the shape of the cavity. As oxygen is paramagnetic, it interacts with the spin probe or label through exchange and

dipolar mechanisms, resulting in a broadening of spectrum lines.
The oxygen must be removed from the sample either by the freeze-
thaw method or by bubbling nitrogen or argon gas through the sample. Another point concerns the spin concentration effects which
come from spin-spin interactions and result in a broadening of
lines. Thus a concentration dependence of the results must be
checked in order to avoid errors in interpretation of line width.
With nitroxide radicals, probe concentration about 4.10^{-4} M is
usely convenient.

When recording E.S.R. spectra, it is important to choose a
field modulation amplitude and a microwave power in such a way
that no disturbing effects occur. In particular, as the modulation amplitude increases, the lines first increase in height,
then broaden and finally become distorted. For accurate line shape
studies, it is necessary to determine if a decrease of modulation
amplitude has an effect on the line shape. As regards the microwave power, it must be chosen in order to avoid saturation effects
which reduce the peak-to-peak amplitude I_{pp}. Below saturation
I_{pp} varies linearly with the power. As saturation sets in, I_{pp}
increases at a lower rate and eventually either flattens or
decreases as the power is increased.

III - THE g-FACTOR

In the case of a free-electron the magnetic moment leads to
the resonance condition given by (II.4). If each unpaired electron in a paramagnetic molecule could be considered as a free-
electron, all resonances would be expected to occur at the same
value of the field. However, this is notfound to be the case for
the magnetic moment of paramagnetic molecules is not simply built
up from free spin electron moments. Indeed, in a molecule or an
atom, there is usually a contribution to the magnetic moment from
the electronic orbital motion.

III.1 - Case of a free atom

An atom will possess a permanent magnetic dipole moment only
if it has nonzero electronic angular moment $\vec{J}$

$$\vec{J} = \vec{S} + \vec{L}$$

$\vec{S}$ = spin angular momentum = $\sum_i \vec{s}_i$

$\vec{L}$ = orbital angular momentum = $\sum_i \vec{\ell}_i$

The magnetic moment of an atom in an electronic state specified
by the quantum numbers L, S and J is :

$$\vec{\mu}_J = g_J \, \beta_e \, \vec{J}$$

$$(III.1) \quad g_J = 1 + \frac{J(J+1) + S(S+1) - L(L+1)}{2\,J(J+1)}$$

g_J is the Lande factor for the atom. In a magnetic field $\vec{H}$ of intensity H_o, the corresponding Hamiltonian is :

$$H = - \vec{\mu}_J \cdot \vec{H} = g_J\,\beta_e\,\vec{J} \cdot \vec{H}$$

and the eigenvalues are

$$E_{M_J} = g_J\,\beta_e\,H_o\,M_J \qquad M_J = -J,\ldots,+J$$

Thus there are $2J + 1$ equally spaced levels. The allowed transitions correspond to $\Delta M_j = \pm 1$ and the resonance condition leading to energy absorption by the atom from the field is met when

$$(III.2) \quad H_{res} = h\nu/g_J\,\beta_e$$

For example, in the case of hydrogen and nitrogen, their ground states are ^2S_o and ^4S_o respectively (using notation $^{2S+1}L_J$), $L = 0$ and there is no orbital contribution to the magnetic moments, thus the g-factor is g_e. For oxygen, the ground state is $^3P_2$ and $g_J = 1.5$. The observed resonances for these gases at low pressure agree with such g values.

III.2 - Molecules and condensed systems

It can be shown that a magnetic moment caused by the orbital motion can result only in the case of atoms, linear molecules or systems with orbitally degenerate electronic states (as p, d... orbitals for an atom). For isolated atoms such an orbital contribution is observed, as afore mentioned for gases. However, for atoms or ions in molecular or crystalline environments, the electron in degenerate orbital is then subjected to the electrostatic interactions arising from other electrons and nuclei composing the molecule or the crystal, in such a way that the primitive degeneracy is lifted and there is no orbital angular momentum. Under these circumstances, it is said that the orbital angular momemtum is "quenched". Of course, for any non-linear, non-degenerate molecule there is no contribution of the orbital angular momentum.

In actual fact, even when the orbital angular momentum is zero or quenched, there are small residual orbital contributions still remaining, due to the magnetic interaction between the spin moment and the magnetic field produced by the orbital motion of the charged particle. Such an interaction is called spin-orbit coupling.

When the paramagnetic molecule is subjected to the action of a magnetic field, the combined effect of the field which lifts the degeneracy and of the spin-orbit coupling reinstates small contributions of orbital angular momentum into the ground state of the system. In this way, the magnetic properties of the system are not due only to the spin angular momentum. Nevertheless, it is convenient to continue to regard the resulting magnetic moment as produced by a pure spin, an effective spin $\vec{S}$, defining consequently an effective g-factor for the system :

$$\vec{\mu} = - g \beta_e \vec{S}$$

Deviations of g from g_e are about 1 ‰ for aromatic free radicals, but they can be 30 % for molecules containing transition metal ions. Furthermore, it is clear that the magnitude of such corrections depends on the orientation of the magnetic field with respect to the molecular axis system. Thus, the g-factor of a molecule cannot be represented by a scalar, but by a second-rank tensor, $\bar{g}$.

Thus, we can write the Zeeman Hamiltonian for a paramagnetic molecule in the form :

$$(III.3) \qquad H_z = \beta_e \vec{H} \cdot \bar{g} \cdot \vec{S}$$

Because of the anisotropy in $\bar{g}$, the resonant field for a particular microwave frequency will depend on the orientation of the sample in the field.

Of course, the specification of the components of the g-tensor depends on the choice of the coordinate axis system. However, the physical observable corresponding to the Hamiltonian operator is the energy, which cannot depend on the reference frame, so the choice of a reference frame is a completely arbitrary one. Nevertheless, it is convenient to consider the principal molecular axis system (p, q, r) which put the g-tensor in a diagonal form. Thus, the Zeeman Hamiltonian becomes :

$$(III.4) \qquad H_z = \beta_e (g_p H_p S_p + g_q H_q S_q + g_r H_r S_r)$$

g_p is written instead of g_{pp} and similarly for g_q, g_r. H_p, H_q, H_r are the components of H in the molecular axis system.

If the g-tensor has an axial symmetry, one gets :

$$(III.5) \quad g_p = g_p = g_\perp \quad ; \quad g_r = g_\parallel$$

The last point concerns the fact that the trace of a tensor is unvariant under any rotational transformation of the axis system. This invariance has the following consequence. Consider a paramagnetic molecule in highly fluid solution, in which the tumbling of the molecule can be sufficiently rapid and random that all orientations are covered in a time short compared to the reciprocal of the frequency-spread of the transitions corresponding to extrema in the anisotropic g-value. For such a case, the position of the resonance is determined by the average of the diagonal elements of the g-tensor :

$$(III.6) \quad H_{res} = h\nu_o / g^o \, \beta_e$$
$$g^o = (1/3) \; Trace \; (\overline{g})$$

It can be convenient, in some cases, to separate this isotropic part of the g-tensor by writing the components in the form :

$$(III.7) \quad g_i = g_i' + g^o \qquad i = p, \; q, \; r$$

Thus, the $\overline{g}'$ tensor is traceless.

IV - HYPERFINE COUPLING

Experimentally, it is often observed that an E.S.R. spectrum consists of a number of lines which are disposed in a centrosymmetric pattern (see, for example Fig. 6). Such conspicuous multiplet structure arises from the interaction of the electron spin magnetic moment with the nuclear magnetic moments of nuclei possessing nonvanishing nuclear spin angular momentum, I, as hydrogen ($I_H = 1/2$), nitrogen ($I_N = 1$). These structures are called hyperfine structures.

Many nuclei possess spins, $\vec{I}_N$, and the associated magnetic moments μ_N are colinear :

$$\vec{\mu}_N = g_N \, \beta_N \, \vec{I}_N$$

where g_N is the nuclear g-factor, β_N the nuclear magneton ($e\hbar/2 \, M_p \, c$), M_p the proton mass (1838 times the electron mass). Thus, the nuclear moments are 10^3 times smaller than the electron moments so that in a magnetic field the nuclear spin states are much more nearly equally populated than are the electron spin states. For this reason, the nuclear levels can be neglected in discussing usual E.S.R..

When a magnetic field is applied, the nuclear moments take (2I + 1) allowable orientations relatively to the direction of the magnetic field. The hyperfine structure comes from the fact that the electron spin magnetic moment interacting with the nucleus feels a different total field depending on which of the

(2 I + 1) allowable orientation is taken by the nuclear spin in the static magnetic field. So, it is evident that the hyperfine coupling will depend on the orientation of the paramagnetic molecule relatively to the static magnetic field, as it is for the spin-orbit coupling involved in the g-factor (see III.2).

The classical interaction between two magnetic moments $\vec{\mu}_1$, $\vec{\mu}_2$, separated by a vector $\vec{r}$, is obtained by considering the energy of one moment in the magnetic field of the other and leads to :

$$E = - (3(\vec{\mu}_1 \cdot \vec{r})(\vec{\mu}_2 \cdot \vec{r}))/r^5 + (\vec{\mu}_1 \cdot \vec{\mu}_2)/r^3$$

The hamiltonian for the magnetic interaction of a nuclear and an electron spin is obtained by using the operators

$$\vec{\mu}_1 = g_N \beta_N \vec{I} \quad \text{and} \quad \vec{\mu}_2 = - g_e \beta_e \vec{S}$$

$$H = - g_N \beta_N g_e \beta_e \{-3(\vec{I}.\vec{r})(\vec{S}.\vec{r})/r^5 + (\vec{I}.\vec{S})/r^3\}$$

The spin Hamiltonian, H_{hf}, acting on spin variables only, which is required for E.S.R., can be obtained by integration of H over the electron spatial coordinates, i.e. by averaging over the electron probability distribution $\psi^2(r)$ corresponding to the ground electronic state. When doing that, a difficulty arises because the integral becomes infinite at r = 0. For this reason, it is interesting to rewrite H as :

$$(IV.1) \quad H = - g_N \beta_N g_e \beta_e \{((\vec{I}.\vec{S})r^2 - 3(\vec{I}.\vec{r})(\vec{S}.\vec{r}))/r^5 - (\vec{I}.\vec{S})\delta(\vec{r}))\}$$

where $\vec{r}$ is the electron-nucleus distance vector and $\delta(\vec{r})$ is the Dirac delta function. The first term appearing in (IV.1) describes the electron nucleus dipolar interaction. The second term, called the Fermi contact coupling exists only when the electron probability at the position of the nucleus does not vanish, as in the case of an s-orbital.

The spin Hamiltonian for the dipolar coupling can be expressed :

$$(IV.2) \quad H_1 = \vec{I} \cdot \vec{A'} \cdot \vec{S}$$

the dipolar coupling tensor A' is a symmetric second rank tensor, with elements given by :

$$(IV.3) \quad A'_{ij} = - g_N \beta_N g_e \beta_e < (r^2\delta_{ij} - 3 ij)r^{-5} > \quad ; \quad i,j= x,y,z$$

where angular brackets denote an expectation value over the electron wave function. It is evident from (IV.3) that the trace of A' is zero. Furthermore, due to the r^3 dependence of the dipolar interaction, only nuclei close to the unpaired electron orbital will lead to significant hyperfine couplings.

Fermi has shown that the spin Hamiltonian for the contact coupling is given by :

$$(IV.4) \quad H_2 = a \, \vec{S} \cdot \vec{I}$$

where a is a scalar, called isotropic hyperfine coupling constant, defined by :

$$(IV.5) \quad a = (8 \, \pi/3) g_N \, \beta_N \, g_e \, \beta_e \, \psi^2(0)$$

$\psi^2(0)$ is the value assumed by the unpaired electron wavefunction at the nuclear position.

Finally, the complete hyperfine Hamiltonian obtained by summing H_1 and H_2, can be written in a compact form :

$$(IV.6) \quad H_{hf} = \vec{I} \cdot \overline{A} \cdot \vec{S}$$

where $\overline{A}$ is the hyperfine tensor with components

$$(IV.7) \quad A_{ij} = A'_{ij} + a \, \delta_{ij}$$

Some consequences of the above considerations must be stressed. First, it is clear that if an unpaired electron is localized in an s-orbital, the dipolar term vanishes and only the a term remains. On the contrary, for unpaired electron in a p, d, f... state there is no contact term and only the dipolar interaction manifests itself. If the electron is described by a wave function that is a mixture of s and p states, both dipolar and contact interactions are present simultaneously. In actual fact, in most organic free radicals, in particular in nitroxide radicals, the unpaired electron distribution is described by p-type orbitals. A second interesting consequence comes from the fact that the A' tensor being traceless, for a radical undergoing rapid tumbling in highly fluid solution, there is no contribution to the hyperfine splitting from the dipolar coupling. The required condition is that the tumbling must be so rapid and random that all orientations are covered in a time short compared to the reciprocal of the difference between the extreme values of the dipolar splitting expressed in frequency units. For such a case the hyperfine splitting will be given by :

$$(IV.8) \quad A^\circ = (1/3) \, \text{Trace} \, \overline{A} = a$$

Finally, we must point out that the hyperfine tensor being symmetric, it can be diagonalised, for instance with the principal molecular axis (p, q, r) of the hyperfine tensor, the spin Hamiltonian assumes the following form :

$$(IV.9) \quad H_{hf} = A_p I_p S_p + A_q I_q S_q + A_r I_r S_r$$

If the A-tensor has an axial symmetry, one gets :

$$(IV.10) \quad A_p = A_q = A_\perp \quad ; \quad A_r = A_\parallel$$

V - SPIN HAMILTONIAN AND ALLOWABLE TRANSITIONS

V.1 - Isotropic Hamiltonian

We consider radicals in highly fluid solutions, so that the random tumbling is sufficiently rapid to average away the aniso-tropic interactions and it remains only the isotropic contributions g^o and a as defined by (III.5) and (IV.8,5) respectively. Furthermore, as nuclear magnetic moments are smaller than electron magnetic moments by a factor 10^3, their contribution to the Hamiltonian will be neglected. Thus, for a single unpaired electron spin interacting via hyperfine coupling with N nuclei of individual nuclear spin quantum number I_k, the Hamiltonian is

$$(V.1) \quad H = g^o \vec{H} \cdot \vec{S} + \sum_{k=1}^{N} a_k \vec{S} \cdot \vec{I}_k$$

Since the hyperfine interactions are always much smaller than the Zeeman interaction energy, a perturbation procedure can be used for computing the energy levels which are given by :

$$E = g^o \beta_e H_o m_s + \sum_k a_k m_s M_k$$

where $m_s = \pm 1/2$ and $M_k = - I_k, \ldots, + I_k$. Due to the effect of the static magnetic field H_o and of the hyperfine coupling, a number of sublevels are generated. Some of these levels will have the same energy if the hyperfine interaction is the same for two or more nuclei (in this case the nuclei are said magnetically equivalent and they can be recognized from the molecular geometry).

Applying a microwave field to the system induces transitions among the sublevels. It can be shown that the states involved in the transitions must satisfy the conditions :

$$\Delta m_s = \pm 1 \quad ; \quad \Delta M_k = 0 \quad \text{for all nuclei.}$$

For simplicity, we can assume that the unpaired electron interacts with a single nucleus with spin I = 1. The corresponding energy diagram is shown in Figure 4 as the allowed transitions. The resulting E.S.R. spectrum will consist of three lines corres-ponding to M = - 1, 0, + 1. Such a case is encountered in nitro-xide radicals in which the unpaired electron interacts with the ^{14}N nucleus of spin 1 (see VI.1).

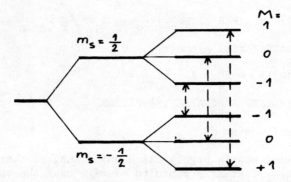

Fig. 4 – Energy level diagram for an electron spin in a static
         magnetic field, interacting with a nucleus of spin 1.
         The coupling constant is assumed to be positive

Now, if we assume that the electron spin is surrounded by N
magnetically equivalent nuclei of spin I_k, the energy of the hyperfine levels is then :

$$(V.2) \quad E(m_S, M) = g^O \beta_e H_o m_S + a m_S M$$

where $M = \sum_i M_k^{(i)}$. The energy difference of the levels involved in
the allowed transitions is

$$(V.3) \quad \Delta E(M) = g^O \beta_e H_o + a M$$

The E.S.R. spectrum will consist of a number of lines equal to the
$(2 N I_k + 1)$ permissible values of M with intensity proportional
to the degeneracy of each nuclear M state, i.e., to the number
of ways in which the individual spin components M_k can be combined
to give the resultant M. If we consider hydrogen atoms ($I_k = 1/2$),
the spectrum will have $(N+1)$ lines with intensities proportional
to the coefficients of the binomial expansion $(1+x)^N$.

V.2 - General spin Hamiltonian

When dealing with viscous solutions, or solid systems, the
paramagnetic molecule is no longer able to perform rapid and random tumbling for averaging away the anisotropic interactions. In
these cases, the complete form of the spin Hamiltonian must be
considered. Thus, for a single unpaired electron spin interacting
with N nuclei of individual spin I_k, we have

$$(V.4) \quad H = \beta_e \vec{H} \bar{g} \vec{S} + \sum_{k=1}^{N} \vec{S} \cdot \bar{A}_{(k)} \cdot \vec{I}$$

where $\bar{g}$ and $\bar{A}$ tensors are defined by (III.3) and (IV.7, 3).

It is interesting to write down this Hamiltonian in order to separate the isotropic and anisotropic parts of $\bar{g}$ and $\bar{A}$ tensors. Thus, we obtain :

$$(V.5) \quad H = (g^0 \beta_e \ \vec{H}.\vec{S} + \sum_k a_k \ \vec{S}.\vec{I}_k) + (\beta_e \ \vec{H}.\vec{g}'.\vec{S} + \sum_k \vec{S}.\vec{A}'_{(k)}.\vec{I}_k)$$

where $\bar{g}'$ and $\bar{A}'$ are traceless tensors defined by (III.7) and (IV.3).

In the case of viscous solutions, in which the paramagnetic molecule undergoes slow motions, the g' and A' tensors will be time dependent. Indeed, it has been pointed out that both spin-orbit coupling and hyperfine coupling depend on the position of the radical relatively to the static magnetic field so that the motions of the radical will modify these interactions. Thus, the corresponding Hamiltonian can be expressed as :

$$(V.6) \quad H = H^0 + H_1(t)$$

Depending on the motional model assumed for the radical, different analytical expressions will be used for evaluating the time average of $H_1(t)$.

Finally, if we assume that the principal molecular axis systems for g and A coincide, in the case of a single unpaired electron interacting with a single nucleus, the corresponding Hamiltonian is the sum of the spin-orbit coupling Hamiltonian (III.4) and the hyperfine Hamiltonian (IV.9):

$$(V.7) \quad H = \beta_e(g_p H_p S_p + g_q H_q S_q + g_r H_r S_r)$$
$$+ A_p I_p S_p + A_q I_q S_q + A_r H_r S_r$$

VI - E.S.R. SPECTRA OF NITROXIDE FREE RADICALS IN DIFFERENT STATES. DETERMINATION OF THE PARAMETERS OF g AND A TENSORS

In studies on molecular motions of polymer solutions or bulk polymers, the most used radicals are nitroxide radicals, with the paramagnetic group :

The main advantages of such radicals are their great stability and chemical inertness due to the protective effect of the four methyl groups. The principal molecular axis system is chosen as shown :

$$\underset{q}{\overset{r}{>}}N\!-\!O\longrightarrow p$$

and it is the same for $\overline{g}$ and $\overline{A}$ tensors.

Due to the nitrogen spin (I = 1) the E.S.R. spectrum contains three main lines corresponding to M = - 1, 0, + 1.

VI.1 - Solution spectra

As above mentioned (see V.1), in highly fluid solutions , only the isotropic contributions g^o and A^o = a are present. Theoretical calculations lead to a positive value of a for nitroxide radicals. The energy diagram shown in Fig. 4 indicates the three allowed transitions. In an E.S.R. spectrometer (see II.3) the microwave frequency ν_o remains constant and the magnetic field is scanned to meet the resonance conditions (V.3). Energy of the various levels increases linearly with H_o (V.2), as represented in Fig. 5 in such a way that when H_o is increased, the first

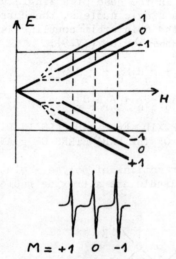

Fig. 5 - Magnetic field strength dependence of the energy levels.
        Positive coupling constant a.

transition appearing (low field transition) corresponds to M= +1, the last one (high field transition) occuring for M= - 1. The spectrum, shown Fig. 6, is composed of three sharp lines. The

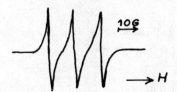

Fig. 6 - X band spectrum of nitroxide free radical in solution
         in toluene

position of the central one allows to determine g^o by using equation (III.6) and the splitting, ΔH, of the two other lines from
the central one leads to a; through the relation :

$$a = g^o \beta_e \Delta H$$

We must point out that for nitroxide radicals in very fluid
solvents as carbon tetrachloride or chloroform, each main line
corresponding to a M value is decomposed in 13 lines among which
11 are easily observable. This multiplet structure comes from
additional hyperfine interactions of the unpaired electron with
the 12 hydrogen atoms of the four neighbouring methyl groups.
Such a structure disappears in more viscous solvents.

Finally, under high magnification satellite lines occur due
to the hyperfine interaction with $^{13}C(I = 1/2)$ and $^{15}N(I = 1/2)$
at their natural abundances, 1.1 % and 0.36 % respectively.

VI.2 - Spectra in oriented crystals

In order to determine the anisotropic components of $\bar{g}$ and $\bar{A}$
tensor, the only possibility consists to introduce the paramagnetic molecule into a single crystal of an host molecule and to
record the E.S.R. spectra for various orientationsof the crystal
relatively to the static magnetic field. In the case of nitroxide
radicals, when one of the principal molecular axis is directed
along the magnetic field, the resonance conditions are given by :

$$h\nu_o = \beta_e g_i H + A_i M \qquad i = p, q, r$$

Thus the various g_i and A_i can be directly obtained from the position of the central line and from the splitting of the other lines
(M = ± 1) in spectra recorded at the convenient orientations. Such
spectra are shown in Fig. 7, and the corresponding values are
reported in Table 1 in the case of the radical :
$(CH_3)_3 - C - NO - C - (CH_3)_3$. Values for other nitroxide radicals
can be found in App. II of (3) and in (4).

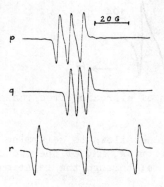

Fig. 7 - X band spectra of nitroxide free radical in a crystal
with the field along the principal axes of the g and
A tensors. a/ p , b/ q ,c/ r

g_p	2.0089	$A_p(A_x)$	7.1 G
g_q	2.0061	$A_q(A_y)$	5.6 G
g_r	2.0027	$A_r(A_z)$	32 G
g^o	2.0059	a	14.9 G

Table 1 - Magnetic parameters for the nitroxide free radical
indicated in the text. Another equivalent notation is
used with $x \equiv p$, $y \equiv q$, $z \equiv r$

VI.3 - Spectra of nonoriented systems

In polycrystalline samples or in glassy matrices, the principal axes of the radical molecules are randomly oriented relatively to the direction of the magnetic field. Thus, the E.S.R.
spectrum is a superposition of the spectra corresponding to all
possible orientations. Nevertheless, though the radicals are randomly oriented in the sample the distribution of the absorption
intensity is not uniform. Indeed, if we consider a paramagnetic
molecule with an axially symmetric g tensor, due to the random
orientation, the number n of molecules with an angle θ between
the symmetry axis, r, and the field direction is proportional to
sin θ, so that there are more molecules with their r axis perpendicular to the field direction than with their r axis parallel.
Thus, there will be more paramagnets absorbing at the resonance
field determined by $g_\perp$ than those absorbing in the field domain
determined by $g_\parallel$. This will result in maxima or minima occuring
in the central line of the E.S.R. spectrum.

Furthermore, as in nitroxide radicals the hyperfine coupling
A_r (or A_z in another notation) is much larger than the two other

ones, this hyperfine splitting can be directly determined from
the extreme peaks of the E.S.R. spectrum recorded in a glassy
solution at sufficiently low temperatures (around 100° K)

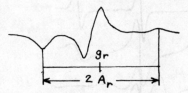

Fig. 8 - X band rigid-limit spectrum of a glassy solution of
         nitroxide free radical at 100°K

for avoiding any motion; it is the rigid-limit spectrum, an example of which is shown in Fig. 8. The middle between the two
extreme peaks leads to a determination of g_r.

        If an axial symmetry is assumed for the $\bar{g}$ and $\bar{A}$ tensors, the
spectrum recorded in a very fluid solution leads to g^o and
A^o = a, the spectrum obtained in the rigid-limit provides $g_{\parallel}$ and
$A_{\parallel}$ and the two other quantities $g_{\perp}$ and $A_{\parallel}$ can be derived by using
relations :

$$g^o = (1/3)(g_{\parallel} + 2\ g_{\perp})$$

$$A^o = a = (1/3)(A_{\parallel} + 2\ A_{\perp})$$

VI.4 - Evolution of E.S.R. spectra with the mobility

        We have above considered the two limit cases of a highly
fluid solution and a rigid medium. In actual fact, when the viscosity of the medium is gradually changed and the radical motion
is progressively slown down till a motionless situation, the
E.S.R. spectrum undergoes a smooth evolution between these two
extreme cases. As an example, spectra obtained with a nitroxide
radical in a polyethylene oxide matrix is shown in Fig. 9 in a
temperature range from the melt to liquid nitrogen temperature.

VII - MOLECULAR MOTIONS AND E.S.R. LINE SHAPES

        From spectra shown in Fig. 9, it is evident that the line
shape depends on the molecular motions. In the case of a rapid
tumbling of the paramagnetic molecule the E.S.R. lines are sharp,
whereas very broad lines are obtained in highly viscous medium
and in the rigid-limit case.

        As pointed out in V.2, the general spin Hamiltonian can be
decomposed (V.6) in an isotropic term H_o and an anisotropic
term $H_1(t)$ which depends on time since the spin-orbit and hyper-

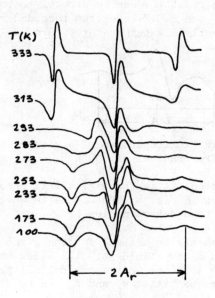

Fig. 9 - X band spectra of a nitroxide radical probe in polyethylene oxide at various temperatures

fine interactions are function of the position of the paramagnetic molecule relatively to the magnetic field direction.

As mentioned in II.2 and in Fig. 2, the linewidth of an
E.S.R. spectrum is directly related to the spin-spin relaxation
time T_2. Thus, the calculation of the E.S.R. line shape implies
the calculation of the time dependence of the macroscopic magnetisation $M_x(t)$. It can be achieved by considering the density
matrix and the relaxation matrix which involves the Fourier transform of the correlation function of $H_1(t)$. Such derivations
require quantum mechanical calculations which are out of the purpose of this paper. Detailed developments can be found in (2,6,7)
and in the original papers quoted hereafter; the principal features of the calculations are given in chapters 2 and 3 of (3).
In this paper, we will only report the main results, showing how
the correlation times of motions can be derived from the experimental results.

VII.1 - Rapid isotropic rotations

It has been pointed out that in the case of rapid tumbling
of the radical molecule, the positions of the E.S.R. lines are
determined by the isotropic part of the g and A tensors. However

even if the anisotropic terms do not contribute to the magnetic
parameters obtained in solution, they play an important role
in the line broadening. Indeed, the molecular tumbling makes $H_1(t)$
a random function of time in such a way that there is a random
modulation of the energy levels and of the transition frequencies
which leads to a broadening of the absorption lines.

In order to characterize a motion as rapid or not, it is
necessary to compare the correlation time τ of the motion with
the frequency shift Δ arising from the anisotropic interactions
(in angular frequency units), mostly the hyperfine interactions
which are the dominant ones.

A motion is considered to be fast if $\Delta \cdot \tau_c < 1$. In the case
of nitroxide radicals, this condition is satisfied for
$\tau_c \leqslant 10^{-9}$ sec.

Line shape theory of rapid isotropic motions has been developed by Kivelson (8) and Freed (9) in assuming an isotropic motion
of the paramagnetic molecule, characterized by a correlation time
$\tau_R = (6\ D)^{-1}$ where D is the rotational diffusion constant. The
two theories lead to the same expression. In the case of a single
unpaired electron interacting with a single nucleus, as for
nitroxide radicals, the line width of E.S.R. spectra is given by :

(VII.1) $T_2^{-1}(M) = \tau_R[(3/20)b^2 + (4/45)(\Delta \gamma H_o)^2$

$$- (4/15)b\Delta \gamma H_o M + (1/8)b^2 M^2] + X$$

where $\Delta \gamma = - (\beta_e / \hbar)[g_r - (1/2)(g_p + g_q)] = - (3\ \beta_e / 2\ \hbar)(g_r - g^o)$

$b = (2/3)[A_r - (1/2)(A_p + A_q)] = (A_r - A^o)$

the g_i' s and A_i' s magnetic parameters are defined in III.2 and IV.

ω_o is the Larmor frequency of the electron in the applied
magnetic field H_o.

M is the nuclear quantum number, which is equal to 0, ± 1 in
the case of nitrogen

X is the intrinsic line width due to other relaxation mechanisms and independent of M.

The A_i' s are expressed in cycle.sec^{-1} and $T_2(M)$ in sec.

This expression is valid under the following conditions :

a/ $T_2^{-1} \tau_R \ll 1$ in such a way that the motion could be considered as
rapid and the line shape be nearly Lorentzian

b/ $A_o^2 \ll \omega_o^2$ in order to be able to consider the hyperfine interaction as a perturbation of the Zeeman Hamiltonian

c/ $g_i^2 \ll g_o^2$ which means that the spin coupling anisotropy is small.

d/ $\omega_o^2 \tau_c^2 \ll 1$ for neglecting the non-secular terms, which implies that $\tau_c \geqslant 5.10^{-11}$ sec.

Expression (VII.1) can be written as :

$$(VII.2) \quad T_2^{-1}(M) = A - BM + CM^2$$

The $T_2(M)$ values are measured from either the peak-to-peak width, $\Delta H_{pp} = (2/\sqrt{3})T_2$, or the peak-to-peak amplitude, $I_{pp} = (3\sqrt{3}/4 \ \pi)T_2^2$, shown in Fig. 2. Two other quantities are frequently used in the case of nitroxide radicals, in which $M = 0, \pm 1 : R_+ = T_2(0)/T_2(+1) ; R_- = T_2(0)/T_2(-1)$

and the following relations can be derived :

$$(VII.3) \quad (R_+ + R_- - 2)T_2^{-1}(0) = 2 \ C$$

$$(VII.4) \quad (R_+ - 1) \ T_2^{-1}(0) = C + B$$

$$(VII.5) \quad (R_- - 1) \ T_2^{-1}(0) = C - B$$

$$(VII.6) \quad (R_+ + R_- - 2)/(R_+ - R_-) = C/B$$

with $C = b^2 \tau_R/8$ and $B = (4/15)b\Delta\gamma H_o \ \tau_R$

For nitroxide radicals, B is always positive, due to the smaller value of g_r compared to $(1/2)(g_p + g_q)$. Thus, for E.S.R. spectra obtained in experimental conditions in which the radical motion is rapid and isotropic, we expect an increase of the line width and consequently a decrease of the peak-to-peak amplitude when going from the low field line (M = + 1) to the central one (M = 0) and to the high field one (M = - 1). Nevertheless, this effect is fairly small for paramagnetic molecules in fluid solutions or for spin-labelled polymers in fluid solutions. In these cases, the same correlation time τ_R is obtained independtly of the expression chosen for doing the calculation (VII.3, 4 or 5).

In order to get a more accurate value of τ_R, it is necessary to take into account the influence of the multiplet structure due to the interaction of the unpaired electron with the 12 hydrogen atoms of the four methyl groups close to the nitroxide radical. For this purpose, a simulation of the complete E.S.R. spectrum, based on rapid motion theory, is required and τ_R is deri-

-ved from the fitting with the experimental spectrum. The resulting correction of τ_R is in the range of 10 %.

VII.2 - RAPID ANISOTROPIC ROTATIONS

In some cases of more viscous solvents or in polymer melts,
the condition required for rapid motion is satisfied
($\tau \leqslant 10^{-9}$ sec) but nevertheless the width of the various lines
(M = 0, ± 1 for nitroxide radicals) does not agree with the expectation from the isotropic motion theory. In particular, expressions (VII.3, 4 and 5) do not lead to the same correlation time.
Such discrepancies can originate from an anisotropic motion of the
paramagnetic molecule. A theoretical treatment has been developed
by Freed (10).

If we consider the nitroxide radical as an ellipsoid, the
effect of anisotropic diffusion on line widths depends on the
orientation of the main axis of rotation (which defines the fastest rotation time and largest diffusion constant, i.e. $D_{\parallel}$ in
Fig. 10) with respect to the orientation of the p_z axis of the

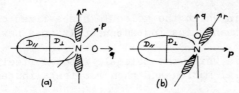

(a)   (b)

Fig. 10 - Relative orientations of the diffusion and nitroxide
          axes in two typical cases.

nitrogen atom, which is the axis of the orbital which contains the
unpaired electron. Nordio (11) has shown that anisotropic diffusion gives a difference in line widths between the low-field and
centre-field lines. In case (a) of Fig. 10, the low-field line is
narrower and more intense than the centre-field line. It is the
reverse in case (b). Consequently, the ratio C/B as derived from
expression (VII.6) has not the same value as for a nitroxide radical in fluid solution and this ratio vary with the contribution
of the anisotropic motion, in particular with temperature and viscosity. In these cases, if an isotropic correlation time, τ_c, is
derived from expression (VII.3), it is defined by :

$$(VII.7) \quad \tau_c = (6 \, \bar{D})^{-1} \quad \text{with} \quad \bar{D} = (D_3 \, D_1)^{1/2}$$

For an ellipsoid molecule, the two correlation times which
can be derived from a E.S.R. spectrum, are defined by :

$$(VII.8) \quad \tau_0 = 6 \, D_{\perp}^{-1} \quad \text{and} \quad \tau_2 = [6 \, D_{\perp}^{-1} + 4(D_{\parallel} - D_{\perp})]^{-1}$$

In order to determine τ_o and τ_2 it is necessary to simulate the spectrum by applying the rapid motion theory with anisotropic rotational diffusion.

VII.3 - Slow isotropic rotations

In the fast rotational motion region, all rotational reorientation processes yield to Lorentzian lines. At the opposite, in the very slow rotational motion region, all models will tend to the rigid-limit spectrum related to the distribution of the absorption intensity (see VI.3). In the intermediate slow rotational region, the E.S.R. line shapes are sensitive to the type of the molecular reorientation process.

Three motional models have been used for describing slow isotropic rotations.

a/ Brownian rotational diffusion, which corresponds to an infinitesimal reorientation of the molecule with each collision with the surrounding particles.

b/ jump diffusion, in which the molecule has a fixed orientation for the time τ and then jumps instantaneously to a new direction

c/ free diffusion, in which the molecule rotates freely for a time τ due to inertial motion and then reorients instantaneously to a new direction.

The application of these models to E.S.R. line widths has been performed by Freed and coworkers (12, 13) and a computer program for spectra simulation is given in appendix B, chap. 3 in (3). The values of the corresponding correlation time are derived from the simulation fitting of the experimental curves. For a spherical radical molecule, with a rotational diffusion coefficient D, the correlation times derived from the motional models above described are :

$$(VII.9) \quad \tau_c^{-1} = 6\ D \qquad \text{for Brownian diffusion}$$

$$(VII.10) \quad \tau_c^{-1} = 6\ D/7 \qquad \text{for jump diffusion}$$

$$(VII.11) \quad \tau_c^{-1} = 6\ D/\sqrt{7} \qquad \text{for free diffusion}$$

and the time τ involved in models b/ and c/ is such that $\tau D = 1$.

The slow motion theory can be applied for correlation times in the range :

$$10^{-9}\ \text{sec} \leqslant \tau_c \leqslant 3.10^{-7}\ \text{sec}$$

If computer simulation of experimental spectra is the best method for treating slow motions of isotropic molecules, nevertheless it is a rather heavy way. For this reason, simplified methods of estimating τ_c have been developed, based on quantities which can be directly obtained from recorded spectra.

1/ Separation of outer hyperfine extrema

In the case of nitroxide radicals, as above pointed out (see VI.3), in the rigid-limit spectrum there are two well-separated outer hyperfine extrema, the separation of which is equal to 2 A_r (or 2 A_z depending on notation) as shown in Fig. 9 and 11. It has been found (13, 14) that a useful parameter is $S = A_r'/A_r$

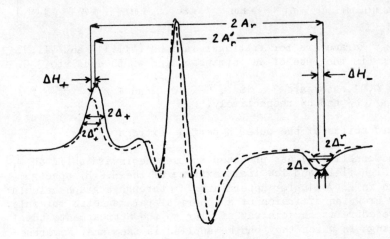

Fig. 11 - Nitroxide spectra in the case of rigid-limit (------)
         and slow motions (————). Measurements required for
         determining τ_c are indicated.

where 2 A_r' is the separation of the two outer hyperfine extrema in the slow motion region. It is evident from Table 1 that, in nitroxide radicals, A_p, A_q and g_p, g_q, g_r contribute only to the central region in the rigid limit spectrum, so that S is insensitive to changes in their values. Furthermore, changes in the magnitude of A_r, in the range 27-40 G encountered with nitroxides, do weakly affect S (less than 3 %). The correlation time τ_c can be obtained from S by the relation :

$$(VII.12) \quad \tau_c = a(1 - S)^b$$

The a and b coefficients depend on the intrinsic line width δ, which can be determined from the half-width at half-heights, $\Delta^r(M)$, for the outer extrema of the rigid-limit spectrum, as shown in Fig. 11 :

$$2 \Delta^r_{(+1)} = 1.59 \ \delta \quad ; \quad 2 \Delta^r_{(-1)} = 1.81 \ \delta$$

In the case of nitroxide probes or labels in polymers, δ is found to be around 3 G. These coefficients depend on the nature of the motional model, the corresponding values (15) are given in Table 2. The longest τ_c 's which can be determined by using the S method are given by the values corresponding to

Diffusion model	$a \times 10^{10}$ sec	b	$a'_{+1} \times 10^8$ sec	b'_{+1}	$a'_{-1} \times 10^8$ sec	b'_{-1}
a/ Brownian diffusion	5.4	- 1.36	1.15	0.943	2.12	0.778
b/ Jump diffusion	25.5	- 0.615				
c/ Free diffusion	1.1	- 1.01	1.29	1.033	1.96	1.062

Table 2 - Parameters for fitting relations (VII.11) and VII.13)
           in the case of an intrinsic line width equal to 3 G.

$(1-S) = 0.01$, they are 3×10^{-7}, 5×10^{-8} and 1×10^{-7} sec for models a/, b/ and c/ respectively.

2/ Inward shifts of the outer hyperfine extrema

        Mc Connell (16) has proposed to use the inward shift $\Delta H_{\pm 1}$ of the high field and low field extrema of the E.S.R. spectrum, as shown in Fig. 11, for determining τ_c through a graph calculated for Brownian diffusion of a spherical paramagnetic molecule. This procedure is in some way similar to the S treatment, the τ_c value range in which they can be applied is the same. Nevertheless, the parameters $\Delta H_{\pm 1}$ are interesting for Kuznetsov (17) has shown that the quantity $R = \Delta H_{-1} / \Delta H_{+1}$ has a dependence on ΔH_{+1} which is a function of the motional model. Thus, from the results shown in Fig. 12 for Brownian diffusion and jump diffusion, it is possible to determine the type of motion undergone by the nitroxide radical in the experimental conditions considered.

3/ Half-widths of the outer hyperfine extrema

        As pointed out, the accuracy of the above methods falls off near $\tau_r = 10^{-7}$ sec. For this reason, Freed (18) has developed a method based on the measure of the half-width at half-height, $\Delta_{\pm 1}$, of the outer hyperfine extrema, as shown in Fig. 11. The considered parameter is $W_{\pm 1} = \Delta_{\pm 1} / \Delta^r_{\pm 1}$, where the sup-index r refers to the rigid-limit spectrum. The correlation time, τ_c, is derived by the relation :

$$(VII.13) \quad \tau_c = a'_{\pm 1} (W_{\pm 1} - 1)^{- b'_{\pm 1}}$$

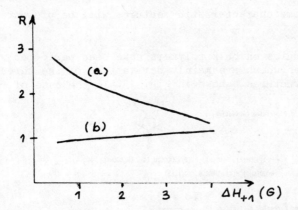

-Fig. 12 - The parameter R as a function of the shift ΔH_{+1} for
a/ Brownian diffusion, b/ jump diffusion

where the a' and b' values, reported in table 2, depend on the
motional model. This method can be used till $\tau_c \simeq 10^{-6}$ sec.

VII.4 - Slow anisotropic rotations

In the case of an axially symmetric radical molecule, the
theory of slow rotations has been extended (13). The two involved
correlation times must be derived from simulation of experimental
spectra by using the computer program listed in appendix B of
chapter 3 in (3). The three types of motional models described in
VII.3 can be used.

When the approximation methods with S or W parameters are
applied to anisotropic rotations, if $D_{\|} > D_{\perp}$ the derived τ_c values
correspond to $(6 D_{\perp})^{-1}$. For small anisotropies about the axes,
i.e. $R_{\|} = 3 R_{\perp}$, one obtains $\tau_c = (1/6)(R_{\|} R_{\perp})^{-1/2}$.

VII.5 - Very slow motions

For correlation times longer than 10^{-6} sec, the line-width
of E.S.R. spectra are no longer sensitive to the motion. During
the last years, experimental techniques have been developped for
studying motions in the τ_c range 10^{-7} sec $< \tau_c < 10^{-3}$ sec. They
are based on nonlinear electron spin response:

- Saturation transfer (19 - 24)
- Electron double resonance (ELDOR) (25)

VIII - E.S.R. APPLICATION TO MOLECULAR MOTIONS IN BULK POLYMERS

Experimental results on molecular motions in bulk polymers,
as studied by the E.S.R. technique, have been recently reviewed
by Törmälä (5) in an extensive way (154 references). For this

reason, only some characteristic features will be presented and
discussed.

Most a studies on bulk polymers have been performed with
nitroxide probes which are mainly derivatives of the three nitroxide cyclic structures, shown in Fig. 13. In the case of reactive

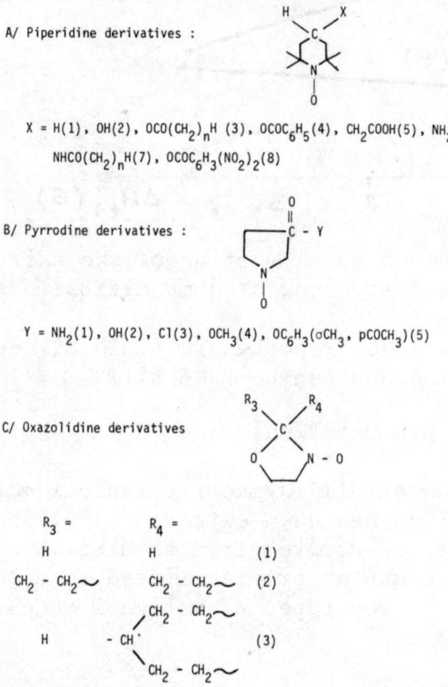

A/ Piperidine derivatives :

X = H(1), OH(2), OCO(CH$_2$)$_n$H (3), OCOC$_6$H$_5$(4), CH$_2$COOH(5), NH$_2$(6),

NHCO(CH$_2$)$_n$H(7), OCOC$_6$H$_3$(NO$_2$)$_2$(8)

B/ Pyrrodine derivatives :

Y = NH$_2$(1), OH(2), Cl(3), OCH$_3$(4), OC$_6$H$_3$(oCH$_3$, pCOCH$_3$)(5)

C/ Oxazolidine derivatives

R_3 =	R_4 =	
H	H	(1)
CH$_2$ - CH$_2$~	CH$_2$ - CH$_2$~	(2)
H	- CH $\backslash$ CH$_2$ - CH$_2$~ / CH$_2$ - CH$_2$~	(3)

Fig. 13 - Spin label and probe nitroxide radicals

derivatives as A2, A5, A6, B3 the nitroxide group can be covalently
bound (labelled) to some polymer chains.

VIII.1 - Tg and E.S.R. parameter T$_{50\ G}$

When the E.S.R. spectra of probe or label radicals in polymers are recorded in a large temperature range, the separation
of the outer hyperfine extrema 2 A'$_r$, as defined in Fig. 11, exhibits a rapid decrease in a rather narrow temperature range. The
temperature region in which 2 A' drops depends on the radical and
the polymer, as shown in Fig. 14. It has been proposed to characterized such a behaviour by the temperature at which 2 A'$_r$ = 50 G,
called T$_{50\ G}$. It has been shown (26) that T$_{50\ G}$ levels when the
radical molecular weight increases, the limit behaviour is reached in a MW range of 250 - 280 g.mol^{-1}. This T$_{50\ G}$ paramater has
been tentatively correlated with the glass-rubber transition temperature of the polymer, but no general relation has been obtained.

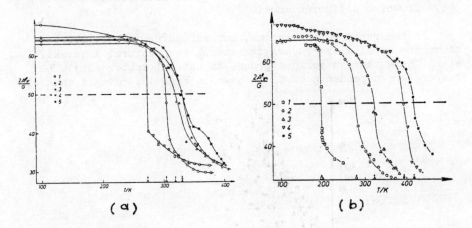

Fig. 14 – Temperature evolution of 2 Λ'_r
a/ Polyisobutylene with various probes : A 2 (1),
B 1 (2), A 7 (3), B 5 (4)
b/ Probe A 7 in various polymers : polydimethylsiloxane (1), polyethylene (2), polyisobutylene (3),
polyvinylchloride (4), polycarbonate (5) (From
ref. (5)).

In actual fact, $T_{50\ G}$ corresponds to the temperature at which
the outer lines clearly converge to the motionally narrowed
spectrum, the corresponding τ_c values lie between 4.10^{-9} to
10^{-8} sec, which is the region between slow and rapid motions.
Thus, the effective frequency ($= (2\,\pi\,\tau_c)^{-1}$) associated to $T_{50\ G}$
would be f = $(3 \pm 1).10^7$ Hz.

The $T_{50\ G}$ parameter must be only considered as a qualitative
means of looking at the onset of rapid motions, but further investigation on the molecular behaviour of polymers using this parameter has no real physical meaning.

VIII.2 –Correlation times of radicals in polymers

The various theoretical developments and the corresponding
expressions of the correlation times, presented in VII, provide
a way of quantitatively studying the motion of probe or label
nitroxides in bulk polymers. A limited number of studies have
dealt with the effect on radical relaxation times of parameters
like the temperature, the nature of polymer chain or the radical
size.

In the region of slow motions, it has been clearly established (27) that the small radical behaviour can be expressed in

terms of jump diffusion while the large probe or label behaviour
obeys Brownian diffusion model.

In the rapid motion regime, corresponding to polymers far
above the usual glass-rubber transition temperature, Kovarskii
et al. have studied various probes in natural rubber (28). The
temperature dependence of τ_R , shown in Fig. 15, reveals that all

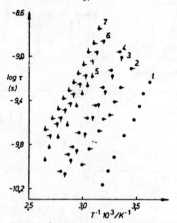

Fig. 15 - Log τ_R vs T^{-1} plots of various probes in natural rubber:
A 1 (1), A 2 (2), A 7 with n = 2 (3), A 3 with n = 2
(4), A 7 with n = 15 (5), A 8 (7) (From ref (5))

studied probes have the same apparent activation energy, but the
correlation times increase with the size of the radical. Though
these results are very interesting, it is delicate to go into
further investigation of molecular behaviour for some of probe
molecules undergo an anisotropic diffusion motion, as pointed out
by the authors, in such a way that a computer simulation of experimental spectra would be required for deriving the actual correlation times (see VII.2).

Another interesting point in the case of natural rubber studies (28) is that the effective frequencies derived from τ_R agree
fairly well with the frequency plot of the glass-rubber transition
obtained from dielectric, mechanical and N.M.R. data, as shown in
Fig. 16. Thus, it appears that these probes reflect local polymer
motions involved in the glass-rubber relaxation processes.

In the case of crystalline polymers, the nitroxide radical,
either free or covalently bound to the chain, is located in the
amorphous phases and enables to look at the molecular motions
occuring in these phases. As an example, Fig. 17 shows the temperature dependence of τ_c for a polyethylene oxide (29) end-label-
led with the nitroxide B 3. For slow motions, the Brownian diffu-

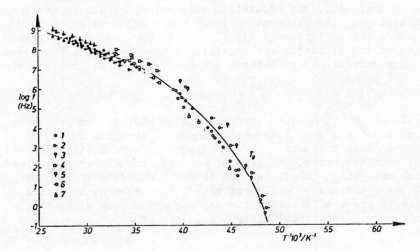

Fig. 16 - Log F vs T^{-1} for natural rubber
1/ Dielectric data; 2/ and 3/ mechanical data;
4/ to 7/ N.M.R. data; E.S.R. data of the probes used
in Fig. 15 with the same symbols. (From reference (5))

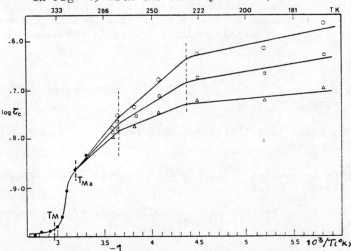

Fig. 17 - Log τ_c vs T^{-1} for labelled polyethylene oxide. Various
motional models are used : ● , rapid motions;
○ , Brownian diffusion; □ , free diffusion;
△ , jump diffusion. (From reference (29))

-sion model should be the most appropriate owing to the fact that
the nitroxide is bounded to the polymer chain. Five regions of
different mobility can be considered. Taking into account that
E.S.R. measurements correspond to high frequency experiments,
comparison with the solid transitions observed on polyethylene

oxide by dielectric and mechanical relaxations, leads to assign
the mobility in the temperature range 227 - 273°K to the γ process, this one in the range 275 - 315°K to the β process corresponding to the glass-rubber-transition of the amorphous phase and
the large decrease of τ_c between 315 and 333°K to the melting zone.

IX - CONCLUSION

The E.S.R. technique, using nitroxide radicals as probes or
labels, provide a very interesting means of studying molecular
motions in bulk polymers either amorphous or crystalline. Indeed,
the size of the probe, the type of bounding to polymer chain (end,
side or main chain labelling) can be vary, allowing to look at
local chain motions of different sizes. Furthermore, theoretical
treatments are available for the various situations of isotropic
or anisotropic rotational diffusion in the rapid or slow motion
regimes.

At the present time, E.S.R. studies performed on bulk polymers have been mostly qualitative, based on the empirical $T_{50\ G}$
parameter. Quantitative studies, based on determination of
isotropic and anisotropic correlation times through computer simulation of experimental spectra, should lead to a significant
improvement of knowledge on molecular motions in bulk polymers.
Such an approach is presently under progress in our laboratory.

References

1. Randby, B., Rabek, J.F.: 1977, "E.S.R. Spectroscopy in Polymer
Research", Springer-Verlag.

2. Atherton, N.M.: 1973, "Electron Spin Resonance. Theory,Applica-
tions", Wiley.

3. Berlinger, L.J., ed.: 1976, "Spin Labelling. Theory and Applications", Academic Press.

4. Gordy, W.: 1980, "Theory and Applications of Electron Spin
Resonance", Wiley.

5. Törmälä, P.: 1979, J. Macromol. Sci.- Rev. Macromol. Chem.,
C 17, pp 297-357.

6. Abragham, A.: 1961, "The principles of Nuclear Magnetism",
Oxford.

7. Slichter, C.P.: 1963, "Principles of Magnetic Resonance",
Harper.

8. Kivelson, D.: 1960, J. Chem. Phys., 33, pp 1094-1106.

9. Freed, J.H., Fraenkel, G.K.: 1963, J. Chem. Phys., 39, pp 326-348.

10. Freed, J.H.: 1964, J. Chem. Phys., 41, pp 2077-2083.

11. Nordio, P.L.: 1970, Chem. Phys. Lett., 6, pp 250-252.

12. Freed, J.H., Bruno, G.V., Polnaszek, C.F.: 1971, J. Phys. Chem., 75, pp 3385-3399.

13. Goldman, S.A., Bruno, G.V., Polnaszek, C.F., Freed,J.H.: 1972, J. Chem. Phys., 56, pp 716-734.

14. Mc Calley, R.C., Shimshick, E.J., Mc Connell, H.M.: 1972, Chem. Phys. Lett., 13, pp 115-119.

15. Goldman, S.A., Bruno, G.V., Freed, J.H.: 1972, J. Phys. Chem., 76, pp 1858-1860.

16. Mc Calley, R.C., Shimshick, E.J., Mc Connell, H.M.: 1972, Chem. Phys. Lett., 13, pp 115-119.

17. Kuznetsov, A.N., Ebert, B.: 1974, Chem. Phys. Lett., 25, pp 342-345.

18. Mason, R.P., Freed, J.H.: 1974, J. Phys. Chem., 78, pp 1321-1323.

19. Hyde, J.S., Dalton, L.R.: 1972, Chem. Phys. Lett., 16, pp 568-572.

20. Hyde, J.S., Smigel, M.D., Dalton, L.R., Dalton, L.A.: 1975, J. Chem. Phys., 62, pp 1655.

21. Robinson, B.H., Dalton, L.R.: 1979, Chem. Phys., 36, pp 207-237.

22. Robinson, B.H., Dalton, L.R.: 1980, J. Chem. Phys., 72, pp 1312-1324.

23. Goldman, S.A., Bruno, G.V., Freed, J.H.: 1973, J. Chem. Phys., 59, pp 3071.

24. Thomas, D.D., Mc Connell, H.M.: 1974, Chem. Phys. Lett., 25, pp 470-475.

25. Bruno, G.V., Freed, J.H.: 1974, Chem. Phys. Lett., 25, pp 328-332.

26. Törmälä, P., Weber, G.: 1978, Polymer, pp 1026 - 1030.

27. Vasserman, A.M., Alexandrova, T.A., Buchachenko, A.L.: 1976, Eur. Polym. J., 12, pp 691-695.

28. Kovarskii,A.L., Vasserman, A.M., Buchachenko, A.L.: 1971, Vysokomol. Soyed. A 13, pp 1647-1653.

29. Lang, M.C., Noël, C., Legrand, A.: 1977, J. Polym. Sci., Polym. Phys. Ed., 15, pp 1329-1338.

QUASI-ELASTIC LIGHT SCATTERING FROM POLYMER SYSTEMS

J M Vaughan

Royal Signals and Radar Establishment
St Andrews Road, Great Malvern, Worcs, UK

The theory and practice of quasi-elastic light scattering
are briefly outlined. The use of wave analysers, intensity
correlators and Fabry-Perot interferometers is described.
Measurements during the past decade of Rayleigh and Brillouin
scattering from a wide range of polymer systems, including amorphous polymers and polymer gels, are reviewed and discussed.

The term 'quasi-elastic' is usually taken to imply light
scattering with frequency shift in the range of a few Hz to a few
tens of GHz. As such it includes the domains of Rayleigh and
Brillouin scattering but excludes Raman scattering where the
frequency shifts are invariably greater than 1 cm^{-1} (30 GHz).
The aims of the present lectures are firstly to give a basic
account of both theory and practice of quasi-elastic laser light
scattering and secondly to discuss and assess the progress that
has been made in applying the techniques to polymer systems. In
regard to the first aim, a simple introduction to the theory of
light scattering is given with reference to many excellent
detailed papers and reviews [eg 1-9]. The available range of
energy and momentum transfer is discussed and the potential this
offers for studying polymer systems. The most widely used techniques of photon correlation and multipass interferometry are
described in a little more detail; these comparatively recent
methods may be unfamiliar to most polymer scientists and certainly
have the reputation of being difficult. It is hoped that the
experimental discussion will help to rectify this situation. With
regard to the second aim many investigations of Rayleigh and
Brillouin scattering from polymer systems have recently been
reported in the literature. Rayleigh scattering in amorphous

305

R. A. Pethrick and R. W. Richards (eds.), Static and Dynamic Properties of the Polymeric Solid State, 305–347.

polymers and polymer gels has given information on relaxation, cross linking, elastic moduli and critical type behaviour and fluctuations. Brillouin scattering and hypersonic viscoelastic measurements have been made on a wide range of systems including gels and biological materials. These investigations are discussed and referenced.

It is useful to appreciate the relative frequency ranges involved in light scattering. Visible (green) light of wavelength ~ 0.5 μm has a frequency ν of ~ 6 x 10^{14} Hz. No detectors are available fast enough to follow this field; what one does have are firstly optical filters that operate directly on the light field and separate it into its constituent frequencies. For such pre-detection analysis the highest resolution is given by a classical instrument - the Fabry-Perot interferometer. This has however a resolving power limited to ~ 10^8 so that it becomes difficult or impossible to analyse frequency components closer than a few MHz. With the enormous expansion of light scattering following the introduction of comparatively inexpensive, reliable lasers it became clear that many phenomena were of much lower frequency than this. In consequence post detection methods were rapidly developed; in fact such methods generally become easier for lower frequencies. In this regime the technique is to expose a (square law) detector to the field and from the resultant measured intensity extract the requisite information about the field. The principle methods rely on either extracting frequency content in the electrical signal from the detector with a scanning wave analyser or alternatively forming the time delayed auto-correlation function of this signal. However before discussing these techniques in detail a simple basis of light scattering is outlined.

1   QUASI-ELASTIC LIGHT SCATTERING

It was long appreciated during the late 19th and early 20th century that a truly homogeneous medium would not scatter light. This may easily be seen by considering a small wavelet emerging from a region in a homogeneous material according to the Kirchoff formulation of light propagation. Then, in any direction other than the forward direction, another region may be found whose corresponding wavelet differs in phase by π and thus cancels out, by destructive interference, the first wavelet. This is true for all directions other than the forward direction, and hence scattering should not occur. Nevertheless even in the most care-fully purified liquids it was clear that appreciable scattering did take place. In 1908 von Smoluchowski suggested that the scattering arose from spontaneous, thermally induced fluctuations occurring on a spatial scale of the order of a wavelength λ. This suggestion was taken up by Einstein in 1910 who provided a

quantitative calculation for the total scattered intensity from
density fluctuations proportional to

$$k_B \cdot T \cdot \rho (\partial \rho / \partial p)_T$$

where k_B is the Boltzmann constant, T the absolute temperature,
ρ the density and the expression in brackets is the isothermal
compressibility. This expression was physically satisfactory
since it showed the strength of thermal fluctuations proportional
to temperature and also accounted for critical scattering since
the isothermal compressibility would diverge sharply on the
approach to critical points.

These considerations were extended independently by
Mandelstam and Brillouin who introduced the idea of propagating
fluctuations - thermally maintained sound waves which were formally pressure waves at constant entropy - given by

$$\Delta \rho = (\partial \rho / \partial p)_S \Delta p$$

They showed that this gave rise to a Doppler shifted doublet of
frequency

$$\Omega = \pm 2n(U/c)\omega_0 \sin \theta/2$$

where n is the medium refractive index, U the speed of the sound
wave, c the velocity of light in vacuo, ω_0 the light wave
frequency and θ the scattering angle. The linewidth of the components was determined by the lifetime of the scattering sound
waves - hence by the effective attenuation, α.

This Mandelstam-Brillouin doublet was first observed for
simple liquids by Gross in 1930. However in addition to the
doublet a central component could not be eliminated and was
evidently not due to elastic scattering from impurities -
suspended particles etc. This was explained by Landau and
Placzek in 1930 who pointed out that an additional density fluctuation would arise from non propagating entropy or temperature
fluctuations at constant pressure given by

$$\Delta \rho = (\partial \rho / \partial S)_p \Delta S$$

Such a temperature fluctuation would be expected to decay away
diffusively giving the unshifted (but broadened by the effective
thermal diffusivity) Rayleigh line. Landau and Placzek also
showed that the intensity ratio of Rayleigh and Brillouin components was given by

$$I_R/2I_{M-B} = \left[(c_p/c_V) - 1\right]$$

where c_p and c_V are the specific heats.

This treatment based on fluctuations was greatly extended in the 1960s notably by Van Hove and Komarov and Fisher who emphasized the idea of correlation of refractive index fluctuations. This will be discussed in greater detail in the next section but in essence derives from the fact that the scattered field $E(R,t)$ from the elements within a scattering volume V is given by

$$E(R,t) \propto \int_V dV \cdot \delta n(r,t) \cdot (\text{phase term})$$

where δn represents the fluctuation in refractive index in the volume elements at r, and the phase term represents the different phase shifts over the path from the source to scattering element to detector. From this the field autocorrelation function (given by $<E(R,t)E*(R,t + \tau)>$) is proportional to

$$\int_V dV <\delta n(r,t)\ \delta n(r',t + \tau)>$$

Applying the Wiener-Khintchine theorem (see footnote) to this

---

Footnote: The Wiener-Khintchine theorem states a relationship between the power spectrum and correlation function of a random process (see eg 10). Suppose the process is given by X(t) which is a real, random function of t (so that it is an ensemble of real functions of t, for which a probability measure is defined. Then in any large finite time interval this can be Fourier analysed. All the Fourier components between ω and $\omega + \Delta\omega$ can be combined to form a function of t denoted by $\chi(t)_{\omega,\omega+\Delta\omega}$. Then the time average of the square of this is the spectrum and is given by

$$\overline{\left|\chi(t)_{\omega,\omega+\Delta\omega}\right|^2} = (\Delta\omega/2\pi) \int_{-\infty}^{+\infty} <X(t + \tau)X(t)>\ e^{i\omega\tau}\ d\tau$$

From this it is for example simple to show that a correlation function of exponential form $\exp(-\tau/\tau_c)$ has a Lorentzian spectrum of half width at half its intensity given by $\Delta\omega$ equal to τ_c^{-1}.

expression the spectrum of the scattered light is then given by

$$S(\underline{K},\omega) \propto \int dV \int d\tau \; <\delta n(0,0) \; \delta n(\underline{r},\tau)> \; e^{-i\omega t}$$

following a small change of axes. The point is that the spectrum of the scattered light is derived from the Fourier transform of the expression in <-> brackets which is the space and time correlation function of fluctuations in the scatterer.

Many detailed aspects of the theory of light scattering may be found in the literature [eg 1-9]. For the present purposes it is probably useful to relate the picture of Brillouin scattering, considered as Bragg reflections off moving sound waves, to a particle like view of scattering with creation or destruction of a phonon. In relation to figure 1 momenta are given by $\hbar\underline{k}$ and energy by $\hbar\omega_o$. Then by momentum conservation

$$\hbar\underline{k}_s = \hbar\underline{k}_i + \hbar\underline{K}$$

and the scattering vector

$$\underline{K} = \underline{k}_s - \underline{k}_i$$

By energy conservation

$$\hbar\omega_i \pm \hbar\Omega = \hbar\omega_s$$

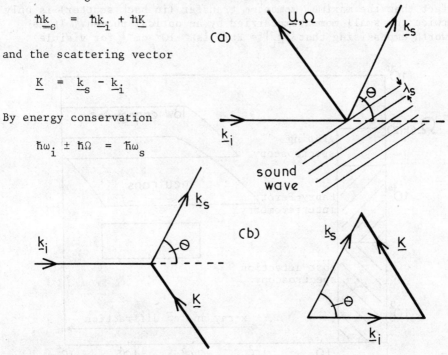

Figure 1.   Light scattering: a) Bragg reflection from a moving sound wave, b) scattering with annihilation of a phonon of momentum $\hbar\underline{K}$.

where the plus and minus refer to annihilation or creation of a phonon. Since $\Omega \ll \omega$ it follows that $\omega_s \approx \omega_i$ and in consequence that $|\underline{k}_s| = |\underline{k}_i|$. Thus the scattering vector has magnitude

$$|\underline{K}| = 2|\underline{k}| \sin \theta/2$$

and the frequency shift is given by the phonon 'frequency' which is $U \cdot |\underline{K}|$ where U is the sound speed. Thus

$$\Omega = \pm 2n(U/c)\omega_o \sin \theta/2 \quad .$$

This conveniently serves to introduce the range of energy and momentum transfer that is available in light scattering experiments. This is illustrated in figure 2 where the various regimes of grating, Fabry-Perot and post-detection spectroscopy are illustrated. The most immediately obvious feature of this diagram is the large energy but small momentum range available compared with other techniques. This of course arises from the fact that the maximum momentum transfer (in back scatter) is only twice the small momentum carried by an optical photon. It is worth emphasising that $|\underline{k}_i| = 2\pi/\lambda$ is $\sim 10^5$ cm^{-1} for visible

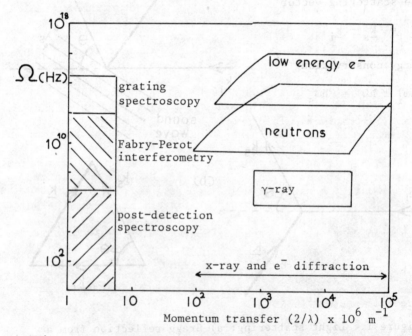

Figure 2.   Frequency and momentum transfer for various domains of scattering. The shaded area indicates the quasi-elastic region.

light, which is much less than the reciprocal of interatomic spacing. Thus quasi-elastic light scattering is effectively probing fluctuations of a cooperative nature occurring on a much larger spatial scale. Some of the potential advantages of light scattering are however not brought out by figure 2. These may be listed as: (a) good sensitivity from high efficiency and noiseless detectors; (b) small sample dimensions (from fine focussing of the laser beam); (c) sample preservation and freedom from damage except in extreme circumstances of high laser power and strongly absorbing materials; (d) high frequency and angular resolution and (e) the possibility of observing rapid and transients effects with time resolution of order 100 nsec.

The simple account given earlier was illustrated by light scattering from a simple fluid. It is worth asking how this relates in practice to polymer materials. It is possible to have fluctuations that are not described in such simple terms. For example one may have molecules in a liquid or material which possess kinetic energy of translation and energy of internal vibration. Fluctuations may occur in their relative contribution to the total energy of the system and which do not correspond to local equilibrium. Such fluctuations may not propagate and would then give rise to scattered light with frequency centred on the incident frequency and line width of the order of the reciprocal of the relaxation time of the fluctuating process. One of the most important questions is how much light is scattered by the particular active or dynamic mechanism that is of interest in the material. The answer of course is that this depends on the fluctuation of polarizability that this mechanism induces on the requisite spatial scale, and how this compares with other possible scattering mechanisms. In practice materials may have large static inhomogeneities that give strong elastic scattering, or, other dynamic mechanisms may also give strong quasi-elastic scattering. The feasibility of studying the mechanism of interest then depends on the strength of scattering and its comparative time or frequency scale - as discussed in the following instrumental sections. It has to be recognised that light scattering techniques and subsequent data interpretation are rendered particularly difficult in anything less than extremely pure, well characterised, samples.

Before discussing the available instruments it is worth considering the form and nature of the scattered light in a little more detail (see eg 11).

1.1 Rayleigh Scattering

If we consider the scattering from a large number of scattering elements the resultant scattered field is given by

$$E_s = \sum_j A_j(t) \, e^{i\underline{k}.\underline{r}_j(t)} \, e^{-i\omega_o t}$$

From this various functions accessible to measurement can be found. These include

a)   The average intensity

$$I_s = <|E_s|^2> = \sum_j \sum_{j'} A_j A_{j'} \, e^{i\underline{k}.(\underline{r}_j - \underline{r}_{j'})}$$

b)   The field autocorrelation function

$$G^{(1)}(\tau) = <E_s^*(t) \, E_s(t + \tau)>$$

c)   The optical spectrum

$$I(\omega) = (1/2\pi) \int G^{(1)}(t) \, e^{i\omega t} \, d\tau$$

d)   The intensity autocorrelation function

$$G^{(2)}(\tau) = <E_s^*(t) \, E_s(t) \, E_s^*(t + \tau) \, E_s(t + \tau)>$$

For a large number of effectively independent scattering elements the interference corresponds to a random walk of many steps in the complex plane - and hence a Gaussian probability distribution for the resultant E_s given by

$$P(E) = (1/\pi<|E|^2>) \exp(-|E|^2/<|E|^2>)$$

In this case the scattered field is a Gaussian random process and the Siegert relation, relating the normalised 1st and 2nd order correlation functions, holds and is given by:

$$g^{(2)}(\tau) = 1 + |g^{(1)}(\tau)|^2$$

where the normalised functions are given by

$$g^{(1)}(\tau) = G^{(1)}(\tau)/G^{(1)}(0)$$

and $g^{(2)}(\tau) = G^{(2)}(\tau)/G^{(2)}(0)$

When the number of scattering elements is not large Gaussian statistics will no longer be found (see eg Pusey in   ).

If the temporal evolution of the scattering process is simply diffusional in character the resultant first order correlation function has the form

$$|g^{(1)}(\tau)| \propto \exp(-\Gamma\tau)$$

where the associated decay constant Γ is determined by

$$\Gamma = D \cdot K^2$$

where K is the scattering vector and D the diffusion constant. (As an example - for a system of diffusing independent particles undergoing Brownian motion D is equal to $(k_B \cdot T/6\pi\eta R_H)$ where k is Boltzmann's constant, T the temperature, η the viscosity and R_{II} the hydrodynamic radius - see eg refs 11-14).

Thus from the Siegert relationship, forming the second order intensity correlation function for a pure intensity fluctuation* process would give

$$g^{(2)}(\tau) = 1 + F \exp(-2\Gamma/\tau)$$

where F (of order unity and to be discussed later) is an experimental factor depending on optical geometry and the fraction of light scattered in the active mechanism. In the event of a strong elastic component at the laser frequency, so that the active scattering is beating with the unshifted light (in the homodyne mode) the intensity correlation function becomes

$$g^{(2)}(\tau) = 1 + F \cdot (2<I_s>/I) \exp(-\Gamma\tau)$$

where $<I_s>$ and I are the (mean) intensity for the active and

---

Footnote: Unfortunately considerable confusion in nomenclature exists within the literature. Throughout this paper the term 'intensity fluctuation mode' means a system where the analysed light is predominantly due to scattering from some active mechanism; 'homodyne' means beating of such light with laser light derived from the unshifted laser beam (which may come from static scatterers within the medium, or be arranged optically), and 'heterodyne' means beating with a frequency shifted laser beam.

elastic scattering. In this case of course the intercept $[g^{(2)}(0) - 1]$ becomes very small for I much greater than $<I_s>$.

Perhaps the two most important points of this discussion are firstly the necessity for establishing very clearly whether the experiment is operating in the intensity fluctuation mode – with temporal relationship $\sim \exp(- 2\Gamma\tau)$, or homodyne mode – with temporal relationship $\sim \exp(- \Gamma\tau)$. The second point is that if the active process is simply diffusional in character it should exhibit an angular dependence given by $\Gamma \propto K^2$. Clearly the angular dependence of Γ is an important diagnostic. The next section shows that dynamic Rayleigh scattering in the frequency range $\lesssim 10^6$ Hz can be studied with best efficiency with intensity correlators (to determine the autocorrelation function) or wave analysers to determine the spectrum of the input electrical signal from the detector.

1.2 Brillouin Scattering

Turning now to Brillouin scattering from hypersonic acoustic waves we see from the earlier discussion that the Fourier component of fluctuation is effectively selected by the wavelength of the light λ and the chosen scattering vector. The measured line width is given by

$$\Delta = \alpha U/\pi \quad \text{Hz}$$

where α is the attenuation coefficient. The complex elastic modulus then becomes

$$M^* = M' + iM''$$

$$= \rho U^2 + i(2\rho U^3 \alpha)\omega_s$$

where ω_s is the sound wave angular frequency given by $2\pi\Omega$. Detailed discussions of Brillouin scattering may be found in several reviews (eg references 1-9).

For the present purposes it is worth extracting a few commonly employed expressions. In the case of a simple medium with a single relaxation mechanism of relaxation time τ_r the frequency dependent viscosity $\eta(\omega)$ is given by

$$\eta(\omega) = \rho_o \left\{ \frac{(U_o^2 - U_\infty^2)\tau_r}{1 + \omega^2 \tau_r^2} \right\}$$

and M' the real part of the longitudinal modulus is

$$M'(\omega) = \rho_0 \left\{ \frac{U_0^2 + U_\infty^2 \omega^2 \tau_r^2}{1 + \omega^2 \tau_r^2} \right\}$$

where U_0 and U_∞ are the limiting zero and infinite frequency sound velocities.

In solids shear modes can also be supported and give rise to scattering which can often be preferentially observed with suitable choice of light polarisation. For anisotropic media the situation becomes quite complex. For a transversely anisotropic system such as often occurs in polymers and biological materials there are five elastic constants and in general three scattering wave modes: two of (quasi) transverse polarisation – T_1 and T_2 and one of (quasi) longitudinal polarisation – L. The hypersound speed is given by

$$U^2(\theta)_{L,T_1} = \frac{1}{2\rho} \left[c_{11} \sin^2\theta + c_{33} \cos^2\theta + c_{44} \pm \left\{ \left[(c_{11} - c_{44}) \sin^2\theta + (c_{44} - c_{33})\cos^2\theta \right]^2 + 4(c_{13} + c_{44})^2 \sin^2\theta \cos^2\theta \right\}^{\frac{1}{2}} \right]$$

and

$$U^2(\theta)_{T_2} = \frac{1}{2\rho} \left[\frac{1}{2}(c_{11} - c_{12}) \sin^2\theta + c_{44} \cos^2\theta \right]$$

where $U(\theta)$ is the speed for a propagation vector at an angle θ to the axis of the transversely anisotropic medium, and the c's are the elastic constants.

In summary Brillouin scattering provides a tool for high frequency (hypersonic) viscoelastic measurements and for thermo-dynamic and elastic studies, with some potential for structural determination in the size range 0.1-20 μm. It is however worth commenting that it should not strictly be thought of as simply ultrasonics at a higher frequency. This point has been made clearly by Evans and Powles [15] who discuss the interpretation of hypersonic and ultrasonic 'velocities' in fluids. They emphasize that in a light scattering experiment a real spatial wave must exist to give a finite scattering intensity, so that the scattering vector K must be real. The amplitude of the wave

varies with time because it is a component of a thermal fluctuation and since it has a limited lifetime its frequency is ill-
defined which corresponds to a complex value for ω_s. For the usual
ultrasonic experiment however a transducer is excited at a fixed
frequency ω_s that is real and well defined. However due to
attenuation the wave is limited in space so that $\underline{K}$ is ill-defined
and complex. (See the extended discussion in reference   ).

2   INTENSITY FLUCTUATION SPECTROSCOPY

This section will deal almost exclusively with the technique
of photon correlation spectroscopy which is generally much more
efficient in the use of signal than wave analyser methods. A
strong signal from the detector can simply be fed into a frequency
analyser; such an instrument takes a sample of the signal, filters
out one frequency component, squares it, takes another sample and
filters, squares, and so on over the set range of frequencies.
After cycling and averaging the spectrum is formed. However since
only one frequency interval is usually examined at a time there is
a loss of efficiency. It is in fact rather more easy to achieve
multi-channel operation in the time domain with correlation
methods. Before discussing these in detail it is worth discussing
the temporal fluctuations in light signals in a qualitative way.
Consider first a detector exposed to a constant light field derived
from an ideal laser. The ideal waveform of the field is a steady
sine wave and the intensity envelope is constant and unvarying.
Photodetections derived from such a field are then Poisson distributed - they are quite random in time and totally uncorrelated
one with the other. The statistical distribution of n photodetection within a sampling time T is given by

$$P(n) = \frac{(\langle n \rangle)^n}{n!} \exp(-\langle n \rangle)$$

where $\langle n \rangle$ is the mean photocount within the sample time T and is
given by

$$\langle n \rangle = \sum_{n=0}^{\infty} n\, P(n)$$

(and is equal to $\beta \cdot I \cdot T$ where I is the constant intensity and β is
the efficiency of the detector).

The temporal autocorrelation function for the photodetections
$\langle n(t)\, n(t + \tau) \rangle$ is accordingly in this case featureless and without structure. Consider however the light from a scattering

process of spectral bandwidth $\Delta\nu$; from the Wiener-Khintchine theorem the associated temporal fluctuations in light intensity occur on a time scale $\Delta\tau_c \sim (\Delta\nu)^{-1}$. The photodetection rate is modulated by the fluctuating intensity. The various frequency components in the optical spectrum may be considered to beat together to give a characteristic coherence period of $\sim (\Delta\nu)^{-1}$. Accordingly as indicated in figure 3 photodetections are now bunched for time scales $< \tau_c$; the intensity correlation is thus larger for short time intervals $(\lesssim \tau_c)$ than for long time intervals $(> \tau_c)$.

A simple extension of the Poisson distribution to the situation where the intensity fluctuates gives the so called Mandel relationship

$$P(n) = \int_0^\infty \frac{(\beta I T)^n}{n!} \exp(-\beta I T) \, P(I) \, dI$$

where $P(I)$ is now the probability distribution for the intensity (and hence $\langle n \rangle = \beta \langle I \rangle T$). This supposes that $T \ll T_c$ (the coherence

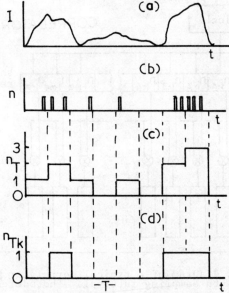

Figure 3.   Optical detection [11]: a) intensity as a function of time; b) corresponding quantised detections; c) number of photodetections within a sampling time T. d) the corresponding clipped signal with the clipping level set equal to 1.

time of the light) so that the intensity is effectively constant throughout the sample period. The distribution of the photo-detections in the fluctuating case is thus the Poisson transform of the intensity distribution. However it can be shown that the corresponding normalised correlation function for the photo-detections is the same as the (normalised) intensity correlation function [see eg 11]. Typically in an experiment with several million photodetections a large amount of information must be processed. This may be performed with relatively small dedicated computers but is commonly carried out with hard wired correlators - particularly at higher speeds. Such a correlator is shown in figure 4. The correlation function is derived by sampling the signal for set periods T (as shown in figure 3c); it is thus calculated for discrete values of the delay τ given by T, 2T, 3T ... mT where m is the number of counters or channels in the instrument. The time regulation of the correlator is provided by a crystal-controlled clock; with fast modern electronic circuitry minimum sampling times of 10 ns are readily obtained. One input channel goes to an accumulator gated at the clock frequency. This component simply accumulates incoming pulses over one sample time T. At the end of the sample time the contents of the accumulator are transferred to the first stage of a multistage shift register, and

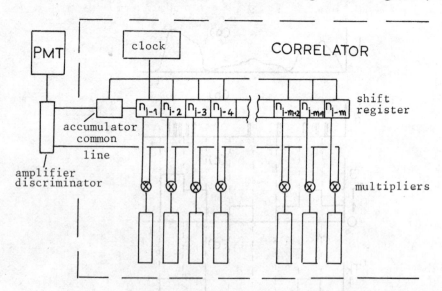

Figure 4.   Block diagram of a typical correlator [11] showing the i-th sampling interval. Photodetections from the photomultiplier (PMT) are standardised by the amplifier discriminator and emitted in two channels. The accumulator counts up the number of pulses in the set sampling time.

the accumulator is reset to zero. Simultaneously the contents of
the first stage are shifted to the second stage etc. The shift
register thus contains the past history of photocount numbers.
Figure 4 shows the correlator during sample interval i. The
first stage of the shift register contains the number of photocounts n_{i-1} registered in the previous interval and so on. Each
stage of the shift register is connected to one input of a
multiplier, the other input of which is connected to a 'common
line' from the discriminator/shaper. The outputs of the multipliers are connected to counters. As pulses pass down the common-
line in real time they are multiplied by the contents of the
shift-register stages and accumulated in the appropriate counters.
During interval i, therefore, $n_i n_{i-1}$ counts will have been added
to counter 1, $n_i n_{i-p}$ counts to counter p, etc. After running for
N sample times, therefore, the contents c_p of counter p will be

$$c_p = \sum_{i=1}^{N} n_i n_{i-p}$$

If the experimental run time NT is much greater than the characteristic fluctuation time T_c, ie the experiment covers many
fluctuations of the light under study, c will approximate its
average value. Thus

$$<c_p> = N<n_T(t)\ n_T(t + pT)>$$

where $n_T(t)$ is the number of photocounts accumulated in the sample
interval of duration T centred at time t. The contents of the
counters are, within a certain statistical error, proportional to
the desired photocount correlation function.

While such full correlators have been built, the real-time
operating speed is limited by the large number of multiplications
necessary. A major advance in the development of real-time
methods came by recognizing the advantages of the technique of
'clipping' as illustrated in figure 3d. This is achieved by
counting the number of detections $n_T(t)$ and choosing a level k;
for $n_T(t) > k$ the clipped signal is set equal to one and otherwise
is set equal to zero. The signals proceeding down the shift
register are thus either zeros or ones; therefore a one-bit per
stage register can be used, with coincidence gates instead of
multipliers. For Gaussian light the single clipped (single, since
clipping is carried out only in one line only - the shift register)
correlation function is related to the first order field
correlation function $g^{(1)}(\tau)$ by:

$$\langle n_{Tk}(t)\, n_T(t + \tau)\rangle \;=\; \langle n_{Tk}\rangle\, \langle n_T\rangle \left[1 + F \cdot \frac{(1 + k)}{(1 + \langle n_T\rangle)} \left| g^{(1)}(\tau) \right|^2 \right]$$

Most importantly the form of correlation function is unchanged and the time dependent part is merely multiplied by the clipping factor $(1 + k)/(1 + \langle n_T\rangle)$. However for fields having non-Gaussian statistics this is no longer true [see eg 13]. In this case the technique of 'scaling' may be applied where the train of pulses passing into the shift register is scaled by a factor S. Every S'th pulse is picked out and S is chosen high enough so that the probability of more than one scaled pulse per sample time is negligible. Thus the advantage of a one-bit delayed signal (down the shift register) is maintained.

Before outlining the component parts of a correlation experiment it is worth commenting that the measured intensity autocorrelation function of Gaussian light contains much information that is often neglected in practice. Most particularly the experimental values of background and intercept after normalisation should approach 1 and 2 for a single Gaussian component (that is $g^{(2)}(\tau) \to 1$ and 2 for $\tau \to \infty$ and 0). The value found in practice for long time delay ($\tau \to \infty$) gives a measure of the overall source and scatterer constancy during the course of an experiment. An intercept close to 2 at $\tau \to 0$ is only found for short sample time T compared with T_c and where the fluctuating intensity is detected over an aperture very small compared with one coherence area. This concept of coherence area has been extensively discussed (see eg [ ]). Essentially if a group of scatterers are contained in a region of cross section ℓ_1 they will produce a scattering pattern at a distance R for which coherence will exist over distances ℓ_2 apart where

$$\left\{ \ell_1 \ell_2 / R\lambda \right\} \lesssim 1$$

The area $\ell_2^2 = (R\lambda/\ell_1)^2$ can be considered as a coherence area in the sense that the electric fields detected at two points separated by less than $R\lambda/\ell_1$ are correlated and show similar fluctuations, whereas for $\ell_2 \gg R\lambda/\ell_1$ they are uncorrelated. There is little advantage in an intensity fluctuation experiment in using a detector aperture of area A much greater than one coherence area A_c. Values of the expected intercept $F(A/A_c)$ for different detector areas have been given [13]. In practice, with a given experimental arrangement, it is often valuable to measure the intercept for a scattering sample known to give a single scattering component (eg mono-disperse latex spheres in suspension). This calibration value then gives a guide for interpretating other unknown specimens. For example if the scatterer gives $<n_S>$ counts per sample interval due to the particular spectral component under study and $<n_B>$ due to some broad-band background scattering then the intercept at $\tau = 0$ of $[g^{(2)}(\tau) - 1]$ is reduced by $[<n_S>/<n_S> + <n_B>]^2$. Thus careful measurements in a well characterised system can indicate the proportion of the light scattered within the particular spectral component under study. Suitable arrangements for establishing the coherence condition are shown in figure 5. The arrangement in figure 5b is particularly convenient; by moving the detector and aperture ℓ_2 along an optical bench the ratio A/A_c may readily be varied without refocussing.

A block diagram of a complete equipment required for a photon correlation experiment is shown in figure 6. It is worth making a few comments on the choice and specification of the various components. Many of these points are considered in detail in the valuable guide by Oliver [in 13]

(a) The *laser* must have good power stability and freedom from eg plasma oscillations which might give rise to spurious correlations. The power level and wavelength range need to be matched to the samples eg their scattering strength and power absorption etc.

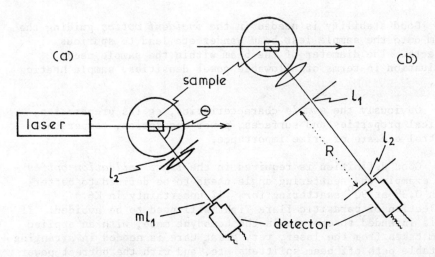

Figure 5.   Optical arrangements for achieving a single coherence area condition (see text): a) the lens images the scattering volume onto the detector aperture $m\ell_1$ with magnification m; b) the lens forms an image of the scatterer on the aperture ℓ_1.

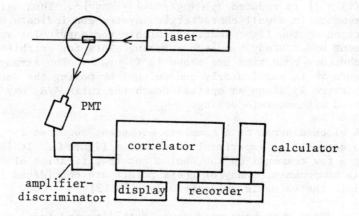

Figure 6.   Block diagram of a complete equipment for a photon
              correlation experiment [11].

(b) Good stability is needed in the *incident optics* guiding the
beam onto the sample lest beam wander etc lead to spurious
effects. The diameter of the beam within the sample needs
evaluation in terms of allowable power densities, sample heating
etc.

(c) Obviously the *sample* characteristics, eg its preparation,
optical properties and surfaces, power absorption, temperature
control etc are of prime importance.

(d) Good definition is required in the *light collection optics* -
for example the scattering angle needs to be defined to better
than 0.3° at 90° scattering for $< \frac{1}{2}\%$ uncertainty in K^2.
Collection of parasitic flare light will need to be avoided. If
it is intended to operate in the homodyne mode, with an applied
beam taken from the laser, very great care is needed in arranging
a stable path off beam splitters etc, and with the correct power
level.

(e) The *photomultiplier* itself needs to be very well characterised.
Any type of correlated afterpulsing may lead to spurious correlations. Ideally the dark count rate should be low - less than
100 counts s^{-1} especially for weakly scattering samples.

(f) In the *correlator* the appropriate choices of sample time, clipping or scaling level, length of experiment and pattern of experiment (eg few long or many short) need to be made.

(g) Finally, in the *analysis of data*, questions of normalisation and intercept must be considered. Of great importance is the problem of evaluation of the experimental correlation function - what model form it is to be compared with and how are parameters of that form to be extracted. These are after all the final output of the experiment - the parameters of some assumed theoretical model. Large computers are not always necessary, fits to models containing several parameters can be performed quite rapidly using currently available programmable desk calculators.

In conclusion it is worth emphasizing the importance in practice of every possible precaution against spurious correlation. These include examination of elastic laser scattering - as a diagnostic of laser instabilities or plasma oscillations, study of stable white light sources - as a guide to photomultiplier afterpulsing etc, together with investigation of dummy samples and variation of scattering angle, sample time and laser power etc as a guard against other spurious effects. With these precautions at least major blunders should be avoided.

3   FABRY-PEROT INTERFEROMETRY

The Fabry-Perot interferometer is basically an optical cavity that serves as a spectral filter (see eg Vaughan in [13]). In one form plane etalon plates with a high reflective coating (typically $R \sim 0.90$ to 0.98) are arranged to be accurately parallel to one another. An incoming light beam at angle θ to the normal is multiply reflected between the plates. Successive reflected beams are retarded optically by $2\mu d \cos \theta$ (giving a phase lag of $\phi = 2\pi(2\mu d \cos \theta)/\lambda$) where μ is the refractive index and d the distance between the plates. Summation of these reflections with a fringe forming lens shows that the beam is transmitted if the usual constructive interference condition $n\lambda = 2\mu d \cos \theta$ is obeyed. From the cylindrical symmetry circular fringes are formed. The plane instrument is thus an angle-dependent optical filter which may be used in different ways. The fringes may be photographed or recorded on a channel plate, or more commonly centre scanning may be used in which a small pinhole is placed at the centre of the fringe pattern. The fringes are then scanned through the pinhole onto a detector either by varying the refractive index μ (by changing the gas pressure) or by changing the plate separation d (usually by piezoelectric scanning). For monochromatic light the fringes are of the form

$$I(\phi) = \frac{T^2}{(1 - R)^2} \times \frac{I_o}{1 + \left[4R/(1 - R)^2\right] \sin^2 \phi/2}$$

where T is the plate transmission. The form of the fringes is shown sketched in figure 7. From this expression the following points may be noted.

(a)   The mode separation, that is the smallest frequency separation such that the components exactly overlap is given by $c/2\mu d$ Hz. This is known as the free spectral range (the FSR or one order of interference).

(b)   The fringes are narrow for high reflectivity coatings. The ratio FSR to fringe width at half height is called the finesse. The theoretical reflectivity finesse F_R is given by $\pi\sqrt{R}/(1 - R)$ but this is reduced in practice by many factors including plate imperfections, lack of parallelism and the effect of the pinhole aperture function. With an effective finesse F the resolving limit is $(c/2\mu d\ F)$ Hz.

(c)   The transmission of the instrument is high; for $T = 1 - R$, that is for no absorption or scatter, there is no loss of intensity in the fringes.

(d)   The fringe contrast or extinction, that is the ratio of peak to trough is quite low and given by $[1 + 4R/(1 - R)^2]$. In practice this may be additionally reduced by scattering or parasitic light and is usually $\sim 10^3$. Thus the fringe system due to a weak spectral feature may be completely obscured by the background due to a strong component.

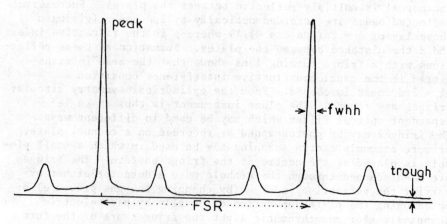

Figure 7.   Fabry-Perot fringes showing two successive orders of interference for a typical Rayleigh-Brillouin triplet.

This last point may be particularly important for polymer systems where strong elastic scattering and the Rayleigh line may be many orders of magnitude greater than the Brillouin components. This problem may be overcome by operating the interferometer in a multipass manner in which the light in a restricted beam is passed back through adjacent parts of the plates with corner-cube retroreflectors. If in single pass the contrast is C_1, the contrast in n pass should approach $(C_1)^n$. In practice contrasts approaching 10^7 to 10^8 are readily achieved in triple pass, and five pass instruments have been employed giving even higher extinction.

Fabry-Perot interferometry is now made relatively straight-forward by commercial instruments of good stability and precision. Spectral line shifts can usually be measured with an accuracy of better than a few percent. Accurate line widths require some form of deconvolution of the instrumental function. Precision of about 10% is usually attained fairly easily but is often hard to improve on.

EXPERIMENTAL REVIEW

In the following sections light scattering from polymer materials is reviewed in the four broad categories of Rayleigh and Brillouin scattering from both solid and gel. Many examples are outlined and discussed in varying degrees of detail in order to provide an overview of progress in the field. However there is presently a great deal of activity in many areas and this review does not set out to provide a complete account of work during the last few years. It should serve nevertheless as a general intro-duction to topics of particular current interest.

4   RAYLEIGH SCATTERING FROM BULK POLYMERS

One of the earliest studies with photon correlation spectro-scopy was conducted by Jackson et al [16] on poly (methylmetha-crylate). Measurements were made on a sample of PMMA over the range $20^{\circ}C$ to $120^{\circ}C$, just above the glass transition temperature T_g. About 40 mW at $4880A^{\circ}$ from a single-mode argon-ion laser was focussed into the sample; horizontally scattered light at 90° was collected from a single coherence area via an aperture and lens and analysed on a 50 channel correlator. With vertically polarised input (VV + VH) weak correlations were found over a wide range of sampling times. Typical correlations functions are shown in figure 8. The intercept of the measured functions was quite small-typically $[g^{(2)}(0) - 1]$ was about 0.01 - and experimental periods of 30 min or more were required to accumulate recordings of suffi-cient statistical accuracy to enable correlation times to be extracted from the data. In this situation of course it is clear that the scattered light arising from the active mechanism is

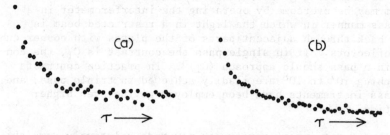

Figure 8.   Intensity correlation functions for PMMA at 20°C
            (from Jackson et al [16]). Sample time for curve (a)
            was 0.4ms per channel and shows the fast process
            dominating. For curve (b) the sample time was 1ms
            and the long tail of the slower process is evident.

only a small fraction of the total scattered light which is
largely inelastic scattering at the incident frequency. This
scattering arises from static inhomogeneities and thus provides
a homodyne reference beam so that the experiment is conducted in
the homodyne mode. It was found that the correlation functions
could be approximately represented by the sum of two exponential
with decay times varying by more than an order of magnitude.
Various sample times were employed at each temperature in order
to determine the decay times.

These early results were interpreted on the following lines.
It was supposed that scattering fluctuations were associated with
motion of the polar side chain which is prominent in dielectric
relaxation. Above the glass temperature a slower motion of the
main chain occurs which is evident in mechanical and nuclear
magnetic relaxation. The frequencies of these two reorientational processes increase very rapidly with temperature and tend
to converge into one process. For light scattering, since the
reorienting entities have anisotropic polarizability the scattered
light would be expected to be 75% depolarised. Thus if the active
scattering were due to reorientational motion of such units,
spectra would be expected in both the polarised and depolarised
light. However in this early work correlation was found only
in the polarised light. The fastest component had a frequency
and temperature dependence which corresponded quite well to the
motion of the side chain which carries the anisotropically
polarisable acrylate unit. This suggested that this active
scattered light arises from density inhomogeneities moving with
characteristic frequencies related to thermally excited motion of
the side chain. However for the lower frequency mode, which
appeared to be independent of temperature, Jackson et al had no

explanation. They commented that it could not be simply related
to main chain motion since this should be suppressed below the
glass temperature.

Following this initial report a further study was undertaken by Cohen et al in 1977 [17]. Two samples of PMMA from
different sources were examined over a temperature range from
6 to 165°C and, valuably, over a range of scattering angles from
30 to 130°. Broadly similar correlation functions were recorded
by Cohen et al and also analysed in terms of two exponential
decays. In summary the results for the higher frequency relaxation mode were found to be independent of angle and scattering
vector. When plotted against temperature these results fell on
two straight lines of different activation energy, indicating the
presence of two coupled relaxation mechanisms. The low frequency
relaxation results were quite sensitive to inhomogeneities of
unannealed samples and had irregular angular dependences. The
angular dependence disappeared for samples close to T_g and for
annealed samples. Below T_g the relaxation process had a fairly
constant width of about 3 Hz independent of temperature. However,
above T_g this increased very rapidly with temperature to reach
130 Hz at 165°C.

These results are shown schematically in figure 9 and were
interpreted by Cohen et al as follows. For the two parts of the
high frequency relaxation mode the high temperature line of
larger slope (region A - figure 9) corresponds to an activation

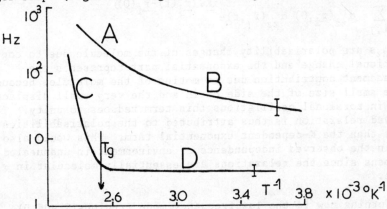

Figure 9.   Representation of the results of Cohen et al [17] for
             PMMA: relaxation frequency plotted against reciprocal
             temperature.

energy of ~8 kcal/mole; the lower frequency line (of smaller
slope - region B) corresponds to ~1 kcal/mole. However the
activation energy from dielectric and NMR measurements, assigned
to side chain (the $-CO_2 \, CH_3$ group) molecular reorientation is

notably higher than 8 kcal/mole. Cohen et al suggest that the
lower value found from light scattering is due to a coupling to
a low activation energy process. This latter they attribute to
torsional oscillations of the chain segments about their equilibrium positions rather than a segmental reorientation (and would
not be observed by dielectric measurements of NMR). This interpretation is supported by light scattering work on polypropylene
glycol (PPG) by Jones and Wang [18]. These latter authors
assign the relaxation time they observe to the backbone segmental
motion of the molecule and obtain an activation energy of ~4
kcal/mole. On the other hand dielectric measurements on PPG give
11.7 kcal/mole. Jones and Wang suggest this discrepancy arises
since in addition to segmental reorientation, torsional oscillations of much lower activation energy are also observed by
light scattering (and similarly would not be observed by dielectric measurements). The light scattering measurements (region
B) are taken to confirm that in PMMA there is a low energy
activation mechanism coupled to side chain reorientation relaxation; accordingly at lower temperatures, when this latter has
lower frequency, the low activation mechanism is observed. As
the temperature is increased the reorientation mechanism
increases in frequency and becomes predominat. The lack of
angular dependence for the light scattering is explained by
Cohen et al on the basis of polarisability fluctuations domina-
ting over displacement contributions. Formally the correlation
function is given as

$$< \sum_{i,j} \alpha_{zz}(r_i,0) \; \alpha_{zz}(r_j,t) e^{i\underline{K} \cdot (\underline{r}_j(t) - \underline{r}_i(0))} >$$

The α_{zz}s are polarisability changes of the molecule due to conformational change and the exponential part represents the
displacement contribution due to motion of the molecule. Because
of the small size of the side chain and the very small displacements in torsional oscillations this term reduces to unity. The
observed relaxation is thus attributed to the polarisabilities
rather than the K-dependent exponential term. This would also
explain the observed independence of environment in unannealed
specimens since the relaxations are essentially molecular in
nature.

Turning now to the low frequency mode (regions C and D),
this was found to be very sensitive to inhomogeneities of the
unannealed samples. It was suggested that, since the relaxation
seems to depend on the overall structure of the scattering volume,
it may represent a rearrangement of free volume or 'configurational entropy'. The constant frequency below T_g argues for a
mechanism with no activation energy; if an excess free volume is
present at and below T_g it should be able to redistribute with
no energy and thus be temperature independent. In terms of the

configurational entropy model the relaxation time would be
given by τ_r^{-1} equal to A exp[$-B/TS_c$] where A and B are constants,
T the temperature and S_c the excess configurational entropy.
Below T_g S_c is essentially constant and T is also constant since
the glassy specimen is locked in a non equilibrium state; thus
τ_r remains constant below T_g. Above T_g the relaxation frequency
(region C) increases very rapidly, with an activation energy of
50 kcal/mole, and follows the behaviour of the main chain motion
(α-relaxation) observed by other techniques. However there
remain unanswered questions about the angle dependence. The
model for free volume or configurational rearrangement is
diffusive in character. Accordingly the α-relaxation mode above
T_g, for such a model for molecular glasses, is due to molecular
diffusion and would be K-dependent. The angular dependence in
the region C was not investigated by Cohen et al. However
according to the model the diffusive character (due to free
volume diffusion) would persist below T_g and also be K-dependent.
The light scattering results (region D) while showing irregular
dependence for unannealed samples gave essentially an independence of K closer to T_g and for annealed samples. The authors
suggest an explanation on the basis that for polymeric glasses
the α relaxation is the longest frequency mode of internal
motion of the macromolecule; as such it is related to diffusion
of the whole molecule but is nevertheless independent of K.

This earlier work has been outlined in some detail to
illustrate the experimental difficulties and the complexity and
uncertainty of interpretation. Very recent work [19-21] on polystyrene, poly (ethylmethacrylate) and poly (n-butyl)methacrylate
has emphasised the prime importance of working with samples of
the highest optical purity. Only with such samples can any
interpretation be put on a reliable base. In their study of
polystyrene, Lee et al [19a,b] prepared prisms by in situ polymerisation of scrupulously clean monomer. For PEMA Patterson
et al [20] annealed their sample at 150°C overnight (T_g is ~65°C)
to remove strains. The spectrum of scattered light for this
latter sample was measured with a Fabry Perot spectrometer and
the intensity of the central (Rayleigh) peak found to be only
3 times that due to Brillouin scattering. In this material the
Rayleigh scattering arises very largely from the slowly relaxing
density fluctuations; in the absence of scattering from inclusions etc the remainder would be due to rapid entropy fluctuations.
In consequence the maximum value of the short time correlation
function [$g^2(\tau) - 1$] that may be expected is 0.56 - equivalent to
$(0.75)^2$ - see section 2. In practice values approaching this
were found, and were used by the authors as a valuable diagnostic.
For these recent studies the simple two-exponential fit to
experimental correlation functions near the glass transition
turns out to be quite inadequate. Correlation functions are
now found to extend over many decades for these bulk polymers

[eg, figure 10]. By comparison with previous dielectric
relaxational studies, a model with continuous distribution of
relaxation times may be adopted for which the Williams-Watt
function [21] is very useful. In this form the field correlation function $\phi_r(t)$, derived from the normalised intensity
correlation function found experimentally (see section 2) is
expressed in the usual nomenclature as

$$\phi_r(t) = \alpha \exp[-(t/\tau_o)^\beta]$$

where α, τ_o and β are adjustable parameters and β lies between
0 and 1. For a single experimental β equals 1, and decreases as
the distribution becomes broader. Parameters may be obtained
[see eg 19a] by expressing the equation as

$$\log\{-\ln[\phi(t)/\alpha]\} = \beta \log(t/\tau_o)$$

and plotting the l.h.s. versus log t; accurate normalisation is
required. The data can also be characterised by an average
relaxation $\langle\tau\rangle$ equal to $(\tau/\beta)\,\Gamma(1/\beta)$ where $\Gamma(1/\beta)$ is the gamma
function.

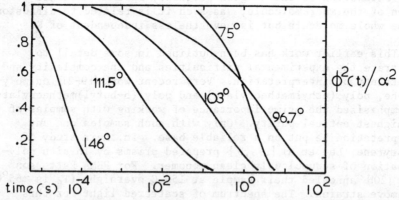

Figure 10. Correlation functions at different temperatures for
atactic polystyrene (from Lee et al [18]).

On considering the materials in turn, for polystyrene the
light scattering is dominated by fluctuations of the anisotropy;
good agreement is found for the results of the two independent
investigations by Lee et al [19] and Patterson et al [20a]. Best
fit to the photon correlation data [see eg figure 10] is provided
by two Williams-Watt terms indicating two relaxation mechanisms.
Above T_g a slow process dominates with a wide distribution of
relaxation time and $\beta_1 \approx 0.4$. The mean relaxation frequencies
are found to be in good agreement with values for the α-
relaxation found in the literature and may thus be assigned to
the main chain glass transition mechanism. This slow relaxation

mode is $\underline{K}$ independent and 'freezes in' at T_g. In addition a
fast decay of small amplitude and approximately single exponential form ($\beta_2 \approx 1$) is found both above and below T_g. Below T_g
this mode is K^2-dependent; it would thus arise from a diffusion
mechanism and may be related to the β-relaxation mechanism.
Various possibilities are considered by the authors including
side chain motion, defect diffusion and transport of dissolved
impurities. Valuable guides to detailed theoretical interpretation of light scattering are provided by the authors.

As noted PEMA is particularly interesting since it has very
weak optical anisotropy and light scattering is largely due to
density fluctuations. In the measurements of Patterson et al
[20b] there was no active correlation for $\tau > 10^{-6}$ at higher
temperatures ($150^{\circ}C$), so that only very fast process must exist
in this range. As the temperature was reduced intensity
correlations became apparent; at $120^{\circ}C$ the average relaxation
time was $\sim 10^{-5}$ sec with a $\beta \approx 0.4$. At $100^{\circ}C$ the value of β
required was below 0.3 and at $70^{\circ}C$ this had fallen to ~ 0.16.
At this temperature the relaxation function stretched over
eight decades in time with an apparent activation energy near
T_g of over 100 kcal/mole. Patterson et al suggest the most
likely explanation for these results is that there are two
processes contributing to the relaxation function. The two
processes are considered to merge at higher temperature, and
to separate as the sample is cooled thus reducing the value
of β.

The results found for poly(n-butyl) methacrylate [21] are
similar to those for polystyrene with a slow broad process and
a fast single exponential decay. However Jamieson et al report
on anomalously large and sudden increase in the amplitude of
the fast mechanism over a narrow temperature range at $\sim T_g + 30^{\circ}$.
Various possibilities, including the coupling of side chain
motion with backbone motion in a resonant fashion, are discussed.

In conclusion it is clear that dynamic light scattering
methods are beginning to make a useful contribution to the study
of hard amorphous polymers. The experiments are difficult -
largely due to sample preparation, the wide dynamic range and
also in some cases to the small fraction of the scattered light
arising from active mechanisms. These constraint appears less
severe in a number of recent investigations of polymer gels
considered in the next section.

5   RAYLEIGH SCATTERING FROM GELS

During the past decade there have been many studies of gels
by dynamic light scattering. These may be classified under three
broad headings: 1) scattering from standing waves,

2) fluctuational studies from gels and semi-dilute solutions, mostly in recent years employing the theoretical background initiated by de Gennes, and 3) the elegant study of critical fluctuations and collapse in gels largely due to Tanaka and colleagues. These related aspects are briefly outlined. There is now a very extensive literature in this field; the present review does not attempt a complete guide or list of references.

5.1 Scattering from Standing Waves

Broad resonances in the frequency spectrum of two dilute gels were first observed by Prins et al [23] in 1972. Measurements with a wave analyser on a 1% agarose in water gel and 5% poly (vinyl alcohol) gel showed a resonance centred at a few kilohertz. The gels were contained within rectangular light scattering cells of 2 mm optical path. It was established that the frequency shifted scattering was not depolarised; spurious effects were eliminated by showing that scattering from an arbitrary static scatterer gave a flat spectrum. The authors attributed the resonances to underdamped oscillatory behaviour of visco-elastic microscopic heterogeneities in the structural makeup of the gels, this was also discussed by Wun & Carlson [24]. In a later publication Wun et al [25] observed the resonance with the onset of oscillatory correlation functions of the scattered light. They reported that the oscillations persisted for several hours but eventually disappeared as the gelation process was completed. They accordingly reinterpreted the phenomena as being associated with mass flow taking place during the process of gelation.

This rather confusing situation was clarified experimentally by Brenner et al [26]. Soft gels of agarose and collagen were examined in various cuvettes with both a correlation spectrometer and a real time spectrum analyser. Complex oscillatory correlation functions were often found; on occasion these could be reduced to a pure cosine form by damping the cuvette with a slight touch or alternatively placing a low frequency mechanical vibrator on the table nearby. For a given cuvette various discrete frequencies could be excited; moreover the lowest of these frequencies for a particular (1 x 1 x 4 cm) cuvette was shown to be independent of the scattering angle. It was suggested that the scattered light came from refractive index modulations arising from microscopic standing displacement waves with mode frequencies depending on the cuvette dimensions. Nossal and Brenner [27], showed that the observable frequencies were given by

$$\omega = \pi c_{\text{long}} \left(\frac{p^2}{a^2} + \frac{q^2}{b^2} + \frac{\gamma^2}{c^2} \right)^{\frac{1}{2}}$$

where a, b, and c are the dimensions of the cuvette (with c
normal to the scattering plane), p and q are integers and γ
is zero or half integral depending on whether the gel is fully
enclosed or not (ie on the boundary conditions). C_{long} was
taken to be the longitudinal sound speed in the network equal
to $[(K + 4G/3)\rho]^{\frac{1}{2}}$ where K and G are the compressibility and
shear modulus. Alternatively C_{long} can be expressed in terms
of Youngs modulus E and the Poisson ratio σ. For different
samples of agarose values of sound speed were obtained increasing by two orders of magnitude over a range of concentration
from 0.1 to 1%. Most importantly Brenner et al maintained that
the modulus being measured was essentially a property, the
longitudinal modulus, of the polymer lattice. More recently
this view has been reinterpreted by Nossal [28]. In a detailed
study he investigates the effects of friction between polymer
strands and solvent, of internal energy dissipation by the
polymer lattice and different boundary conditions with freely
moving surfaces along or normal to the cell walls. Nossal con-
cludes that the standing wave resonance observations in fact
provide a measure of the shear modulus of the bulk gel as a
whole. Independent lattice motions are so strongly inhibited
by friction between polymer molecules and the fluid that such
resonances would be broadened and unobservable. Moreover the
scattered light is considered to arise from inherent microscopic inhomogeneities in the gel structure as considered in
[24] and not from continuously varying refractive index modulation.

5.2 Diffusional Studies of Gels

One of the earliest studies of light scattered from a
visco-elastic gel was described in a seminal paper by Tanaka,
Hocker and Benedek [29] in 1973. In a detailed theoretical and
experimental investigation they showed that the measurement of
the intensity and spectrum of the light scattered by gels gives
information on the elastic properties of the gel lattice and on
its viscous interaction with the gel fluid. In the modes they
considered the fibre or lattice network mass against the gel
liquid so that the viscous properties are determined by the
friction between them. In a detailed evaluation of the amplitude and time dependence of thermally excited displacements of
the gel network.Tanaka et al showed that the temporal auto-
correlation of the scattered light field from longitudinal
fluctuations was given by

$$\langle E_{pol}(\underset{\sim}{K},t)E_{pol}(\underset{\sim}{K},0)\rangle = \frac{I_o}{c}\left(\frac{\omega_o}{c}\right)^4 \frac{\sin^2\phi}{4\pi R^2}\left(\frac{\partial\varepsilon}{\partial\rho}\right)_T^2 \rho^2 \frac{FkT}{(K+4\mu/3)}$$

$$\times \exp\left(-\frac{(K+4\mu/3)|\underset{\sim}{K}|^2 t}{\gamma}\right)$$

for the polarised scattered light where the terms have their usual meaning, F is a geometrical factor, ϕ is the angle betwe between the polarisation of the incident light and $\underline{k}_s$, ε are the diagonal elements of the dielectric tensor and γ is the friction constant. Since $\varepsilon = n^2$ the term $(\partial\varepsilon/\partial\rho)_T$ can be obtained by determining the index of refraction n for different gel concentrations ρ. A somewhat similar expression is found for depolarised scattering due to transverse fluctuations with the temporal component replaced by $\exp(-\mu|K|^2 t/\gamma)$, that is containing only the shear modulus. Light scattering measurements on 5% and 2.5% polyocrylamide gels showed that the correlation functions were of single exponential form and the decay rate Γ was proportional to $|K|^2$ as anticipated. From 10° to 70^oC the diffusion constant $D = \bar{\Gamma}/2|K|^2$ increased from ~ 1.5 to $\sim 6.5 \times 10^{-7}$ cm^2 s^{-1}. From the decay constant Γ a value of $[(K + 4/3\mu)/\gamma]$ may be found; in principle, from the measured intensity of (polarised) scattering, $(K + 4/3\mu)$ can also be obtained (similarly μ/γ and μ for depolarised scattering). For the network the Poisson ratio is nearly zero and the longitudinal modulus E_L equal to $(K + 4\mu/3)$ is 2μ. From these considerations Tanaka et al measured both μ and γ by macroscopic mechanical means and showed good agreement between the macroscopic ratio μ/γ and the value derived from the light scattering decay constant.

In other early work both networks and polymer solutions were examined by correlation methods [eg 30, 31]. McAdam et al in 1974 compared results for polystyrene chains in tetralin for cross linked gels and concentrated solutions. Modified diffusion constants in the gel were roughly twice those in solution and for large values of $\underline{K}$ the autocorrelation function was not a single exponential. Recently there has been a great upsurge of work in dynamic light scattering from gels and semi-dilute solutions. Following de Gennes [32] the two systems should differ only in their low frequency behaviour: in the gels long lived bonds connect different polymer coils, while in solution entanglements of characteristic lifetime T_r are formed. Thus for periods shorter than T_r the solution exhibits gel-like properties; the other parameter controlling the behaviour is a coherence or screening length ξ representing some average distance between entanglements and beyond which no correlation of chains exists. For both solution and cross linked network (swollen to equilibrium) in good solvents, ξ is expected from scaling to vary with polymer concentration c as $c^{-0.75}$ and the dynamic elastic modulus (for solutions at high enough frequency) is proportional to $1/\xi^3$. Furthermore the co-operative diffusion coefficient is given by $D_c = k_BT/6\pi\eta\xi$ where η is the solvent viscosity, so that D_c is proportional to $c^{0.75}$.

In a comprehensive series of papers Munch et al [33a,b,34] investigated model networks of polystyrene swollen by benzene. E_L was determined as a function of the molecular weight of the elastic chains, and the frictional constants were determined [33b,34]. The scaling laws were investigated with measurement of the cooperative diffusion constant, on gels of different concentration. In one experiment Candau et al [35] discarded, by filtering, slow variations in scattered light which they attributed to motion of macroscopic inhomogeneities. They found for fully swollen gels in the equilibrium state good agreement for the behaviour of E_L determined from mechanical measurements.

Polyacrylamide gels have been investigated under a wide variety of conditions and in considerable detail by Hecht and Geissler [36-42]. In one study [37] D_c was found to vary as $c^{0.65\pm0.03}$; the difference between this exponent and that for the inverse screening length was interpreted as consistent with a free volume behaviour of the solvent viscosity. In a later paper [39] they concluded that solvent viscosity did not depend on polymer concentration and the scaling exponents could only be deduced after correction for solvent displacement and residual hydrodynamic interactions. The Poisson ratio was investigated in good and poor solvent [40a,b] by comparing values of E_L and μ - the former measured by dynamic light scattering and mechanical means, the latter also by observation of standing waves (section 5.1). Most recently [42] values of ξ obtained by small angle neutron scattering have been reported in good agreement with light scattering measurements.

5.3 Critical Fluctuations in Gels

In an experiment reported in 1977 Tanaka et al[43] described critical type behaviour of polyacrylamide gels observed by light scattering. As the temperature was decreased both the intensity and correlation time of the scattered light increased by a factor of more than 200 and appeared to diverge at a certain temperature (figure 11a). As noted earlier the intensity is proportional to $kT/(K + 4\mu/3)$ and the reciprocal correlation time is given by $\Gamma = (K + 4\mu/3)|\underline{K}|^2/\gamma$. The bulk modulus K can be related to the osmotic pressure in the gel which itself derives from the three forces acting upon it - the rubber elasticity (of the strands themselves), the polymer - polymer affinity (due to interaction between polymer strand and solvent) and the hydrogen ion pressure (associated with the electrical state of the network).Tanaka et al

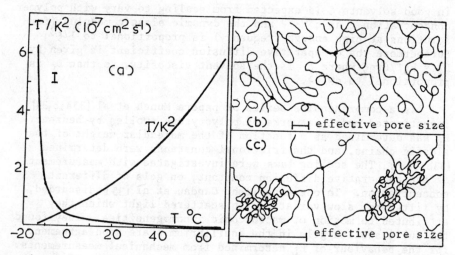

Figure 11.   Critical behaviour in gels. a) Intensity and decay
             rate versus temperature (after Tanaka et al [43]).
             b) Configuration of the network away from the critical
             point and c) near the critical point (after Tanaka
             [44c]).

showed that the longitudinal modulus E_L (equal to $K + 4\mu/3$) was
proportional to $(T - T_s)$ and would thus vanish at temperature
T_s. The frictional coefficient γ was obtained from the measurements of intensity and Γ; since it does not increase on approaching T_s the slowing down of network density fluctuations is thus
attributable to the reduction of E_L. In fact γ itself becomes
very small close to T_s. In a paper on the dynamics of the fluctuations [44a] Tanaka derives the scaling relations $E_L \propto \xi^{-2}$ and
$\gamma \propto \xi^{-1}$ where ξ is the correlation length of concentration
fluctuations. Thus on approaching T_s the effective 'pore size'
in the gel increases and appears to diverge (figure 11c). The
critical behaviour is thus associated with a phase separation of
the binary mixture of the network and the fluid medium, leading
to shrinkage of the network.

In later publications [44b,c] factors other than temperature
are shown to give rise to gel collapse, eg altering the solvent
composition or its ionic strength. This is very evident on
immersing fully hydrolysed gels in acetone-water mixtures. In
the critical region an arbitrarily small increase in acetone
concentrations leads to a dramatic change in volume. Full three
dimensional phase diagrams plotting gel volume, ionisation, and
temperature or concentration were obtained with identification
of a critical end point. Most interestingly Tanaka relates the
study to physiological phenomena such as vitreous collapse in
retinal detachment (see eg reference [44c]).

In concluding this section on gels and networks it is worth emphasizing the many new insights that dynamic light scattering is now providing. For a greater appreciation of their interest and value the reader should consult the original papers.

6   BRILLOUIN SCATTERING FROM SOLID POLYMERS

In the first study of Brillouin scattering in a glassy amorphous polymer, Peticolas et al [45] reported an extremely sharp discontinuity at T_g in the intensity ratio $I_c/2I_B$ of the Rayleigh to Brillouin scattering in polyethyl methacrylate (PEMA). For the temperature range $90^{\circ}C$ down to T_g ($61^{\circ}C$) the ratio was ~10; at T_g it increased suddenly to ~13. With further decrease in temperature to $20^{\circ}C$ $I_c/2I_B$ increased to ~19. However Friedman et al [46] reported that annealing at 175ºC lowered the overall ratio by a factor of 2 and eliminated the discontinuity. As in their earlier work on PMMA [47] they found that the change in slope of the hypersonic speed – temperature curve above and below T_g was simply related to the change in the coefficient ot thermal expansion (and hence the density ρ – see section 1.2). These results were considered by Stevens et al [48] in terms of scattering from stationary random strains, which would be temperature dependent, and inclusions (ie dust) which would not be. They suggested that a small discontinuity might still be present in the results of Friedman et al and related this to strains introduced on polymerisation that are only partly relieved on annealing. This explanation was reinforced by the results of Jackson et al [49] on hypersonic attenuation in PVC and PMMA. They too found a change in slope of the hypersonic speed – temperature curves at T_g, and the ratio of central peak to Brillouin scattering was large – of order 20 and 30 for PVC and PMMA respectively. However for both materials the attenuation per wavelength, $\alpha\lambda_s$, was almost independent of temperature below T_g and had values very similar to those found by ultrasonic measurements. Thus the attenuation α was approximately proportional to ω_s over a frequency range of order 2000:1. Various explanations of such an unusual dependence were discussed including the possibility of a structural relaxation. It was concluded that the most likely explanation was attenuation of the hypersound waves due to scattering by spherical strain fields around small imperfections in the materials.

Further studies by Stevens et al [50] on poly (n-butyl) methacrylate (PBMA) showed T_g was up to $18^{\circ}C$ lower for well annealed samples (compared with poorly annealed samples), where T_g was judged by the change in temperature dependence of the speed, attenuation and intensity ratio. This was taken as further evidence of scattering of strain fields around defects. In other investigations Lindsay et al [51] measured the anisotropy of the high frequency elastic modulus in strained PMMA; the variation of sound speed for different directions relative to the draw axis was determined. Cohen et al [52] considered in detail the

hydrodynamics of amorphous solids with particular application
to light scattering.

In recent studies by the Guelph group [53, 54] particular
attention has been paid to sample preparation and the measurement of the fullest range of parameters including depolarisation
ratio and intensities. Measurements on clean styrene [53b] were
made over a period of 200 hours as polymerization proceeded to
completion at 90°C. The intensity ratio $I_c/2I_B$ increased rapidly
once polymerization was initiated and levelled off to a value of
~5. The Brillouin shift and hypersonic speed approximately
doubled during the process. In comparison with PBMA, studies

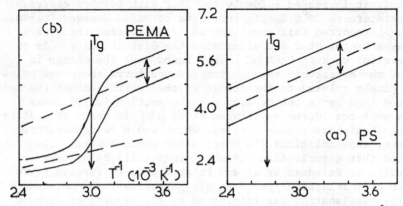

Figure 12. Rayleigh/Brillouin intensity ratio versus T^{-1},
(a) atactic polystyrene (derived from Coakley et al[53a])
(b) atactic PEMA (from Mahler et al [54]). The dashed lines
are $\propto T^{-1}$.

of the intensity ratio for polystyrene (PS) showed no discontinuity around the glass transition (figure 12a) and its value
was continuously proportional to 1/T. This together with the
observations that annealing had no effect on the hypersonic
properties suggested that for PS there is at most a very small
concentration of spherically symmetric strain fields. These
studies were in noted contrast with later work [ ] on PEMA.
Here the thermal history of annealing was crucial and annealing
at 135°C was essential for reproducable results. The temperature
dependence of the intensity ratio and the discontinuity around T_g
is as indicated in figure 12b. Mahler et al suggest that in
PEMA centers of entanglement exist at room temperature with long
relaxation time. These are able to relax over the temperature
range between 40 and 70°C and give the discontinuity in the
intensity ratio versus 1/T curve in that region (figure 12b).
The result for PS implies that PS molecules do not form centers
of entanglement under these conditions. This the authors attri-
but to the shorter and less mobile side chains of PS compared

with PBMA and PEMA. In this regard it would seem interesting to compare attenuation data in these materials and in particular its frequency dependence. The interpretation of attenuation as largely due to scattering would seem to be much less likely in PS and would provide a useful check.

7   BRILLOUIN SCATTERING FROM GELS AND FIBRES

The first measurements of Brillouin scattering in a gel were conducted by Bedborough and Jackson in 1976 [55]. They investigated the Brillouin spectrum of gelatin gel over a range of concentrations up to 0.2 gm cm^{-3} with a double pass Fabry-Perot spectrometer at 5145 Å and 90° scattering. It was found that both the Brillouin shift and line width increased linearly with concentration - the shift by some 10% over the range and the line width by 300%. In the latter case the rate of increase was sensitive to the thermal history of the gels, being greater for gels cooled slowly. Bedborough and Jackson found that an explanation of the results in terms of a relaxation phenomenon was improbable. They suggested that the observations could be explained in terms of the structure of the gel and the number density of crosslinks which is dependent on the concentration. The increase in shift was consistent with the change in bulk properties; the increase in line width was interpreted as due to scattering of the hypersonic phonons by the crosslinks. The model for the gel structure took the crosslinks between the polymer molecules (which are the triple helices of the parent collagen molecule) as cylindrical scattering entities. In order to fit to theoretical expressions these were approximated as spherical; a density of $N \sim 10^{20}$ cm^{-3} giving an inter crosslink distance of ~ 20 Å was found at the highest concentration - in reasonable accord with expectation. The scattering interpretation was further supported by the fact that the line width greatly decreased on melting the gel so that the network was destroyed. Further measurements of Brillouin spectra from polymer fluids and rubber networks were outlined by Lindsay et al [51].

In recent years interferometry and measurement of Brillouin spectra have been extended to natural biological materials including fibres and gels [56-61]. The ultimate object of observing spectra of such materials must be to relate the measurements to their intrinsic molecular properties. The optical properties may be related with the elastic ones and in some instances with the inter and intra molecular forces of the molecules involved. The mechanical properties of biological materials measured by imposing known stresses are of practical value but may not relate to fundamental molecular characteristics because of natural faults and uneven stress. Natural collagen from rat tail tendons was examined in several studies [56,57,59]. In dried material Cusack and Miller observed a propagating shear mode and measure-

ments were made at different angles between the scattering
vector and fibre axis. A fairly complete analysis was conducted
with determination of the five elastic constants. The ratio
$(E_\parallel /E_\perp)$ of Young's modulus for strain along the fibre axis to
that for strain normal to the axis is found to be 1.43 and the
ratio of the shear modulus to $E_\parallel$ is 0.28. In this instance, it
seems likely that Brillouin scattering is probing the intermolecular forces between the rodlike molecules that comprise the
collagen fibrils. It is interesting that the high frequency
Brillouin measurements of Young's modulus are an order of magnitude greater than for static macroscopic measurements. This
may be due to inhomogeneities on a large scale (which could
accomodate the strain) to which the high frequency Brillouin
measurements are insensitive since they are effectively probing
on a microscopic scale. Another suggestion [57] is that the
hypersonic modulus is that of the individual fibril, whereas the
macroscopic modulus is that of the tendon as a whole made up of
a composite of fibrils. A third possibility involves a genuine
viscoelastic relaxation mechanism arising from viscous damping
of the motion of the collagen molecules. Clearly ultrasonic
measurements at intermediate frequencies would be of value in
examining this problem.

In other investigations keratin [59] and oriented DNA
fibres and films [58] have been examined in various states of
hydration. Propagation along the fibre axis is usually markedly
different from propagation across the axis. Shear modes have
not been reported for these materials. Brillouin scattering has
recently been extended [59,60] to lens material which is a gel
composed of globular proteins. In the lens of the vertebrate
eye various globular proteins called crystallins occur in three
predominant classes, α, β and γ. The γ's are monomers while the
α and β crystallins are polymeric. Rather remarkable differences
are found in the concentration of these crystallins in the lenses
of different species. In man and most mammals the weight concentration is about 0.25 at the periphery of the lens increasing to
~ 0.35 at the centre; the lens has a flattened lenticular shape.
In bony fishes the lens is spherical and the concentration rises
to probably greater than 0.85 at the centre and is extremely
hard. These properties also hold true, remarkably, for the rat
and other rodents. Such differences are immediately apparent in
the Brillouin spectra [60,61] as seen in figure 13. These
measurements have been allied with accurate density data and
elastic moduli obtained. The results are similar in form to
those found in gelatin gels [55] and have also been interpreted
in terms of propagation and scattering in the mixed medium of
globular proteins and water. For such a system, of N spherical
particles per unit volume, each of radius a_n, density ρ_n and
compressibility κ_n respectively, within a continuous medium of
density ρ_o and compressibility κ_o, the sound speed is given by

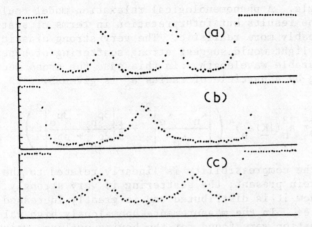

Figure 13 Brillouin spectra recorded with a triple pass
interferometer at different positions in the fish
lens *Sarotherodon mosambicus*: a) periphery, shift
4.47 GHz, width 0.51 GHz; b) half-way in, shift
6.1 GHz, width 1.1 GHz; c) nucleus, shift 8.4 GHz,
width 1.1 GHz (from [60a]).

$U^2 = (\kappa_R \rho_R)^{-1}$. In this expression the mean compressibility is
given by [see eg 62].

$$\kappa_R = \kappa_o + \sum_{n=1}^{N} \frac{4}{3}\pi \, a_n^3 (\kappa_n - \kappa_o)$$

and the effective mean density ρ_R (not the actual mean density)
is given by

$$\rho_R = \left[\frac{1}{\rho_o} + \sum_{n=1}^{N} \frac{4}{3}\pi \, a_n^3 \left(\frac{1}{\rho_n} - \frac{1}{\rho_o} \right) \right]^{-1}$$

By using reasonable values for ρ_o, ρ_n, κ_o and κ_n quite good
agreement for the calculated and observed values of hypersonic
speed are found for lens gels of protein concentration less than
~ 0.4. The model supposes well separated spheres in a continuous
medium so poorer agreement at higher concentration is not too
surprising.

Somewhat similar considerations apply to the measured line
width and attenuation. These were measured over a nearly an
order of magnitude range of scattering vectors by varying the

scattering angle. A phenomenological relaxation model could be
made to fit the results but interpretation in terms of scattering is probably more realistic. The very strong elastic
scattering of light would suggest strong scattering of hypersonic
waves of comparable wavelength. In this case the cross section
for scattering by a particle of radius a_n is given by

$$\sigma_{sc} \approx \frac{4}{9}\pi \, a_n^2 (|\underline{K}|a_n)^4 \left(\left| \frac{\kappa_n - \kappa_o}{\kappa_o} \right|^2 + \frac{1}{3} \left| \frac{3\rho_n - 3\rho_o}{2\rho_n + \rho_o} \right|^2 \right)$$

Whereas the compressibility is linearly related to the
amount of protein present, the scattering is very strongly
dependent on how it is distributed and is greatly increased by
larger particles. In the measurements anomalously high values
for the attenuation were found for the bovine nucleus, which
contrasted notably with those for sheep and foul. The fact that
the attenuation is so strong may be attributed to a relatively
high concentration of large aggregates in the bovine nucleus.
It has been suggested [63-65] that in the intact calf nucleus
aggregates as large as 200 nm may exist a few degrees above the
cold cataract temperature. Such aggregates present in concentrations amounting to less than $\sim 1.4\%$ of the protein material
would give the hypersonic scattering and attenuation derived
from the Brillouin scattering data for the bovine nucleus.

8   CONCLUSIONS

In summary quasi-elastic light scattering offers valuable
information about polymer systems that is often not obtainable
by other techniques. In particular for polymer gels recent
progress has given exciting new insights and it may be hoped
that these will extend to other domains. With increased
theoretical understanding of the underlying processes and their
interaction with light it seems likely that light scattering
techniques will prove increasingly valuable.

ACKNOWLEDGEMENTS

I am indebted to many colleagues for valuable discussions
and in particular to Prof G Williams, Dr E Geissler,
Prof S E Rickert and Mr J Tribone for many useful reprints and
preprints. As always any errors or omissions are the sole
responsibility of the author.

REFERENCES

1) I L Fabelinskii: *Molecular scattering of light*, Plenum
    Press, New York, 1968.
2) I D Landau and E M Liftshitz: (a) *Electrodynamics of
    Continuous Media*, Chapt. XIV, Pergamon Press, London 1960.
    (b) *Statistical Physics*, Chapter XII, Pergamon Press,
    London 1958.
3) B Chu: *Laser Light Scattering*, Academic Press, New York,
    (1975).
4) B J Berne: *Dynamic Light Scattering*, Wiley, New York,
    (1976).
5) G Benedek and T Greytak: Brillouin scattering in liquids,
    Proc. IEEE. 53, 1623 (1965).
6) W L Peticolas: Inelastic laser light scattering from
    biological and synthetic polymers, *Fortschr. Hochpolym.
    Forsh.* 9, 285 (1972).
7) R Vacher and L Boyer: Brillouin scattering - a tool for
    measurement of elastic and photoelastic constants, *Phys.
    Rev.* 6B, 639 (1972).
8) *Light Scattering in Liquids and Macromolecular Solutions*,
    Ed. V Degiorgio, M Corti and M Giglio, Plenum, New York,
    1980.
9) P A Fleury and J P Boon: Laser light scattering in fluid
    systems. *Adv. in Chemical Physics*, 24, 1, (1973).
10) W B Davenport and W L Root: *An Introduction to the Theory
    of Random Signals and Noise*, McGraw-Hill, New York, 1958.
11) P N Pusey and J M Vaughan: Light Scattering and intensity
    fluctuation spectroscopy: Chapter 2 in *Dielectric and
    Related Molecular Processes*, London - The Chemical Society
    SPR2, (1975).
12) E Jakeman and E R Pike in *Advances in Quantum Electronics
    Vol 2*; ed D Goodwin, Academic Press, London 1974.
13) *Photon Correlation and Light Beating Spectroscopy*.
    Proceedings of the NATO ASI, ed H Z Cummins and E R Pike,
    Plenum, New York 1974, and the many articles therein.
14) *Photon Correlation and Laser Velocimetry*, Proceedings of
    the NATO ASI, Ed H Z Cummins and E R Pike, Plenum, New York
    1977, and the many articles therein.
15) W A B Evans and J G Powles: The interpretation of hypersonic
    and ultrasonic 'velocities' in fluids, *J. Phys. A: Math.
    Gen.* 7, 1944 (1974).
16) D A Jackson, E R Pike, J G Powles, J M Vaughan: The
    possibility of detecting slow molecular reorientation in
    polymers by photon correlation spectroscopy, *J. Phys. C:
    Solid State Phys.* 6, L55 (1973).
17) C Cohen, V Sankur, C J Pings: Laser correlation spectro-
    scopy of amorphous polymethylmethacrylate, *J. Chem. Phys.*
    67, 1436, (1977).

18) Dane R Jones and C H Wang: Depolarised Rayleigh scattering and backbone motion of polypropylene glycol, *J. Chem. Phys.* 65, 1835, (1976).

19) H Lee, A M Jamieson, R Simha: a) Laser light scattering as a probe for structure and dynamics in bulk polymers: polystyrene, *Colloid and Polymer Sci.*, 258, 545, (1980). b) *Macromoecules* 12, 329, (1979).

20) G D Patterson, J R Stevens, C P Lindsey: a) Depolarised Rayleigh spectroscopy of polystyrene near the glass-rubber relaxation. *J. Chem. Phys.* 70, 643, (1979). b) Photon correlation spectroscopy of poly (ethyl methacrylate) near the glass transition, Preprint 1981.

21) A M Jamieson, R Simha, H Lee, J Tribone: Laser light scattering studies of bulk polymers in the glass transition regime, Preprint 1981.

22) G D Williams, D C Watts: *Trans. Faraday Soc.*, 66, 80, (1970).

23) W Prins, L Rimai and A J Chompff: An audio frequency resonance in quasielastic light scattering of polymer gels, *Macromolecules* 5, 104, (1972).

24) K L Wun and F D Carlson: Harmonically bound particle model for quasielastic light scattering by gels. *Macromolecules* 8, 190, (1975).

25) K L Wun, G T Feke and W Prins: Laser light scattering by polymer gels, *Faraday Discuss. Chem. Soc.*, 57, 146, (1974).

26) S L Brenner, R A Gelman and R Nossal: Laser light scattering from soft gels, *Macromolecules*, 11, 202 (1978).

27) R Nossal and S L Brenner: Correlation functions for light scattering from soft gels, *Macromolecules*, 11, 207, (1978).

28) a) R Nossal: A theory of quasielastic laser light scattering by polymer gels, *J. Appl. Phys.* 50, 3105, (1979). b) R A Gelman and R Nossal: Laser light scattering from mechanically excited gels, *Macromolecules* 12, 311, (1979).

29) Toyoichi Tanaka, L O Hocker and G B Benedek: Spectrum of light scattered from a viscoelastic gel, *J. Chem. Phys.* 59, 5151, (1973).

30) D G McAdam, T A King and A Knox: Molecular motion in polymer networks and concentrated solutions from photon correlation spectroscopy, *Chem. Phys. Lett.*, 26, 64, (1974).

31) P N Pusey, J M Vaughan and G Williams: Diffusion of polystyrene in solution studied by photon correlation spectroscopy, *J. Chem. Soc., Faraday Trans. II*, 70, 1969 (1974).

32) P G de Gennes: Dynamics of entangled polymer solutions. I The Rouse model, *Macromolecules* 9, 587, (1976).

33) J P Munch, S Candau, R Duplessix, C Picot, J Herz and H Benoit: a) Etude par diffusion inélastique de la lumière des propriétés viscoélastique de gels modèles. *J. de Physique* 35, L-239 (1974). b) Spectrum of light scattered from viscoelastic gels. *J. Polym. Sci* 14, 1097, (1976).

34) J P Munch, S Candau, J Herz and G Hild: Inelastic light
     scattering by gel modes in semi dilute polymer solutions
     and permanent networks at equilibrium swollen state, *J. de
     Physique* 38, 971 (1977).

35) S J Candau, C Y Young, T Tanaka, P Lemarechal and J Bastide:
     Intensity of light scattered from polymeric gels: influence
     of the structure of the networks, *J. Chem. Phys.* 70, 4694,
     (1979).

36) E Geissler and A M Hecht: Dynamic light scattering from
     gels in a poor solvent, *J. de Physique* 39, 955, (1978).

37) A M Hecht and E Geissler: Dynamic light scattering from
     polyacrylamide-water gels, *J. de Physique* 39, 631, (1978).

38) A M Hecht, E Geissler and A Chosson: Dynamic light
     scattering by gels under hydrostatic pressure, *Polymer*, 22,
     877, (1981).

39) E Geissler and A M Hecht: Translational and collective
     diffusion in semi-dilute gels, *J. de Physique*, 40, L-173,
     (1979).

40) E Geissler and A M Hecht: a) The Poisson ratio in polymer
     gels, *Macromolecules* 13, 1276, (1980). b) The Poisson
     ratio in polymer gels 2, *Macromolecules* 14, 185, (1981).

41) A M Hecht and E Geissler: a) Gel deswelling under reverse
     osmosis, *J. Chem. Phys.* 73, 4077, (1980). b) Pressure
     induced deswelling of gels, *Polymer* 21, 1358, (1980).
     c) Gel deswelling under reverse osmosis II, *J. Chem. Phys.*
     to be published.

42) E Geissler, A M Hecht and R Duplessix: Comparison between
     neutron scattering and quasi electric light scattering by
     polyacrylamide gels, *J. Polymer Sci.* to be published.

43) Toyoichi Tanaka, Shinichi Ishiwata and Coe Ishimoto:
     Critical behaviour of density fluctuations in gels, *Phys
     Rev. Lett*, 38, 771, (1977).

44) Toyoichi Tanaka: a) Dynamics of critical concentration
     fluctuations in gels, *Phys. Rev.* 17A, 763 (1978).
     b) Collapse of gels and the critical end point, *Phys. Rev.
     Lett.* 40, 820, (1978). c) Gels: *Scientific American*, 244,
     110, (January 1981).

45) W L Peticolas, G I A Stegeman and B P Stoicheff: Intensity
     ratio of Rayleigh to Brillouin scattering at the glass
     transition in polyethylamethacrylate, *Phys. Rev. Lett.* 18,
     1130, (1967).

46) E A Friedman, A J Ritger, Y Y Huang and R D Andrews:
     Landau-Placzek ratio and hypersonic attenuation in amorphous
     acrylic polymers, *Bull. Am. Phys. Soc.* 15, 282 (1970).

47) E A Friedman, A J Ritger and R D Andrews: Brillouin
     scattering near the glass transition of polymethyla-
     methacrylate, *J. Appl. Phys.* 40, 4243, (1969).

48) J R Stevens, I C Bowell and K J L Hunt: Light scattering
     from isotropic polymeric solids, *J. Appl. Phys.* 43, 4354,
     (1972).

49) D A Jackson, H T A Pentecost and J G Powles: Hypersonic absorption in amorphous polymers by light scattering, *Molecular Physics* 23, 425, (1972).

50) J R Stevens, D A Jackson and J V Champion: Evidence for ordered regions in poly (n-butyl) methacrylate from light scattering studies, *Molec. Phys.* 29, 1893, (1975).

51) S M Lindsay, A J Hartley and I W Shepherd: Multi-pass Fabry-Perot spectroscopy of polymers, *Polymer*, 17, 501, (1976).

52) C Cohen, P D Fleming and J H Gibbs: Hydrodynamics of amorphous solids with application to the light scattering spectrum, *Phys. Rev.* 13B, 866, (1976).

53) R W Coakley, R S Mitchell, J R Stevens and J L Hunt: a) Rayleigh-Brillouin light scattering studies on atactic polystyrene, *J. Appl. Phys.* 47, 4271, (1976). b) Study of polymerising styrene through depolarised light scattering, *J. Macromol. Sci-Phys.* B12, 511, (1976).

54) D S Mahler, R W Coakley, J R Stevens and J L Hunt: Rayleigh: Brillouin light-scattering study of atactic polyethyl-methacrylate (PEMA), *J. Appl. Phys.* 49, 5029, (1978).

55) D S Bedborough and D A Jackson: a) Brillouin scattering study of gelatin gel using a double passed Fabry-Perot spectrometer, *Polymer*, 17, 573, (1976). b) Brillouin scattering from silica colloids, *Phys. Lett.* 60A, 140, (1977).

56) R Harley, D James, A Miller and J W White: Phonons and the elastic moduli of collagen and muscle, *Nature* 267, 285, (1977).

57) S Cusack and A Miller: Determination of the elastic constants of collagen by Brillouin light scattering, *J. Mol. Biol.* 135, 39, (1979).

58) G Maret, R Oldenbourg, G Winterling, K Dransfeld and A Rupprecht: Velocity of high frequency sound waves in oriented DNA fibres and films determined by Brillouin scattering, *Colloid and Polymer Sci.* 257, 1017, (1979).

59) J T Randall and J M Vaughan: Brillouin scattering in systems of biological significance, *Phil. Trans. R. Soc. Lond.* A293, 341, (1979).

60) J M Vaughan and J T Randall: a) Brillouin scattering, density and elastic properties of the lens and cornea of the eye, *Nature* 284, 489, (1980). b) Brillouin scattering and elastic properties of biological materials, *Proc. 4th Nat. Quant. Elect. Conf. Edinburgh 1979*, Wiley Interscience ed. B S Wherrett, p 185, (1980).

61) J T Randall and J M Vaughan: The measurement and interpretation of Brillouin scattering in the lens of the eye. *Proc. Roy. Soc. B* (1981) in press.

62) P M Morse and K U Ingard: *Theoretical Acoustics* Chapter 8,
     McGraw-Hill, New York (1968).

63) Toyoichi Tanaka and G B Benedek: Observation of protein
     diffusivity in intact human and bovine lenses with
     application to cataract, *Investig. Opthalmol.* 14, 449,
     (1975).

64) J A Jedziniak, D F Nicoli, H Baram and G B Benedek:
     Quantitative verification of the existence of high mole-
     cular weight protein aggregates in the intact normal human
     lens by light scattering spectroscopy, *Investig. Opthalmol.*
     17, 51, (1978).

65) M Delaye, J I Clark and G B Benedek: Identification of the
     scattering elements responsible for lens opacification in
     cold cataracts, preprint 1981.

QUASIELASTIC AND INELASTIC NEUTRON SCATTERING

J.S. Higgins

Department of Chemical Engineering and Chemical
Technology, Imperial College, London.

The neutron is a heavy particle compared to photons and
therefore scattering spectroscopy using neutrons covers a unique
range of frequency-spatial correlations. For inelastic experiments the far infra-red spectroscopic range up to ~ 1000 cm^{-1} may
be observed with sizeable wave vector transfers and phonon dispersion curves mapped out. In quasielastic experiments motion
of main chain and side groups may be followed at frequencies from
10^7 Hz upwards, again with simultaneous information about the
spatial characteristics of the motion in question.

The scattering properties of hydrogen and deuterium differ
greatly so that neutron scattering from isotopically-labelled
samples allows inelastic peaks arising from particular side
groups in molecules to be simply identified,while labelling of
whole molecules allows their motion in bulk samples to be followed.

1.   INTRODUCTION

1.1 Neutron Spectroscopy in Context

The neutron cross-section of the hydrogen nucleus is at
least an order of magnitude larger than that of most other nuclei.
Investigations of synthetic polymers for this reason featured
fairly early among inelastic scattering experiments carried out
with neutrons (1, 2, 3). Neutrons leave thermal reactors with a
Maxwellian distribution of wavelengths peaked around 1.4 Å and a
corresponding energy of order 42 meV (340 cm^{-1}). As a result,
energy transfers of order 10-1000 cm^{-1} can conveniently be
measured corresponding to the far infra-red spectroscopic range.

349

R. A. Pethrick and R. W. Richards (eds.), Static and Dynamic Properties of the Polymeric Solid State, 349–381.

Despite the advantage of the large hydrogen cross-section, flux
limitations of even the most powerful reactor sources allow
resolution, $\Delta E/E$, of at best only one or two per cent. Neutron
spectroscopy (4, 5, 6, 7) is always at a disadvantage in resolution terms when compared to optical spectroscopy (8). Given as
well, the inconvenience of using a central facility, rather than
a laboratory-based spectrometer, there have to be strong advantages to encourage the use of neutron spectroscopy for investigating any given sample. From the point of view of polymeric
systems, these advantages are twofold. Firstly, the interaction
of the neutron with deuterium is very different from its interaction with hydrogen, thus offering the possibility of a relatively simple labelling technique which does not grossly perturb
the system. Secondly, the neutron is a relatively heavy particle
having large momentum for modest energies. The wave vector
changes associated with energy transfers are much larger than for
photons and allow, for example, exploration of phonon dispersion
curves away from the range close to zero phonon wave vector in
optical spectroscopy.

Quasielastic scattering arises from effectively non-quantised
motion (translation, rotation) and appears as a Doppler broadening of the incident energy spectrum. Ultra-high resolution techniques such as those described in Section 3 have allowed observation of energies as small as a few nano-electron volts (frequencies around 10^7 Hz) but these are still relatively high frequencies compared to the slow reorganisational motion of molecules in
a polymer melt. There is considerable overlap with results at
these frequencies from NMR relaxation (8, 9), from high frequency
viscosity (10) and from ultrasonic measurements (11) - all techniques which can be conveniently carried out in the laboratory.
As will be seen, it is again the hydrogen-deuterium labelling
possibilities and the explicit information about spatial correlations obtained from the sizeable wave vector changes which mean
that neutron quasielastic experiments can provide unique information about polymer melts and networks.

1.2 Neutron Scattering Principles and the Scattering Laws

Neutron energies in these experiments are very much lower
than the binding energies of nuclei so the scattering from an
isolated nucleus is isotropic and the interaction characterised
by a single parameter - the scattering length, or amplitude, b.
The scattering cross-section, σ, is the ratio of scattered neutrons to the incident flux, and for an isolated stationary nucleus

$$\sigma = 4\pi b^2 \tag{1}$$

The neutron carries spin $\frac{1}{2}$ and therefore has two interaction
states with any nucleus which also contains spin. The magnetic

moment of the neutron may Larmor precess in a magnetic field
(with an important application in the spin-echo technique described in Section 2.5) and it interacts weakly with unpaired electron spins in magnetic crystals. This magnetic effect leads to
an important branch of neutron scattering, but since it has not
so far been of importance for polymer samples, only the strong
nuclear interaction is considered here.

For nuclei with spin the two interaction states with the
neutron lead to two values of the scattering length. In this
case, or even for an isotopically impure sample, the scattering
length thus varies from site to site.

When scattering occurs from stationary nuclei there can be
no energy transfer, but the wave vector $\underline{k}$ of the neutrons changes
($\underline{k}$ is a vector in the direction of travel of magnitude $|\underline{k}| = 2\pi/\lambda$
$= mv/\hbar$). The momentum transfer $\hbar Q$ is defined in terms of the
change in wave vector on scattering

$$\underline{Q} = \underline{k}_i - \underline{k}_f \tag{2}$$

where the subscripts i and f refer to the initial and scattered
beams respectively. However, since there is no energy transfer
$|\underline{k}_i| = |\underline{k}_f|$ and simple trigonometry shows

$$Q = |\underline{Q}| = 4\pi/\lambda \sin \theta/2 \tag{3}$$

where θ is the angle of scatter.

The differential cross-section per atom with respect to solid
angle Ω contains information about the spatial arrangements of the
scattering nuclei. In a sample containing N atoms

$$\frac{d\sigma}{d\Omega} = \frac{1}{N} | \sum_n b_n \exp i(\underline{Q}.\underline{R}_n)|^2 \tag{4}$$

where $\underline{R}_n$ is the position vector of the nth nucleus. This expression can be manipulated (4, 5) to give

$$\frac{d\sigma}{d\Omega} = \{<b^2> - <b>^2\} + <b>^2 | \sum_n \exp i(\underline{Q}.\underline{R}_n)|^2 \tag{5}$$

where <> implies averaging over all sites.

Only the last term in equation (5) contains spatial information and this is the coherent scattering. The coherent scattering cross-section is given by the mean square scattering amplitude, $\sigma_{coh} = 4\pi<b>^2$. The first term is constant in a static
experiment and forms the incoherent background, σ_{inc}. The term
$\sigma_{inc} = 4\pi(<b^2> - <b>^2)$ can also be written as $4\pi<(b - <b>)^2>$.

This explicitly shows that σ_{inc} is the mean square deviation of the scattering lengths from their average value. For isotopically pure samples of nuclei such as ^{12}C, ^{16}O which have zero spin, σ_{inc} itself is zero. For hydrogen it is very large. Table 1 lists values of b, σ_{inc} and σ_{coh}, together with the absorption cross-section σ_{abs} for nuclei commonly found in synthetic polymers.

	b	σ_{coh}	σ_{inc}	σ_{abs} (1.08 Å)
	10^{-12} cm	10^{-24} cm^2	10^{-24} cm^2	10^{-24} cm^2
^1H	-.374	1.76	80	.19
^2D	.667	5.59	2	.0005
^{12}C	.665	5.56	0	.003
^{14}N	.94	11.1	.3	1.1
^{16}O	.58	4.23	0	.0001
^{19}F	.56	3.94	.06	.006
ave 28.06$_{Si}$	.42	2.22	0	.06
^{32}S	.28	.99	0	.28
ave 35.5$_{Cl}$	.96	11.58	3.5	19.5

TABLE 1

In general, if the scattering nuclei are in motion on the time scale of the experiment, $k_i \neq k_f$. It is then the double differential cross-section $d^2\sigma/d\Omega dE_f$ which is measured in the experiment, and the incoherent term is no longer a flat background, since it carries spatially uncorrelated information about motion of the scattering nuclei. Although for convenience the coherent and incoherent terms are expressed and discussed separately, experimentally they may be very difficult to separate. The scattering cross-sections are related to two space-time correlation functions introduced by Van Hove (4, 12), $G(\underline{R}, t)$ and $G_S(\underline{R}, t)$. $G(\underline{R}, t)$ expresses the probability that if there is a nucleus at position $\underline{R}_i$ at time t = 0, there will be another nucleus at position $\underline{R}_j$ at time t. $G_S(\underline{R}, t)$ is the self-correlation function, expressing correlations between the same nucleus at time 0 and t.

$$\frac{d^2\sigma_{coh}}{d\Omega dE_f} = \frac{k_f}{k_i} \frac{<b>^2}{2\pi \hbar} \int\int d\underline{R}dt \exp\{i(\underline{Q}.\underline{R} - \omega t)\}G(\underline{R}, t) \qquad (6)$$

$$\frac{d^2\sigma_{inc}}{d\Omega dE_f} = \frac{k_f}{k_i} \frac{(<b^2>-<b>^2)}{2\pi\hbar} \int\int d\underline{R}dt \exp\{i(\underline{Q}.\underline{R} - \omega t)\}G_S(\underline{R}, t) \qquad (7)$$

where $\quad \hbar\omega = \Delta E = E_f - E_i = \frac{\hbar^2}{2m}(k_f^2 - k_i^2) \qquad (8)$

The integrals in equations (6) and (7) contain all the information about the scattering system and are usually called the scattering laws $S(\underline{Q}, \omega)$ and $S_S(\underline{Q}, \omega)$

$$S(\underline{Q}, \omega) = \frac{1}{2\pi} \int\int \exp\{i(\underline{Q}\cdot\underline{R} - \omega t)\}G(\underline{R}, t)d\underline{R}dt \qquad (9)$$

and there is an analagous definition for $S_S(\underline{Q}, \omega)$ in terms of
$G_S(\underline{R}, t)$. If only the spatial Fourier transforms (in equation
(9) and its incoherent analogue) are performed, the quantities
obtained are the so-called intermediate scattering functions
$S(\underline{Q}, t)$ and $S_S(\underline{Q}, t)$.

2. EXPERIMENTAL TECHNIQUES

2.1 General Introduction

In order to determine the inelastic neutron scattering cross-
sections discussed in the previous section, it is normally necessary to define the incident energy more or less precisely using
some form of monochromator and to then analyse the energy of
the scattered neutrons as a function of scattering angle. To
date, most neutron experiments have used the thermal neutrons
from reactors with a Maxwellian wavelength distribution peaked
at 1.4 Å (in some cases shifted to longer wavelengths by the use
of a cold source, as described below).

A slice of wavelengths is taken out of this distribution
using either mechanical choppers (time of flight and spin-echo
spectrometers use mechanical monochromation) or using Bragg reflection from a suitably-oriented crystal (the triple axis and
back-scattering spectrometers use this method). Analysis of the
scattered neutrons can again use Bragg reflection from a suitable
crystal, or measurement of the time taken to cover a measured
flight path (time of flight analysis) of a pulsed beam.

Recently, a new generation of neutron sources (13-15) is
being constructed where accelerated beams of charged particles
(electrons or protons) are incident on a target of high atomic
number and produce intense bursts of high energy neutrons. The
high energies and short wavelengths of these neutrons make them
unique for investigation of large energy transfers and high wave
vectors. This area is not typically associated with polymer
dynamics; however, the neutrons can be moderated to wavelengths
of the same order as reactors and then it is the extremely high
intensity in the pulse which is exploited. Generally, spectrometers on pulsed sources (15) use the pulsed nature of the beam
together with time of flight analysis to obtain good energy resolution. Many such spectrometers are at the design stage at
present (14) but none has yet been constructed for the low energy

(high resolution) range of interest for studying polymer dynamics.
For this reason, the spectrometers described here are all the
most recently constructed reactor instruments of their type.
They are all at the high flux reactor at the Institut Laue-Langevin (16) in Grenoble which is jointly owned by the scientific
communities of France, Germany and the U.K. All four instruments
are at the end of neutron guides directed at a cold source. The
guides conduct the neutrons to areas away from the reactor where
there is more space and lower background "noise". The cold
source is a container of liquid deuterium close to the reactor
core where the neutrons are scattered many times and lose energy
to finish with a wavelength distribution shifted to long wavelengths (typically 5-10Å).

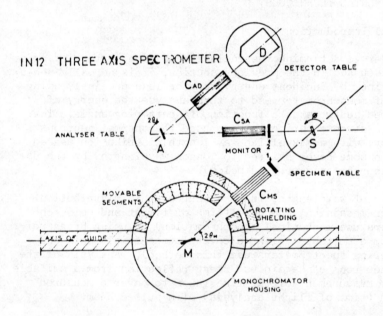

Fig. 1 - The IN12 triple-axis spectrometer at ILL, Grenoble

2.2 The Triple Axis Spectrometer

Figure 1 shows the layout of the triple-axis spectrometer
IN12 (17). The spectrometer uses neutrons from a guide on a
cold source with a useful flux of wavelengths between 2.3 and
6.3Å. Apart from the wavelength range, it is typical of all such
spectrometers (18, 19). The whole spectrometer rotates about the
monochromator (M) which, on IN12, is normally pyrolitic graphite
(spectrometers for shorter incident wavelengths use copper or
germanium monochromator crystals). The analyser and detector arms

rotate about the sample (S) and the detector (D) about the
analyser crystal (A). The analyser can often be the same material
as the monochromator but could also be chosen from any crystal
suitable for covering the desired energy transfer range. The
three axes of the apparatus are thus M, S and A. In between are
three collimators C_{MS}, C_{SA} and C_{AD}. These are well-defined narrow
slots in the beam direction covered in neutron absorbing material
(cadmium or gadolinium) which reduce beam divergence. The three-
axis construction, which allows variation of E_i, E_f and θ, gives
almost unlimited flexibility in choosing $Q = \underline{k}_i - \underline{k}_f$ and $\Delta E =
E_i - E_f$ independently. It is thus possible to map the whole of
$\underline{Q}$ - E space and the spectrometer is particularly suited to the
tracing of phonon dispersion curves. Since the first one designed
by Brockhouse (20), these spectrometers have changed little and
have been one of the basic tools for neutron scattering use in
solid state physics. Since each point is counted separately,
however, this is not a particularly fast spectrometer for covering large areas of $\underline{Q}$ - E space. Such experiments are better
tackled using the time of flight technique. The spectrometer
measures the differential cross-section $d^2\Omega/d\Omega dE_f$ directly and
IN12 has a best resolution of $\Delta E/E$ around 0.5%.

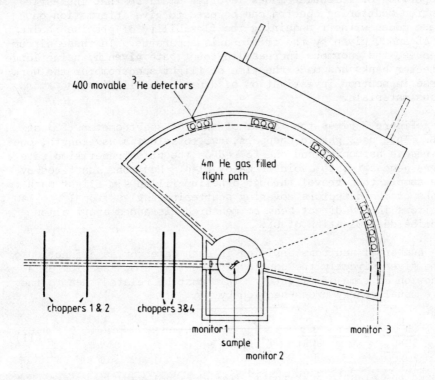

Fig. 2 - The time of flight spectrometer IN5 at ILL, Grenoble

2.3 The Time of Flight Spectrometer

Time of flight spectrometers (21) increase the counting rate by using banks of detectors at fixed scattering angles but lose the flexibility of uncoupled ΔE and Q. There is more variation between spectrometers possible here, but in general, a pulsed monochromatic beam produced either by rotating a crystal monochromator or by phased mechanical choppers is scattered by the sample and the neutron flight time to the detectors is measured. If the flight path d is measured, then time of flight, $\tau = t/d = $ velocity^{-1}. Hence ΔE is known but in each experiment E_i is constant, so that Q is coupled to ΔE

$$\text{since} \quad Q = \sqrt{k_i^2 + k_f^2 - 2k_i k_f \cos \theta} \quad)$$

$$\text{and} \quad \Delta E = \left(\frac{\hbar^2 k_i^2}{2m} - \frac{\hbar^2 k_f^2}{2m} \right) \quad) \tag{10}$$

In quasielastic experiments, ΔE is very small so this coupling to Q may not be important. Similarly, in non-crystalline samples, dispersion of molecular modes is often small so that inelastic neutron scattering spectra can be used to give information on these modes without requiring the flexibility of precise choice of ΔE and Q given by the triple-axis apparatus. In these circumstances, the enormous increase in count rate given by using large detector banks has made the time of flight spectrometer the workhorse in neutron investigation of dynamics of liquids and amorphous materials.

Figure 2 gives the layout of the IN5 spectrometer (22) at ILL. This uses phased choppers, two to select a wavelength, one to remove harmonics and one to reduce the pulse repetition rate where necessary, to avoid frame overlap. Neutrons scattered by the sample then travel through a helium-filled 4 m flight path to banks of ^{3}He detectors covering scattering angles from 1° to 140°. Monitors in the direct beam at measured distances apart allow precise determination of E_i.

The number of neutrons arriving in each detector as a function of time gives directly the double differential cross-section as a function of time of flight τ. This must be related then to the more usual energy cross-section by

$$\frac{\partial^2 \sigma}{\partial \Omega \partial E_f} = \frac{-\tau^3}{m} \frac{\partial^2 \sigma}{\partial \Omega \partial \tau} \tag{11}$$

Around the elastic peak the τ^3 term does not vary very fast, but across the inelastic spectrum this correction has an important effect on relative intensities of different parts of the spectrum.

Figure 3 shows the time of flight spectra for two polymers compared to the corresponding scattering laws.

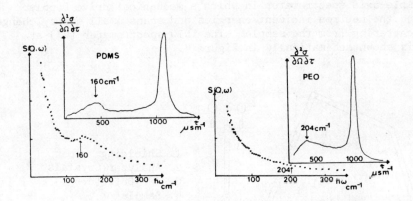

Fig. 3 - Time of flight spectra and corresponding scattering laws for polydimethylsiloxane (PDMS) and polyethylene oxide (PEO)

For PDMS the inelastic peak at 160 cm^{-1} is a genuine feature arising from the torsional vibrations of the methyl groups. The apparent peak at 204 cm^{-1} in the time of flight spectrum of PEO disappears in the scattering law however. It is purely a result of the competition between the τ^3 term and the exponential Boltzmann factor for occupation of energy levels, and does not correspond to any true feature in the spectrum. The energy resolution of such time of flight spectrometers depends both on the rotation rate of the choppers and the value of E_i chosen, but the best value of $\Delta E/E$ for IN5 is 1%. In terms of quasielastic scattering, this leads to a best resolution for 10 Å incident neutrons of around 20 μeV ($\omega \approx 3 \times 10^{10}$ Hz).

2.4   The Back-Scattering Spectrometer

The 10^{10} Hz resolution limit of IN5 represents the best that is likely to be achieved with the time of flight technique, mainly for reasons of flux limitation. It is normally possible to detect motion about one order of magnitude in energy less than the resolution of the spectrometer - thus IN5 allows detection down to ~10^9 Hz. However, the interesting range for polymer backbone motion has been even lower in energy. In order to achieve better resolution the backscattering spectrometer sacrifices the wide window in energy allowed on other spectrometers and concentrates the available flux in a narrow range observed with very high resolution.

The back-scattering technique uses a crystal monochromator set at a 180° Bragg angle, which gives an extremely sharp energy spectrum (~ 1 μeV or 10^9 Hz). Its basic construction is that of a triple-axis spectrometer in which a mechanical drive Doppler shifts the neutron incident energies and scans small energy changes on scattering from the sample. The IN10 spectrometer (23) at ILL is shown schematically in Figure 4.

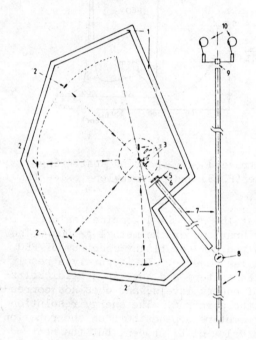

1. Shielding
2. Analyzer Crystals
3. ^{3}He Detectors
4. Sample
5. Monitor
6. Auxiliary Chopper
7. Neutron Guide
8. Graphite Crystal
9. Monochromator Crystal
10. Doppler Drive

Fig. 4 - The back-scattering spectrometer IN10 at ILL, Grenoble

Neutrons arrive in a neutron guide from the reactor and are incident on a monochromator crystal. This is silicon, oriented so that the 111 planes Bragg reflect 6.2 Å neutrons at 180°. Because of this 180° reflection angle, θ, the wavelength spread which is proportional to $\cot(\theta/2)d\theta$ is very narrow. This 6.2 Å beam travels back along its path to a graphite crystal and is there reflected at 45° through a chopper (the function of which will be explained later) to a sample. Neutrons scattered by the sample are incident on silicon crystal analysers oriented with the 111 planes again in back reflection for 6.2 Å neutrons. The analysers

are on curved mountings focussed on detectors just behind the
sample position. If the sample scatters inelastically, the intensity counted will be diminished since some neutrons no longer
reach the analysers with the necessary 6.2 Å wavelength. However,
the monochromator crystal is mounted on a Doppler drive which
imparts a small shift in energy to the 6.2 Å reflected neutrons.
Only if they lose or gain to the sample the energy they have
gained or lost at the Doppler drive, so that they again have
exactly 6.2 Å wavelength, will the neutrons be back-reflected at
the analysers and reach the detector.

(The auxiliary chopper coarsely pulses the neutrons and
time of flight analysis allows discrimination against neutrons
scattered directly into the detectors). In this way, a small
energy scan (± ∿12 μeV) can be made with a very sharp (1 μeV)
resolution. As can be seen, in this case resolution has been won
at the expense of the width of the energy window.

Other limitations concern the Q-range and resolution. In
order to increase the counting flux, the focussing analyser crystal banks are made relatively large, relaxing the angular resolution, and giving a ΔQ/Q of order 10%. These analysers can, of
course, be masked to give better Q resolution but this is only
done for special circumstances because of the severe flux penal-
ties. For similar reasons the lowest angles of scatter used are
limited and the smallest Q value is 0.07 Å^{-1}.

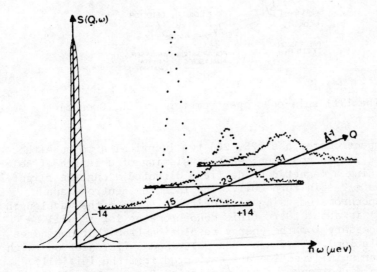

Fig. 5 - Quasielastic peaks obtained on IN10 from a 3% solution
of polytetrahydrofuran (d8) in CS_2 at room temperature. The curve
at zero Q is the resolution function of the apparatus.

Even so, there is a danger of some interference at the lowest Q
values, and $\Delta Q/Q$ becomes very large. Within these limitations
the apparatus has been very successfully used in observation of
tunnelling splittings of order a few μeV and of quasielastic
scattering arising from motion of about 10^9-10^{10} Hz in simple
liquids, macromolecules and liquid crystals. Figure 5 shows the
quasielastic scattering at low Q from a polymer solution.

2.5 The Spin-Echo Spectrometer

This technique originally devised by F. Mezei (24) uses the
precession of the neutron spin in a magnetic guide field as a
counter to measure very small changes in neutron velocity. In
order to explain the method, it is necessary to follow the path
of neutrons through the apparatus, which is shown schematically
in Figure 6.

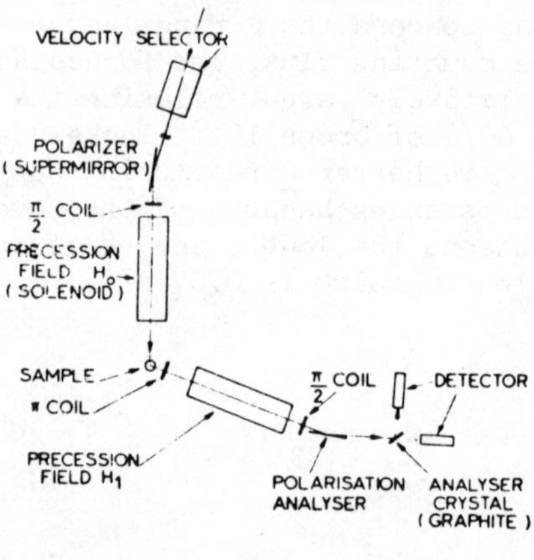

Fig. 6 - The IN11 spin-echo spectrometer at ILL, Grenoble

The spectrometer has (25, 26) two identical arms on either
side of the sample position, each consisting of a length of sole-
noid providing a magnetic guide field directed along the flight
path, and a π/2 spin turn coil. The incoming neutrons are
roughly monochromated using a velocity selector - the wavelength
spread $\Delta\lambda/\lambda$ is about 10%. (This monochromator is, as will be
seen, unnecessary for the _energy_ resolution (26b) but dominates
the Q-resolution). They are then polarised - i.e. neutrons with
a particular spin direction are extracted from the initially
random spin directions, by the reflection from a specially coated
surface - a "super mirror". After the polarizer the neutron
spins are aligned along the flight path. The π/2 spin turn coil
rotates the neutron spin direction which starts precessing about

the guide field. The precession rate of each neutron is the
same - given by the Larmor precession frequency ($= \omega_L = 2\mu n B_o/\hbar$
where μ_N is the magnetic moment and B_o the guide field strength).
However, the number of precessions will be governed by the length
of time taken to traverse the guide field, i.e. the neutron velocity. A beam containing a wavelength spread loses its initial
phase coherence as it travels through the guide field and in
analogy with NMR the spins will have fanned out with respect to
each other. In the π turn coil, located at the sample, the
neutrons make a $180°$ precession about a field perpendicular to
the guide field ($H^\perp$). This has the effect of reflecting the
distribution in the $H^\perp$ axis. Those spins that were furthest
behind in the fan of precession angles are now furthest ahead and
vice versa. Thus, if the length of the second guide field is
adjusted to be identical to the first, all the spins will once
more be aligned at the second $\pi/2$ turn coil and 100% polarisation
in the flight path direction will be observed at the analyser.

To observe quasielastic scattering, the number of precessions
in each guide field is set equal for elastic events so that small
changes in energy at the sample result in a neutron performing
unequal precessions in the two fields and produce a reduction in
the polarisation received at the analyser. The net polarisation
$\langle P_Z \rangle$ is then given by

$$\langle P_Z \rangle = \int_0^\infty I(\lambda)d\lambda \int_{-\infty}^{+\infty} P(\lambda, \delta\lambda)\cos \frac{2\pi N_o \delta\lambda}{\lambda_o} d(\delta\lambda) \qquad (12)$$

where $I(\lambda)d\lambda$ is the initial wavelength spread and $P(\lambda, \delta\lambda)$ is the
probability that a neutron of wavelength λ will be scattered with
a wavelength change $\delta\lambda$. Since $\delta\lambda$ can be transformed into an
energy change $\delta E = \hbar\omega$ this is, of course, just the scattering
law $S(Q, \omega)$

$$\int P(\lambda, \delta\lambda)d(\delta\lambda) = \int S(Q, \omega)d\omega \qquad (13)$$

N_o is the number of precessions made by a neutron of the mean
wavelength λ_o. Now since

$$E = \frac{\hbar^2}{2m} \frac{1}{\lambda^2} \qquad (14)$$

$$\hbar\omega = \delta E = \frac{\hbar^2}{m} \frac{\delta\lambda}{\lambda^3} \text{ or } \delta\lambda = \frac{m\lambda^3}{2\pi\hbar} \omega$$

and
$$\langle P_Z \rangle = \int_0^\infty I(\lambda)d\lambda \int_{-\infty}^\infty S(Q, \omega)\cos \left\{ \left(\frac{N_o \ m\lambda^3}{\hbar\lambda_o} \right) \omega \right\} d\omega \qquad (15)$$

The expression $N_o m \lambda^3 / h \lambda_o$ has the dimensions of time, and can be designated $t(\lambda)$.

P_Z, then, is just the Fourier transform of $S(\underline{Q}, \omega)$, i.e. the time correlation function $S(\underline{Q}, t)$ is observed directly in spin-echo experiments. The experimental time scale $t(\lambda)$ is governed through N_o by the strength of the magnetic guide field. For neutrons of 8Å the time scale covers about two decades from 10^{-9} to 10^{-7} secs. For study of polymer dynamics, direct observation of $S(\underline{Q}, t)$ has considerable advantages as will appear below when analysis of the experimental data is considered. Figure 7 shows the correlation functions $S(\underline{Q}, t)$ obtained on IN11 for the same polymer solution as the data in Figure 5.

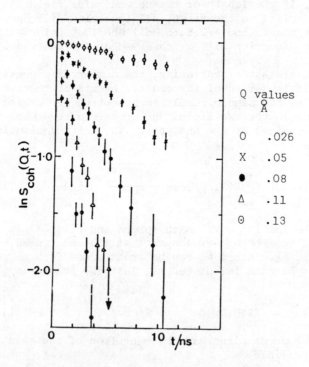

Fig. 7 - Correlation functions $S_{coh}(\underline{Q}, t)$ for the same polymer solution as the data in Figure 5

As with the back-scattering spectrometer, the very high resolution (<0.1 µeV, 10^{-8} Hz) is won at the expense of Q resolution. In this case, the angular definition of Q is very good, since the beam has to be highly collimated within the guide fields. $\Delta\lambda/\lambda$ is, however, quite broad ($\sim 10\%$) in order to increase flux and this imposes a 10% limit on $\Delta Q/Q$. On the other hand, the tight collimation does allow use of small ($\sim 1^{\circ}$) scattering angles and the smallest Q available is ~ 0.025 Å^{-1}.

One further important property of the spin-echo technique is its ability to distinguish coherent from incoherent scattering. Most incoherence arises from the neutron spin and since this machine explicitly follows the spin, it is possible to select only the coherently scattered neutrons. In cases where the coherent and incoherent scattering laws differ, this property is a great advantage. At low Q, also the coherent scattering law has high intensity for macromolecular systems conveniently increasing the signal to noise ratio.

3. INELASTIC SCATTERING FROM VIBRATIONAL MOTION

3.1 Model Scattering Laws

In the classical limit for a molecular vibration of frequency ω_o and amplitude $\underline{u}_o$, which gains or loses energy in a neutron scattering event,

$$\frac{d^2\sigma}{d\Omega dE_f} = \frac{k_f}{k_i} \sum_{n,m} <b_n b_m> \exp\left[i\,\underline{Q}.(\underline{n}-\underline{m})\right] \exp(-2W) \times$$

$$\left[\delta(\hbar\omega) + \frac{1}{2}(\underline{Q}.\underline{u}_o)^2\delta(\hbar\omega - \hbar\omega_o) + \frac{1}{2}(\underline{Q}.\underline{u}_o)^2\delta(\hbar\omega+\hbar\omega_o) + ..\right] \quad (16)$$

where $\underline{n}$ is the position vector of the nth nucleus and $\underline{u}_n$ its displacement. This classical limit shows some of the more important features of scattering from molecular vibrations. The δ functions define an elastic component $\hbar\omega = 0$ and two inelastic components $\hbar\omega = \pm\hbar\omega_o$ corresponding to neutrons gaining or losing a quantum of vibrational energy (the higher terms in the series correspond to multi-quantum effects). The intensity of the inelastic peaks is proportional to $(\underline{Q}.\underline{u}_o)^2$, i.e. it increases with Q and with the amplitude of the vibration observed. In the simplified form of equation (16), both the elastic and inelastic components are attenuated by the term $\exp(-2W)$. For real situations the inelastic factor may be more complex and energy dependent. This attenuation arises because interference effects between atoms at different sites are smoothed out by thermal vibrations. In equation (16) $\exp(-2W) = \exp(-(W_n+W_m))$, where $W_n = \frac{1}{2}<(\underline{Q}.\underline{u}_n)^2>$. The single quantum inelastic scattering in equation (16) can be manipulated to show explicitly the coherent and incoherent terms

$$\frac{d^2\sigma(1)}{d\Omega dE_f} = \frac{k_f}{k_i} \sum_{n,m} <b>^2 \exp(i\underline{Q}.(\underline{n}-\underline{m}))\frac{(\underline{Q}.\underline{u}_o)^2}{2} (\delta(\hbar\omega-\hbar\omega_o) +$$

$$+ \ \delta(\hbar\omega + \hbar\omega_o)) + \frac{k_f}{k_i} \ (<b^2> - <b>^2)N(\underline{\frac{Q \cdot \underline{u}_o)^2}{2}} \ (\delta(\hbar\omega - \hbar\omega_o) + \delta(\hbar\omega + \hbar\omega_o))$$

$$(17)$$

The first term is the coherent scattering and contains a phase factor relating atoms at $\underline{n}$ and $\underline{m}$. This imposes conditions on the $\underline{Q}$ values at which the scattering can take place - conditions which are absent for the second, incoherent, term.

Note that because of the term $\underline{Q} \cdot \underline{u}_o$ both coherent and incoherent scattering from oriented samples will show direction properties with intensity at a maximum when $\underline{Q}$ is parallel to $\underline{u}_o$ and zero when $\underline{Q}$ is perpendicular to $\underline{u}_o$. In unoriented samples averaging over all molecules removes this condition.

In extended arrays of ordered molecules (or atoms) the intermolecular motions of the weakly coupled molecules constitute sets or branches of normal modes which may be represented by travelling waves of phonons. The frequencies of these modes $\omega(\underline{q})$ are characterised by a wave vector $\underline{q}$, and the complex motions reflect the symmetry and periodic properties of the lattice. Because the motion in adjacent unit cells is correlated, the spatially sensitive coherent neutron scattering cross-section is especially useful for observation of phonon modes.

Since both energy and wave vector or momentum are conserved in this scattering process, sharp peaks are observed when either the energy transfer ω or momentum transfer $\underline{Q}$ is scanned and the two conservation conditions are satisfied

$$\underline{Q} \ = \ \underline{g} + \underline{q}$$

$$\omega(\underline{q}) \ = \ E_f - E_i$$

where $\underline{g}$ is a reciprocal lattice vector (the mode frequencies are periodic with the lattice periodicity).

For N molecules there are 3N-6 intermolecular modes possible; considering M molecules per unit cell, these lattice modes are distributed into 6M branches (3 translational and 3 librational). Along high symmetry directions in the crystal a clear distinction may be made between modes parallel (longitudinal) and transverse (displacements $\underline{u}$ perpendicular) to the direction of propagation $\underline{g}$. Analysis of the dispersion of the phonon energy $\omega(\underline{q})$ with $\underline{q}$ leads to detailed information on the intermolecular potential.

The incoherent scattering from a crystal contains much less detailed information. In particular, the δ function in $\underline{Q}$ disappears and with it much directional information. All possible energy transfers will be observed at any given value of Q. Assuming one atom in a cubic unit cell, the incoherent term in (16) becomes

$$\frac{d^2\sigma^{(1)}}{d\Omega dE_f} = N \frac{k_f}{k_i} (<b^2>-<b>^2) \frac{Q^2<u^2>}{2M} \frac{\exp(-2W)}{\sinh(\hbar\omega)/kT} \frac{Z(\omega)}{\omega} \quad (18)$$

where $Z(\omega)$ is the density of phonon states

$$Z(\omega) = \frac{\hbar\omega}{3N} \sum_{q j} \frac{(\overline{\hbar\omega+\hbar\omega_j(q)})}{\omega_j(q)} \quad (19)$$

$<u^2>$ is the mean square amplitude of vibration and M the mass of the vibrating atom. The sinh term in the denominator arises from the Bose-Einstein population factor for the energy levels.

The scattering law $S_S(Q, \omega)$ is obtained from $d^2\sigma/d\Omega dE_f$ by

$$S_S(Q, \omega) = \frac{k_i}{k_f} \frac{\hbar}{N(<b^2>-<b>^2)} \frac{d^2\sigma}{d\Omega dE_f} \quad (20)$$

For comparison with optical spectra obtained at zero Q the experimental scattering laws are sometimes extrapolated to obtain the zero Q amplitude weighted density of states (7, 27) $g(\omega)$ (rather than $Z(\omega) \simeq g(\omega)/<u^2>$).

$$g(\omega) = \omega \sinh \frac{\hbar\omega}{2kT} \lim_{Q\to 0} \left[\frac{S_S(Q, \omega)}{Q^2} \right] \quad (21)$$

The thermal effects of the population factor have been divided out and the Q variation in the $Q^2<u^2>$ term and the Debye Waller factor removed by the extrapolation to zero Q.

It is interesting to note that extraction of $S(Q, \omega)$ from $\partial^2\sigma/\partial\Omega\partial E_f$ (see equations (6-9)) introduces a factor k/k_o which is numerically equal to τ/τ_o so that $S(Q, \omega)$ is effectively multiplied by τ^4 in the time of flight spectrum.

When the density of states is required, as in equation (21), $S(Q, \omega)$ has to be multiplied by $\omega \sinh \hbar\omega/kT$. This factor varies approximately as ω^2 over a limited energy range. Now $\hbar\omega = \Delta E$ which will vary over a limited range roughly as E_f^2 - i.e. as $1/\tau^4$, so the two factors cancel and the time of flight spectrum is often very close in shape to the density of states finally required.

3.2 Torsional Vibrations of Side Groups on Polymer Molecules

Examination of equations (16) and (17) shows that large
amplitude vibrational modes involving nuclei with large cross-
sections will give intense peaks in the inelastic neutron spectra.
The incoherent scattering from hydrogen is an order of magnitude
larger than that from other nuclei (see Table 1). Torsional
vibrations of methyl or phenyl side-chains of polymers with their
large amplitude are, therefore, good candidates for neutron investigation, especially since these bands are often weak and difficult
to identify from infra-red and Raman spectra. Deuteration of the
side group where this is possible reduces the cross-section by a
factor of around twenty and the peak drops out of the spectra
confirming the assignment. Methyl torsion motion has been investigated for polypropylene oxide (28), polymethyl methacrylate (29),
polypropylene (2, 30-32), polydimethyl siloxane (29, 33, 34),
poly-1-alanine (35), polyacetaldehyde (36), poly 4-methyl pentene-
1 (29), polyisobutene (29) and poly vinyl methyl ether (29), and
a brief investigation of the phenyl torsion in polystyrene is mentioned in reference (30). In many of these examples, partial
deuteration of the molecule has allowed an unambiguous assignment
of the torsional mode.

In the case of polydimethyl siloxane (see Fig. 3) deuteration would not help assign the only clearly identifiable peak in
the spectrum of this polymer melt. However, the methyl torsion
is the only low frequency mode expected and discussion (29, 33,
34) has centred not on the assignment but on the relative freedom
of the rotation. Estimates of the barrier to rotation have ranged
from less than 1.6 kcal mol^{-1} by Amaral et al. (34) using total
scattering measurements to 6.8 kcal mol^{-1} from the torsional
frequency (29). The peak is very broad due partly to the low
barrier ($\sim$ 3 kT , thus allowing transitions between many of the
levels and broadening from anharmonicity) and partly to the other
motions of the molecules in the melt which will broaden both the
elastic and inelastic δ-functions in equation (16). Recent NMR
spin lattice relaxation measurements (37) arrive at an activation
energy for methyl rotation in PDMS of 6.7 kcal mol^{-1}.

3.3 Crystalline Samples - Phonon Modes

The observation of travelling waves in polymer crystals offers
the attractive possibility of obtaining both the inter- and infra-
chain force constants. The drawback is the near impossibility of
obtaining single crystals of polymers. Most polymers are only
semi-crystalline with small crystallites embedded in an amorphous
matrix. The scattering from such an unoriented sample essentially
averages out much of the directional information in equation (17).
However, in the coherent scattering, for each value of ω there is
a single allowed $\underline{Q}$ value for a given mode and the ω-$\underline{Q}$ pairs must

obey the dispersion relationship. Thus the dispersion curve of a
particular vibrational mode may be picked out from a series of
time of flight spectra at different scattering angles or followed
on a triple-axis spectrometer by tracing the way a peak moves in
ω-Q space. The coherent scattering from polytetrafluorethylene
(38) and deuterated polyethylene (39) is strong enough to allow
the acoustic modes to be analysed in this way and compared with
calculated dispersion curves. For hydrogenous polymers, however,
the incoherent scattering dominates and only the H amplitude
weighted density of states, g(ω) is observed (2, 40). This may
be compared with a calculated curve (see equations (18) and (19))
but such comparison does not provide a very stringent or direct
test of the force constants used (40, 41). If the polymer samples
are stretch-oriented, then some of the directional information is
recovered.

Figure 8 shows the density of states g(ω) for a stretch-
oriented polyethylene determined using a triple-axis spectrometer
so that in one case, the Q vector was oriented always along the
chain axis (longitudinal) and in the second, Q was perpendicular
to this axis (transverse) (3).

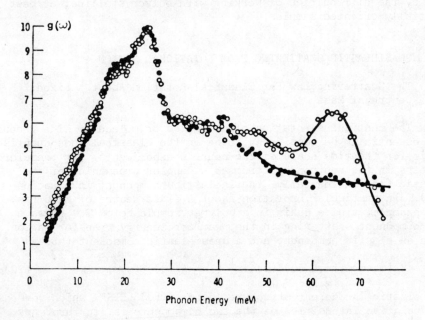

Fig. 8 - Amplitude weighted directional frequency spectra for a
stretch-oriented sample of polyethylene at 100 K reprinted with
permission from reference (3).
O - longitudinal; ● - transverse g(ω)

The two curves have been normalised at 190 cm^{-1}. The two main peaks at 525 and 190 cm^{-1} have been assigned respectively to the longitudinal stretch bend (accordion) mode of the C-C-C skeleton and to the out-of-plane torsion of the methylene groups about the C-C bond. The disappearance of the 525 cm^{-1} band from the Q$^{\perp}$ spectrum confirmed its assignment to the longitudinal mode - in this configuration, there is no way for the neutron to excite the motion. The fact that the transverse mode shows up in both spectra (though without the normalisation mentioned, it is, in fact, reduced in intensity by a factor two in the Q" spectrum) is caused by crystalline field-mixing of the modes.

Stretch-orientation of coherently scattering samples also helps the measurement of dispersion curves and these have been compared with calculations for deuterated polyethylene (42) and PTFE (43), both stretch-oriented. In only one case (44), a large single crystal was grown (of polyoxymethylene) and a number of dispersion curves reported. Where it has been possible to measure the dispersion of the interchain modes, the force constants obtained provide unique information specific to the neutron technique. However, as can be seen, the amount of experimental data is limited by the difficulties of working with polycrystalline, at best partially-oriented samples.

4.   QUASIELASTIC SCATTERING FROM ROTATIONAL MOTION

4.1   The Scattering Law For Classical Rotation About a Fixed
      Centre of Mass

The continuous rotation of molecules or molecular side groups can essentially be considered to be in the classical (non-quantised) limit for the time scale of the neutron experiments. It therefore appears in the neutron spectrum as a Doppler broadening of the elastic (and incidentally, the inelastic) component in equation (16). In particular, rotation about a fixed centre of mass (as, for example, a side chain in a polymer sample below T_g) gives a two-component scattering in the near zero energy transfer region with an elastic component and a quasielastic component (45-47)

$$S_S(\underline{Q}, \omega) = A_o(Q)\delta(\omega) + \frac{1}{\pi} (1 - A_o(Q))F(\omega) \qquad (22)$$

The elastic incoherent structure factor $A_o(Q)$ (EISF) which governs the relative intensities of the two components is, in turn, governed by the shape swept out by the rotating nuclei. For a hydrogen nucleus in a methyl group rotating about a fixed axis between three equivalent sites (45)

$$A_o(Q) = \frac{1}{3} (1 + 2j_o(Qa)) \qquad (23)$$

where a is the radius of gyration.

The broadened component is a Lorentzian in form

$$F(\omega) = \frac{2\tau_r/3}{1 + \omega^2(2\tau_r/3)^2} \tag{24}$$

with a full width at half maximum $4\tau_r/3$, where τ_r is the time between jumps. For more complex motion, various forms for $A_o(Q)$ and $F(\omega)$ have been derived. The general principle of a broadened component governed by the rotational frequency is maintained. $A_o(Q)$ is essentially the spatial Fourier transform of the shape swept out around their fixed centre of gravity by the scattering nuclei. A large rotational volume Fourier transforms to give an $A_o(Q)$ peaked close to zero Q and in the limit as the volume tends to infinity (i.e. we have no fixed centre of mass and the sample becomes liquid) the elastic component disappears at finite Q.

4.2 Rotational Motion of Polymer Side Chains

Separation of the elastic from the quasielastic components in equation (22) depends crucially on the incident energy spread, (the resolution of the apparatus) and on the shape of the resolution function. To some extent, since the rotational motion is an activated process, its frequency can be adjusted to match the available resolution by changing the sample temperature. However, as a polymeric sample is heated above T_g motion of the main chain will broaden the elastic component (see Section 5), and make separation that much more difficult. For this reason it is preferable to observe side chain rotations in samples at around or below their glass transition temperatures. Although there has been extensive study of quasielastic scattering from rotational motion in plastic and liquid crystals (48), there has, to date, been only limited work on polymeric systems.

Polypropylene oxide (49) and very recently, polymethylmethacrylate (50, 51) have been subjected to fairly extensive investigation. The high glass transition temperature of PMMA gives a wider temperature range for studying the methyl rotation while the backbone motion is frozen out. Moreover, there are two methyl groups in the monomer unit with very different torsional frequencies and some evidence that the crystal structure strongly affects at least one of these torsional barriers (29, 52-54). The normal preparation of PMMA produces predominantly syndiotactic sequences and for these samples, the α-methyl torsional band occurs at about 350 cm^{-1} yielding a barrier $V_3 \sim 34$ kJ mol^{-1}. In earlier work, this band was assigned by comparing the α chloro analogue with PMMA itself (29), but deuteration of this group has recently reconfirmed this assignment (50, 51). In a predominantly isotactic

sample this barrier is reduced to 23 kJ mol^{-1}. The ester methyl
group has a torsional band at rather lower energy (around 100
cm^{-1}) leading to a value for V_3 of between 3 and 10 kJ mol^{-1}.
The band is difficult to assign with more precision because it
occurs in a region (25) of other backbone motions which are also
reduced in intensity in the neutron spectra after deuteration of
the methyl group (29, 50, 51). From the inelastic spectra, it is
not possible to determine the effect of tacticity on this barrier.
These barriers are different enough so that even without the aid
of specific deuteration, it is possible to separate the effects
of their rotational motion in the quasielastic spectra for samples
below T_g. At room temperature, the α-methyl group is rotating
at less than 10^9 Hz while the ester methyl moves much faster at
10^{11}Hz. Specific deuteration reduces the elastic scattering from
the other hydrogen nuclei (immobile in the glass on this time
scale, apart from the high frequency vibrations which average out).

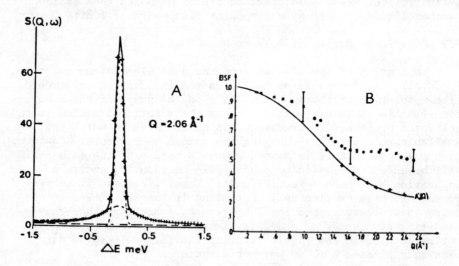

Fig. 9 - S($\underline{Q}$, ω) for ester group in PMMA at 25°C and corresponding
EISF as a function of Q.

Figure 9a shows the quasielastic scattering from a sample of predominantly syndiotactic PMMA in which only the three hydrogens in
the ester methyl group have not been replaced by deuterium. (The
data are obtained from the time of flight spectrometer, IN5).
The peak is divided into a central component with the same shape
as the triangular resolution function, and a Lorentzian broadened
component - in the form given by equation (22). The experimental
A_o(Q) or EISF for the ester methyl is compared in Fig. 9b to the
calculation for a rotating methyl group (solid curve A'(Q)). This

calculation has been adjusted for the residual incoherent scattering, and the discrepancy between experiment and calculation in Fig. 9b arises from coherent scattering.

The X-ray scattering from this polymer shows (55) peaks at around $Q = 0.8 \, Å^{-1}$ and $2.2 \, Å^{-1}$. The neutron diffraction from the same sample as that used for the rotational measurements also showed peaks around these values at the points of maximum deviation of the data from the calculated $A'_0(Q)$. Correction of the data for this coherent elastic contribution brings the data into excellent agreement with the calculations as shown by the crosses in Fig. 9. Similar experiments have been carried out for the α-methyl rotation in a predominantly syndiotactic sample. In both cases, the rotational frequency was obtained as a function of temperature allowing an activation energy E_a to be extracted. If effects of quantum mechanical tunnelling are taken into account (52, 56) then E_a may be compared to values of V_3 estimated from the torsional frequencies, as well as to the NMR activation energies. Table 2 summarises all the values at present available.

Sample	E_a neutrons (V_3 tunnelling)	E_a NMR (V_3 tunnelling)	V_3 neutron inelastic measurements
α-methyl: predominantly syndiotactic	29.1 (31.1)		33
α-methyl: pure sydiotactic		23.3 (30)	33
α-methyl: pure isotactic		15.8 (22.6)	23
ester methyl: predominantly syndiotactic	7.6 (8.8)	10.1	3.2 to 10.5

TABLE 2

Barriers to Rotation in kJ mol^{-1} for PMMA

Completion of the neutron experiments will eventually allow values of E_a to be assigned for both groups in both stereo isomers with much better precision than that allowed by calculation from the rather poorly resolved torsional frequencies (and better than NMR estimates for the very low ester methyl barrier). Preliminary data (50, 51) indicate that the ester barrier also may be affected by tacticity.

5.   QUASIELASTIC SCATTERING FROM MAIN CHAIN MOTION

5.1 Scattering Laws and Correlation Functions

In liquids the elastic component in equations (16) and (22) are completely replaced by Doppler broadened quasielastic peaks. Although a polymer solution or melt is a true liquid in the sense that there is no fixed centre of mass, the experimental Q-value defines whether the motion of all or only part of the molecule is observed. The exponents $(Q.R)$ in equations (5)-(7) ensure that the scattering from self- or pair-correlations over a distance R will be maximum when $Q = 1/R$ (or multiples of 1/R for a periodic array such as a crystal). The Q-value will thus 'pick out' different scales of molecular motion.

Figures 5 and 7 show typical quasielastic data from the high resolution back-scattering and spin-echo spectrometers respectively. Figure 5 shows quite clearly that the width of the instrument resolution function is not negligible when compared to the measured scattering laws. Deconvolution procedures are unreliable under these circumstances and it is the usual procedure to choose a model scattering function, convolute it with the instrument resolution and compare with the data. The Fourier transform of a convolution is a product, so the time correlation functions are directly obtained from the data by a simple division by the resolution function (obtained for both techniques by measuring the scattering from an elastic scatterer).

Complete model laws have been calculated for simple polymer motion in some cases. These are usually characterised by the Q-dependence of a width function in the case of $S(Q, \omega)$ or an inverse correlation time for $S(Q, t)$. The width function is the half-width at half maximum of $S(Q, \omega)$ called $\Delta\omega$. The extraction of this parameter for data such as that in Fig. 5 requires a knowledge of the shape of the scattering laws in order to remove the instrumental resolution. The inverse correlation time may be the slope Γ, of the logarithmic plot of $S(Q, t)$ if a simple exponential decay is expected, or the initial slope of this plot, Ω, when a more complex form results.

$$\Omega = \lim_{t \to 0} \frac{d}{dt} (\ln S(Q, t)) \tag{25}$$

This latter quantity, Ω, may be extracted even if the shape of $S(Q, t)$ is not known exactly and this shows the particular importance of the spin-echo technique results which allow $S(Q, t)$ to be obtained directly.

As already mentioned, the Q-value will determine the type of motion observed in a given experiment. There has recently been considerable success in describing the motion of dilute solutions of macromolecules over the whole range of distances from the large scale (small Q) of centre of mass diffusion through to high Q, local segmental motion. Based on the Rouse (57) bead-spring model, the theoretical developments of Pecora (58), de Gennes (59, 61) and, more recently, Akcasu et al. (61-63) fit very well both light scattering (61) and neutron scattering (64) results. Motion in melts and networks is more complex and only neutron scattering offers the possibility of labelling and observing single molecules. Analytical calculations of scattering laws can only be made for a number of simplified cases.

At high Q local segmental motion might be expected to resemble Brownian motion. The scattering law for such motion is a simple Lorentzian and the corresponding correlation functions shows an exponential decay with

$$\Delta\omega \equiv \Gamma \equiv \Omega = DQ^2 \tag{26}$$

where D is a diffusion coefficient. Recent computer calculations for a "real" polyethylene chain (65) tend to confirm a Q-dependence of this type at higher Q-values. At smaller Q, the internal motion of the molecule is observed as it constantly changes its conformation. This continually changing conformation causes fluctuation in the distances between points on the chain, and the entropy gain as the length of any segment increases from its equilibrium value causes that segment to act as a spring under tension. The model developed by Rouse (57) represents the polymer molecule as a series of such springs of length 'a' connected by beads. The effect of the surrounding solvent molecules is described by a frictional drag on the beads. Zimm (66) modified the model to take account of hydrodynamic coupling between the beads existing in many real situations. Despite the loss of any chemical structure for Rouse chains (the only variable parameter is the spring length 'a') these models have been extremely successful in describing viscoelastic behaviour of polymer solutions excepting only the highest frequencies where local vibrations and rotations are dominant and the chemical structure can no longer be ignored.

Pecora (58) and de Gennes (59, 60) calculated the intermediate scattering laws for Rouse chains with and without hydrodynamic coupling in the region $R_g^{-1} \ll Q \ll a^{-1}$. Both the coherent and the incoherent scattering show an unusual dependence on time.

For a melt, hydrodynamic effects are expected to be screened out, and for this case, the time dependence is $t^{\frac{1}{2}}$ and   .

$$\Delta\omega \propto \Omega \propto \frac{k_B T}{\xi_o} Q^4 \tag{27}$$

where ξ_o is a friction coefficient per bead. At longer distances (small Q) the entanglement effects of the surrounding polymer molecules will be observed.

Recent discussions describe the motion of polymer molecules in a melt in terms of reptation (67, 68) along a 'tube' formed by the effect of the surrounding chains. Graessley (69) has recently calculated the diameter of this tube from experimentally measured quantities for several polymers and found that it is surprisingly large - of the order of 35 Å for polyethylene. It would not be expected, therefore, that entanglement effects would show up over the distance scales explored in the neutron experiments, for which the lowest Q-values are of order .025 Å^{-1}.

The calculations indicate that if this Q-range could be attained, an extremely fast Q^6 dependence would be observed for $\Delta\omega$ and Ω. Two lengths are thus important, a, the Rouse spring and ℓ, the distance between entanglements, and the motion of polymer molecules in the melt is expected to be characterised by three Q-ranges. For $Q > a^{-1}$, local segmental motion will dominate with, possibly, a Q^2 dependence of Ω and $\Delta\omega$; for $a^{-1} > Q > \ell^{-1}$, Rouse modes should be observed with a Q^4 dependence and for $Q < \ell^{-1}$, reptation may be expected with a possible Q^6 dependence.

5.2 Neutron Scattering From Melts and Networks

Figure 10 shows the values of Ω extracted from measurements of $S_{coh}(Q, t)$ made using the IN11 spectrometer for samples of deuterated polytetrahydrofuran containing a small percentage of the hydrogenous polymer (70). The difference in scattering length for H and D gives a large coherent scattering pattern from the differentiated chains (71) so that the dynamics of single molecules in a melt environment can be observed. Also shown in Fig. 10 are the results for the same polymer in dilute solution in CS_2. The data in Fig. 10 have been extended to higher Q by including results for $\Delta\omega$ obtained from measurements on the backscattering spectrometer which does not allow the same separation of coherent and incoherent scattering. Allowance has been made for the inevitable contamination of $S(Q, \omega)$ by incoherent scattering at higher Q where the coherent small angle signal no longer dominates. In a melt, provided the distances explored do not include entanglements, pure Rouse behaviour would be expected to give a Q^4 dependence of Ω as shown in equation (26). When hydrodynamic effects are important (60) a Q^3 dependence should result. Ω shows a faster Q dependence of the bulk data than the solution data in Fig. 10, though neither of the slopes are at the limiting values of Q^3 and Q^4. This is because the distances explored are short (of order 1 or 2 Å) compared to the Rouse spring length and local segmental motion is becoming important.

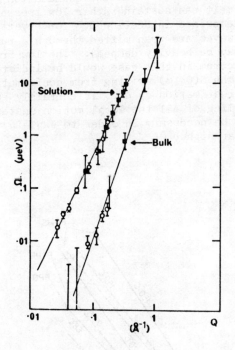

Fig. 10 - Ω against Q for PTDF in 3% solution in CS_2 and for 3% PTHF in PTDF. Reprinted with permission from reference (70).

Results of experiments measuring $S(\underline{Q}, \omega)$ for the incoherent scattering from a number of polymer melts using the back-scattering spectrometer are summarised in Fig. 11 (72). On this double logarithmic plot, the slope is the power of Q variation. For polydimethyl siloxane (73) this is approximately Q^4, but for all the other polymers it is somewhat less, and for polyisobutylene at 73°C, it is not much faster than Q^2. Polypropylene oxide, polytetrahydrofuran and polyisobutylene all have glass transition temperatures around 200 K. The very much slower motion and stiffer unit (if we interpret a low power Q variation as implying $Q > a^{-1}$ and therefore a long spring unit 'a') for polyisobutylene is somewhat surprising if the glass transition is interpreted in terms of the freezing out of local segmental motion. Polydimethyl-

siloxane does have a much lower T_g than the other polymers, so its
faster motion and more flexible structure (in terms of a short
'a' value) fits this interpretation. The Q range in Fig. 11
extends much higher than that in Fig. 10 and the persistance of
power laws greater than Q^2 to such values is somewhat puzzling,
since it is clearly unrealistic to imagine values of 'a' less
than 2 or 3 Å. At present it seems the explanation must lie in
the fact that the melt can sustain rather low frequency transmitted
modes like those described in Section 3.3 for crystalline samples.
(After all, sound waves are transmitted through a rubber or liquid
even though they may be heavily damped). The low frequency density of states spectrum in this case would be indistinguishable
from the wings of the $S(Q, \omega)$ arising from conformational modes
and lead to an overestimation of $\Delta\omega$. It certainly appears that
more careful modelling of all the local motion sustained by the
melt polymers will be necessary in order to explain the behaviour
of $S(\underline{Q}, \omega)$ and $\Delta\omega$ at higher Q.

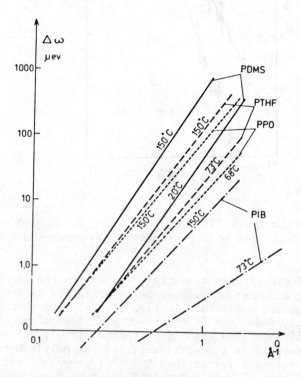

Fig. 11 - $\Delta\omega$ plotted logarithmically against Q for 4 polymers
each at two temperatures.

When a polymer melt is cross-linked to form a rubber, the
macroscopic motion of the molecules is removed and a three-
dimensional network is produced. It is by no means clear that on
the local scale observed in the neutron experiments the same
modifications take place. In fact, experiments on model networks
(74) showed no effects of cross-linking until the links are relatively close (~ 30 monomer units) together. The question then
arises as to whether the resultant slower motion observed as
narrower $S(\underline{Q}, \omega)$ curves is a slower motion of the whole molecule,
or a superposition of slower motion of the junctions onto an
effectively unaltered Rouse spectrum. Comparison with a sample
in which the sections of chains away from junction has been deuterated showed that it is, in fact, just the section around the
junctions which are slowed down (75). Figure 12 shows data plotted
as $\Delta\omega/Q^2$ against Q for such model networks at 70°C.

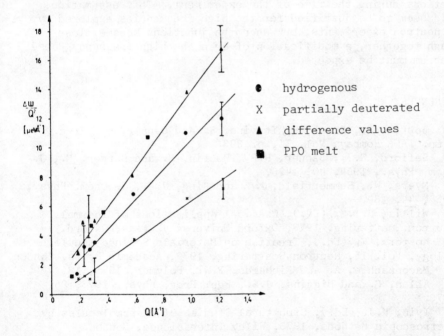

Fig. 12 - $\Delta\omega/Q^2$ against Q for model networks of polypropylene
oxide at 70°C

The curve marked '●' is for a fully hydrogenous model tri-func-
tional network of polypropylene oxide with 30 monomer units between
each junction. The points marked 'x' are data for an identical
network which has been deuterated except for a few units around

each cross-link. Finally, the upper set of data marked '▲' is
the value of Δω from an appropriately weighted difference between the two previous sets of S(Q, ω) data, and therefore represents the motion of the free chain sections away from junctions.
This is indistinguishable from the motion of uncross-linked polypropylene oxide melt at the same temperature indicated by the '■'
symbols in Fig. 12. Calculations (76) of the effect of crosslinking on the Rouse motion of a polymer chain indicate that the
Rouse spectrum is unaltered away from the cross-links and that the
cross-links should reduce their frequency by a factor of 2/N
where N is the functionality of the network. The reduction factor
for the trifunctional networks in Fig. 12 is approximately $\frac{1}{2}$
rather than the 2/3 predicted, but this discrepancy is probably
not surprising since the neutron experiments only explore the
limits of the Rouse model. The calculation assumes that the cross-
link points are infinitely far apart, which corresponds to an
assumption that a given Rouse mode cannot propagate between the
junctions during the time of the experiment. This assumption
would seem to be justified for the high frequencies explored by
the neutron experiments, but when the junctions become close
enough together, a modification of even the high frequency Rouse
spectrum must be expected.

REFERENCES

(1) Boutin, H., Prask, H., Trevino, S. and Danner, H., Proc.
 Symp. IAEA Bombay, Vol. II, p. 393.
(2) Safford, G.J., Danner, H.R., Boutin, H. and Berger, M., J.
 Chem. Phys., 1964, 40, 1426.
(3) Myers, W., Summerfield, G.C. and King, J.S., J. Chem. Phys.,
 1966, 44, 189.
(4) Willis, B.T.M. (Ed.), Chemical Applications of Thermal
 Neutron Scattering, 1973, Oxford University Press, Oxford.
(5) Kostorz, G. (Ed.), Treatise on Materials Science and Technology, Vol. 15, Neutron Scattering, 1979, Academic Press, London.
(6) Maconnachie, A. and Richards, R.W., Polymer, 1978, 19, 739.
(7) Allen, G. and Higgins, J.S., Rep. Prog. Phys., 1973, 36,
 1073.
(8) Ivin, K.J. (Ed.), Structural Studies of Macromolecules by
 Spetroscopic Methods, 1976, Wiley Interscience, London.
(9) McBrierty, V., Polymer, 1974, 15, 503.
(10) Osaki, K. and Schrag, J.L., Polymer Journal, 1971, 2, 541.
(11) Dunbar, J.H., North, A.M., Pethrick, R.A. and Teik, P.B.,
 Polymer, 1980, 21, 764.
(12) Van Hove, L., Phys. Rev., 1954, 95, 249.
(13) Lynn, J.E., Contemporary Physics, 1980, 21, pp 483-500.
(14) Fender, B.E., Hobbis, I.C.W. and Manning, G., Phil. Trans.
 R. Soc. London, 1980, B290, pp 657-672.
(15) Windsor, C., Pulsed Neutron Scattering, 1981, Taylor & Francis.

(16) The Institut Laue-Langevin is owned and run jointly by
France, Germany and the U.K. for the benefit of their scientific
communities. For further information, contact the Scientific
Secretariat, ILL, BP156X Centre de Tri, 38042 Grenoble, France.
(17) Stirling, W.G., Neutron Beam Facilities at the HFR, ILL,
(see Ref. 16).
(18) Iyengar, P.K. in Thermal Neutron Scattering, ed. P.A. Egel-
staff, 1965, Academic Press, London.
(19) Dolling, G. in Dynamical Properties of Solids, eds. G.K.
Horton and A.A. Maradudin, Vol. I, 1974, North Holland Publ.
(20) Brockhouse, B.N. in Inelastic Scattering of Neutrons in
Solids and Liquids, Proc. IAEA Sym., 1961, 113.
(21) Brugger, R. in Thermal Neutron Scattering, ed. P.A. Egel-
staff, 1965, Academic Press, London.
(22) Douchin, F., Lechner, R.E. and Blanc, Y., ILL Internal
Scientific Report, 1973, ITR 26/73 and ITR 12/73.
(23) Birr, M., Heidemann, A. and Alefeld, B., Nucl. Inst. &
Methods, 1971, 95, 435.
(24) Mezei, F., Z. Physik, 1972, 255, 146.
(25) Hayter, J.B. in Neutron Diffraction, Ed. H. Dachs, 1978,
Springer Verlag, Berlin.
(26) a - Dagleish, P., Hayter, J.B. and Mezei, F. in Neutron
Spin-Echo, ed. F. Mezei, 1980, Physics 128, Springer Verlag,
Berlin.
       b - Hayter, J.B., ibid.
(27) Howard, J. and Waddington, T.C., Mol. Spectroscopy with
Neutrons in 'Advances in Infra-red and Raman Spectroscopy',
R.J.H. Clark and R.E. Hester, Eds., Vol. 7, 1980, Heyden, London.
(28) Allen, G., Brier, P.N. and Higgins, J.S., Polymer, 1972,
13, 157.
(29) Allen, G., Wright, C.J. and Higgins, J.S., Polymer, 1975,
15, 319.
(30) Wright, C.J., Chapter 3 in ref. 6.
(31) Yasukawa, T., Kimura, M., Watanabe, N. and Yamada, Y., J.
Chem. Phys., 1971, 55, 983.
(32) Takeuchi, H., Higgins, J.S., Hill, A., Maconnachie, A.,
Allen, G. and Stirling, G.C., Polymer (in press).
(33) Henry, A.W. and Safford, G.J., J. Pol. Sci. A2, 1969, 7,
433.
(34) Amaral, L.W., Vinhas, L.A. and Herdade, S.B., J. Pol. Sci.
Pol. Phys., 1976, 14, 1077.
(35) Drexel, W. and Peticolas, W.L., Biopolymers, 1975, 14, 715.
(36) Longster, G.F. and White, J.W., Mol. Phys. 1969, 17, 1.
(37) Pellow, C., private communication.
(38) Twistleton, J.F. and White, J.W., Polymer, 1972, 13, 41.
(39) Holliday, L. and White, J.W., Pure and Applied Chemistry,
1971, 26, 545.
(40) Lynch, J.E., Summerfield, G.C., Feldkamp, L.A. and King,
J.S., J. Chem. Phys., 1968, 48, 912.
(41) Zerbi, G. and Piseri, L., J. Chem. Phys., 1968, 49, 3840.

(42) Pepy, G. and Grim, H. in Neutron Inelastic Scattering, p. 605, 1978, IAEA, Vienna.

(43) Piseri, L., Powell, B.M. and Dolling, G., J. Chem. Phys., 1973, 58, 158.

(44) White, J.W. in Dynamics of Solids and Liquids by Neutron Scattering, Eds. S.W. Lovesey and T. Springer, 1977, Springer Verlag, Berlin.

(45) Barnes, J.D., J. Chem. Phys., 1973, 58, 5193.

(46) Skold, K., J. Chem. Phys., 1968, 49, 2443.

(47) Hervet, H., Dianoux, A.J., Lechner, R.E. and Volino, F., J. Phys. (Paris), 1976, 37, 587.

(48) Leadbetter, A. and Lechner, R.E. in The Plastically Crystalline State, Ed. J.N. Sherwood, 1979, Wiley & Sons.

(49) Allen, G. and Higgins, J.S., Macromolecules, 1977, 10, 1006.

(50) Ma, K.T., PhD Thesis, 1981, Dept. Chem. Eng. and Chem. Tech., Imperial College, London.

(51) Higgins, J.S. and Ma, K.T., to appear in Polymer.

(52) Stejskal, E.O. and Gutowsky, H.S., J. Chem. Phys., 1958, 28, 388.

(53) Powles, J.G., Strange, J.H. and Sandiford, D.J.H., Polymer, 1963, 4, 401.

(54) Connor, T.M. and Hartland, A., Phys. Lett., 1966, 23, 662.

(55) Lovell, R. and Windle, A.H., Polymer, 1981, 22, 175.

(56) Eisenberg, A. and Reich, S., J. Chem. Phys., 1969, 51, 5706.

(57) Rouse, P. Jr., J. Chem. Phys., 1953, 21, 1272.

(58) Pecora, R., J. Chem. Phys., 1968, 49, 1032.

(59) de Gennes, P.G., Physics, 1967, 3, 37.

(61) Akcasu, A.Z., Benmouna, M. and Han, C.C., Polymer, 1980, 21, 866.

(62) Akcasu, A.Z. and Benmouna, M., Macromolecules, 1978, 11, 1193.

(63) Akcasu, A.Z., Benmouna, M. and Alkhafaji, S., Macromolecules, 1981, 14, 147.

(64) Nicholson, L.K., Higgins, J.S. and Hayter, J.B., Macromolecules, 1981, 14, 836.

(65) Allegra, G. and Gannazzoli, F., J. Chem. Phys., 1981, 74, 1310.

(66) Zimm, B., J. Chem. Phys., 1956, 24, 269.

(67) Doi, M. and Edwards, S.F., J. Chem. Soc. Far. II, 1978, 74, 1789, 1802, 1818.

(68) de Gennes, P.G., J. Chem. Phys., 1971, 55, 572.

(69) Graessley, W.W., J. Pol. Sci. Pol. Phys., 1980, 18, 27.

(70) Higgins, J.S., Nicholson, L.K. and Hayter, J.B., Polymer, 1981, 22, 163.

(71) Maconnachie, A. and Richards, R.W., Polymer, 1978, 19, 739.

(72) Higgins, J.S., Ghosh, R.E., Allen, G. and Maconnachie, A., to be published in J. Chem. Soc. Faraday II.

(73) Higgins, J.S., Ghosh, R.E., Howells, W.S. and Allen, G., J. Chem. Soc. Faraday II, 1977, 73, 40.

(74) Walsh, D.J., Higgins, J.S. and Hall, R.H., Polymer, 1979, 20, 951.

(75) Higgins, J.S., Ma, K. and Hall, R.H., accepted by J.
      Physics A.
(76) Warner, M., accepted by J. Physics A.

LUMINESCENCE POLARIZATION METHODS FOR STUDYING MOLECULAR MOBI-
LITY AND ORIENTATION IN BULK POLYMERS

Lucien MONNERIE

Ecole Supérieure de Physique et de Chimie Industrielles
de Paris
10 rue Vauquelin
75005 Paris - France

ABSTRACT - The basic phenomena of luminescence and luminescence
polarization are briefly presented. A general formalism is developed and successively applied to both mobility in isotropic
systems and orientation in uniaxially symmetric systems. The particular case in which orientation and mobility simultaneously
occur is examined. Typical examples of application of luminescence
polarization in polymer field are indicated.

I - INTRODUCTION

The first study on fluorescence polarization has been performed in 1929 by F. Perrin (1) for measuring the fluorescence lifetime of small molecules in solution. Since this time, this technique has been applied to polymer field for measuring either molecular motions in isotropic systems or chain orientation in stretched materials. In this paper we will first briefly recall the
basic phenomena of luminescence, then we will show how luminescence polarization can lead to information on mobility and orientation. We will try to emphasize the advantages and difficulties
of these techniques when they are applied to polymer field. Some
typical examples will be given.

II - LUMINESCENCE PHENOMENA

Several books deal with luminescence, in particular those
cited in ref. 2 - 6, and hereafter we only summarize the main
features. Using the energy level diagram of the electronic states
of a luminescence molecule, (Fig. 1 and 2) one can easily describe
the electronic transitions occuring in fluorescence and phosphorescence. Thus, light absorption brings the molecule in a singlet
excited state which depends on the absorbed wavelength, then intra-

383

R. A. Pethrick and R. W. Richards (eds.), Static and Dynamic Properties of the Polymeric Solid State, 383–413.
Copyright © 1982 by D. Reidel Publishing Company.

molecular energy conversion process brings it back to the lowest vibrational level of the 1 st singlet excited state. In the case of fluorescence (Fig. 1), the molecule remains in this state for

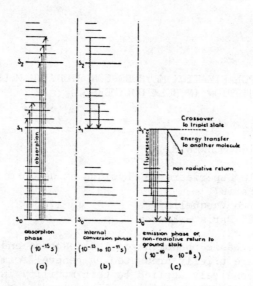

Fig. 1 - Electronic states of a molecule and fluorescence emission between 10^{-10} and 10^{-7} s, then emission occurs. On the contrary, in a phosphorescent molecule (Fig. 2), there is a cross over bet-

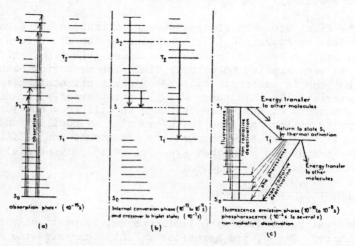

Fig. 2 - Electronic states of a molecule and phosphorescence emission

-ween the 1 st singlet excited state and the vibrational levels
of an excited triplet state, and the molecule can reach the lowest
vibrational level of the 1 st triplet state. For such a state,
return to the ground state by luminous emission (phosphorescence)
is difficult since it is theoretically forbidden. Consequently, it
occurs a fairly long time after excitation (10^{-4} to several
seconds) and there is a considerable competition from non-radiative processes which explain that phosphorescence is mainly observed in rigid or viscous medium at low temperature. Phosphorescence
emission occurs at longer wavelengths than fluorescence.

We should notice that the lifetime of the excited state from
which luminescence occurs (we will call it "luminescent excited
state") is an important quantity, since it is during this time
that the physical phenomena affecting luminescence emission must
occur (non-radiative processes, energy transfers, molecular
motions...). In simplest cases, the fluorescence and phosphorescence intensity decays after the with drawal of the excitation
light is given by :

$$I_f(t) = I_f^o \exp(- t/\tau_f)$$

(I.1) $$I_p(t) = I_p^o \exp(- t/\tau_p)$$

$$I_f^o = \alpha I_o \tau_f \; ; \; I_p^o = \alpha I_o k_{ICS} \tau_f \cdot \tau_p$$

where τ_f and τ_p are the fluorescence and phosphorescence lifetimes respectively, αI_o is the fraction of excitation light absorbed and k_{ICS} is the rate constant for internal energy conversion
from singlet to triplet state. If intensity measurements are performed under continuous excitation, fluorescence intensity is
directly proportional to fluorescence lifetime.

A schematic diagram of the simplest spectroluminescence apparatus is shown in Fig. 3. With such an instrument excitation and

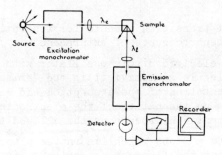

Fig. 3 - Schematic diagram for luminescence spectra recording.

emission spectra for both normal fluorescence and phosphorescence
can be recorded. In order to isolate phosphorescence, a mechanical
device is used which separates the excitation and observation phases and thus eliminates normal fluorescence emission.

When we deal with luminescence polarization, it is very important to consider the transition moments associated with the electronic transitions involved. Indeed, when absorbing a suitable
wavelength, a molecule behaves like an electric dipole oscillator
with a fixed orientation with respect to the geometry of the molecule. Such an equivalent oscillator is called "absorption transition moment". The direction of the transition moment in the molecule depends on the electronic transition under consideration
(Fig. 4). In the same way, for the luminescence emission the mole-

a) Absorption

ground state → 1st singlet state

ground state → 2d singlet state

b) Fluorescence emission

c) Phosphorescence emission

Fig. 4 - Directions of transition moments for absorption, fluorescence and phosphorescence of anthracene

-cule behaves like an electric dipole oscillator with a definite
orientation in the molecule geometry: it is the "emission transition moment". The orientation of the absorption and emission transition moments depends on the nature of the electronic transition
undergone. Thus in aromatic hydrocarbons the allowed transitions
$^1(\pi^\star \leftarrow \pi)$ correspond to moments which lie in the plane of the molecule, in such a way that the fluorescence emission is called "in-
plane" polarized. But the phosphorescence emission is polarized
perpendicular to the plane of rings for it corresponds to a for-

-bidden transition. As an example some transition moments of anthracene are shown in Fig. 4. On the contrary, with molecules where $\pi^\star \leftarrow n$ occurs, as heterocyclic molecules or carbonyl containing molecules, the $\pi^\star \leftarrow n$ singlet transition is out-of plane polarized whereas the phosphorescence from n, $\pi^\star$ triplet state is in-plane polarized. The relationship between the chemical structure of a molecule and the nature of electronic transitions can be found in (3).

III - BASIC EXPRESSIONS OF LUMINESCENCE POLARIZATION

As far as luminescence polarization is concerned, luminescent molecules can be represented by their absorption and emission transition moments $\vec{M}_o$ and $\vec{M}$ (Fig. 5). When a polarized light is

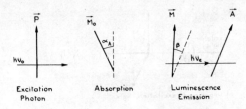

Fig. 5 - Polarized absorption and luminescence emission. $\vec{P}$ = polarizer, $\vec{A}$ = analyzer.

used, the probability of absorption is proportional to $\cos^2 \alpha_A$. In the same way, the luminescence intensity measured through an analyzer $\vec{A}$ is proportional to $\cos^2 \beta$. Thus, for $\vec{P}$ and $\vec{A}$ direction of polarizer and analyzer, the observed luminescence intensity is proportional to $\cos^2 \alpha_A \cdot \cos^2 \beta$.

Due to the lack of phase correlation between excitation and emission lights, luminescence emission can be described as resulting from three independent radiations respectively polarized along OX, OY, OZ with intensities I_x, I_y and I_z. The total luminescence intensity I is equal to :

$$I = I_x + I_y + I_z$$

and the polarization state is characterized by the relative values of I_x, I_y and I_z. The Curie symmetry principle, applied to different polarizations of the excitation light, leads to the results shown Fig. 6.

Luminescence emission is commonly observed either at right angle of the excitation direction or along it. Two quantities are used for characterizing the emission polarization :

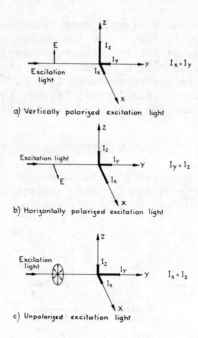

Fig.6 –States of polarization of luminescence emission for various
    polarization conditions of the excitation light.

. the polarization ratio : p

. the emission anisotropy : r which is more convenient for
deriving analytical expression and will be used below.

For simplicity, only polarized excitation light will be considered and these quantities are defined as :

$$(III.1) \qquad \begin{aligned} p &= (I_{\parallel} - I_{\perp})/(I_{\parallel} + I_{\perp}) \\ r &= (I_{\parallel} - I_{\perp})/(I_{\parallel} + 2I_{\perp}) \end{aligned}$$

$I_{\parallel}$ and $I_{\perp}$ correspond to luminescence intensity obtained with an
analyzer direction respectively parallel and perpendicular to that
of the polarizer. The quantity $(I_{\parallel} + 2I_{\perp})$ is the total luminescence intensity.

Both quantities are related by :

$$r = 2 p/(3 - p)$$

III.1 – Emission anisotropy for motionless molecules isotropically
       distributed

Detailed calculations about this topic can be found in (6).

Consider a set of molecules, excited at time t = 0 by an infinitely short light pulse, vertically polarized (Fig. 7a). At time t, the emission transition moment of a molecule i lies along

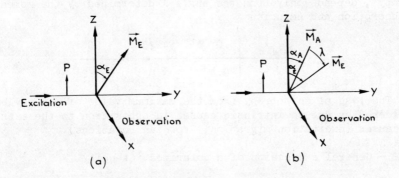

Fig. 7 - Orientation of transition moments in the reference frame. (a) emission moment (b) non-parallel absorption and emission moments

$\vec{M}_E$. Taking into account that the excited molecules are symmetrically distributed around the polarization direction OZ, we obtain:

$$(III.2) \quad r(t) = (3 \overline{\cos^2 \alpha_E(t)} - 1)/2$$

where the bar means an average over the N molecules emitting at time t. Further evaluation requires the calculation of such an average and two cases must be considered.

1 - Parallel absorption and emission moments

The number of excited molecules with moments at an angle to OZ between α and $(\alpha + d\alpha)$ is $\cos^2 \alpha . \sin \alpha . d\alpha$ in such a way that :

$$(III.3) \quad \overline{\cos^2 \alpha} = 3/5 \quad \text{and} \quad r_{o,\ell} = 2/5 = 0.4$$

2 - Non-parallel absorption and emission moments

Such a $r_{o,\ell}$ value is never observed. Indeed, in fluorescence, vibronic coupling between atomic vibrations and electrons leads to some electronic delocalization which results in an average angle between transition moments (Fig. 4a, b). In the case of phosphorescence the transition moments are almost perpendicular (Fig. 4a, c). The corresponding situation is represented in Fig. 7b and leads to .

$$(III.4) \quad (3 \overline{\cos^2 \alpha_E} - 1)/2 = ((3 \overline{\cos^2 \alpha_A} - 1)/2)((3 \cos^2 \lambda - 1)/2)$$

$$(III.5) \quad r_o = 0.4(3 \cos^2 \lambda - 1)/2$$

Notice that this r_o value, called "fundamental emission anisotropy", depends only upon the angle λ determined by the moments of absorption and emission

λ	$0°$	$54.73°$	$90°$
r_o	0.4	0	-0.2

The loss of anisotropy from the maximum value of 0.4 may be considered as due to intrinsic causes in opposition to the extrinsic causes (excitation migration, Brownian rotations).

III.2 - General expression of a polarized intensity

In order to apply further luminescence polarization to the study of molecular mobility and molecular orientation, it is first convenient to consider the most general situation where the luminescent molecules display an orientation distribution in the sample and undergo a motion during the lifetime τ of the luminescent excited state. In such a case, a general expression of a polarized intensity has been derived (7), from which the particular expressions for mobility in isotropic system or orientation in uniaxial symmetry can be deduced.

The general expression is derived under the assumptions that the transition moments in both absorption and emission coincide with a molecular axis M of the molecule and that the polarization of the exciting radiation and luminescence is not affected by the birefringence of the anisotropic sample. The effects arising from delocalization of the transition moments and from birefringence can be considered in each particular case (mobility, orientation).

The luminescent sample can be represented as a set of unit vectors $\vec{M}$ whose direction is specified by the spherical polar angle $\Omega = (\alpha, \beta)$ in the reference frame (Fig. 8). Let us introduce the angular functions :

$N(\Omega_o, t_o)$, the orientation distribution of M vectors at time t_o ($\vec{M}_o$ in Fig. 8)

$P(\Omega, t | \Omega_o, t_o)$, the conditional probability density of finding at position Ω at time t a vector $\vec{M}$ which was at position Ω_o at time t_o.

After illumination by a linearly polarized short pulse light at t_o, the intensity emitted by the sample at time $(t_o + u)$ for $\vec{P}(\gamma_o, \varphi_o)$ and $\vec{A}(\gamma, \varphi)$ directions of polarizer and analyzer, is

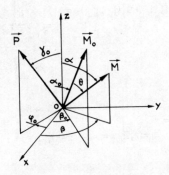

Fig. 8 — Illustration of the angles which define the position P of the polarizer and the orientation of a molecular axis $\vec{M}$ at time t_o and $\vec{M}$ at time t_o + u with respect to the fixed frame $0XYZ$

given by :

$$i(\vec{P},\vec{A},t_o + u) = K \iint N(\Omega_o,t_o)P(\Omega,t_o + u|\Omega_o,t_o)$$
$$\times \cos^2(\vec{P},\vec{M}_o) \cos^2(\vec{A},\vec{M}) e^{- u/\tau} d\Omega_o d\Omega$$

where K is an instrumental constant. In this expression t_o corresponds to the macroscopic evolution of the sample, for example in a rheological experiment, whereas u corresponds to a microscopic reorientational motion at the scale of the luminescence lifetime τ. If we assume that the t_o dependence of N and P functions can be ignored within the time τ, the luminescence intensity under continuous excitation is :

$$i(\vec{P},\vec{A}, t_o, \tau) = \int_o^\infty i(\vec{P},\vec{A}, t_o + u)du$$

In order to derive convenient expressions, it is interesting to do an expansion of the angular functions of Ω_o and Ω in a series of spherical harmonics $Y_\ell^m(\Omega_o)$ and $Y_\ell^m(\Omega)$. Taking into account the properties of spherical harmonics, it results :

$$(III.6) \quad \cos^2(\vec{P},\vec{M}_o) = \sum_{k=0,2} \sum_{m=-k}^{+k} p_k^m(\gamma_o, \varphi_o)\overline{Y_k^m}(\Omega_o)$$

$$(III.7) \quad \cos^2(\vec{A},\vec{M}) = \sum_{\ell=0,2} \sum_{n=-\ell}^{\ell} a_\ell^n(\gamma_o, \varphi_o) Y_\ell^n(\Omega)$$

$$(III.8) \ K \ N(\Omega_o,t_o)P(\Omega_o t_o+u|\Omega_o,t_o) = \sum_{k=0}^{\infty} \sum_{\ell=0}^{\infty} \sum_{m=-k}^{+k} \sum_{n=-\ell}^{+\ell} f_{k\ell}^{mn}(t_o,u)$$

$$x \ Y_k^m(\Omega_o) \ \overline{Y_\ell^n(\Omega)}$$

$$(III.9) \ i(\vec{P},\vec{A},t_o,\omega) = \sum_{k=0,2} \sum_{\ell=0,2} \sum_{m=-k}^{+k} \sum_{n=-\ell}^{+\ell} p_k^m \ a_\ell^n \ f_k^{mn}(t_o,u)$$

$$x \ e^{-u/\tau}$$

$$(III.10) \ \text{with} \ f_{k\ell}^{mn}(t_o,u) = K < Y_k^m(\Omega_o) \ \overline{Y_\ell^n(\Omega)} >$$

(the angular brackets denote an ensemble average)

$$(III.11) \ i(\vec{P},\vec{A},t_o,\tau) = \sum_{k=0,2} \sum_{\ell=0,2} \sum_{m=-k}^{+k} \sum_{n=-\ell}^{+\ell} p_k^m \ a_\ell^n \ f_{k\ell}^{mn} t_o,\tau)$$

$$(III.12) \ \text{where} \ f_{k\ell}^{mn}(t_o,\tau) = \int_o^{\infty} f_{k\ell}^{mn}(t_o,u)e^{-u/\tau} \ du$$

Thus, any fluorescence intensity is expressed as a linear combination of 36 $f_{k\ell}^{mn}$ coefficients, in agreement with Frehland's result (8). All $f_{k\ell}^{mn}$ values are proportional to K, and f_{00}^o is simply

$$(III.13) \ f_{00}^o = K/4 \ \pi$$

When $\vec{P}$ or $\vec{A}$ lies along the fixed frame axes, some coefficients vanish, leading to the following relations :

$$i(Z,Z) = (4 \ \pi/9)[\ f_{00}^o + 2(5)^{-1/2} f_{20}^o + 2(5)^{-1/2} f_{02}^o + 4(5)^{-1} f_{22}^o]$$

$$i(Z,X) = (4 \ \pi/9)[\ f_{00}^o + 2(5)^{-1/2} f_{20}^o - (5)^{-1/2} f_{02}^o - 2(5)^{-1} f_{22}^o]$$

$$(III.14) \quad i(X,Z) = (4\pi/9)[f^o_{00} - (5)^{-1/2}f^o_{20} + 2(5)^{-1/2}f^o_{02} -$$
$$2(5)^{-1} f^o_{22}]$$

$$i(X,X) = (4\pi/9)[f^o_{00} - (5)^{-1/2}f^o_{20} - (5)^{-1/2}f^o_{02} +$$
$$(5)^{-1}f^o_{22} + 8(5)^{-1} f^2_{22}]$$

$$i(X,Y) = (4\pi/9)[f^o_{00} - (5)^{-1/2}f^o_{20} - (5)^{-1/2} f^o_{02} +$$
$$(5)^{-1}f^o_{22} - 8(5)^{-1} f^2_{22}]$$

These expressions will be used to derive the emission anisotropy and the orientation functions from the experiments.

IV - MOLECULAR MOBILITY OF ISOTROPICALLY DISTRIBUTED MOLECULES AS STUDIED BY LUMINESCENCE POLARIZATION

IV.1 - Expressions of the emission anisotropy

First, assume that absorption and emission moments are parallel. In the case of isotropic systems, the number of $f^{mn}_{k\ell}$ coefficients is considerably reduced and the required intensities, derived from (III.14), becomes :

$$i(Z,Z) = (4\pi/9)f^o_{00} + (16\pi/45)f^o_{22}$$
$$i(Z,X) = (4\pi/9)f^o_{00} - (8\pi/45)f^o_{22}$$

which leads to :

$$(IV.1) \quad r = (2/5) f^o_{22}/F^o_{00} = (2/5)(4\pi/K)f^o_{22}$$

With flash pulse light, the instantaneous anisotropy $r(t)$ is obtained by taking $f^o_{22}(t)$ as defined by (III.10), whereas for continuous excitation the mean anisotropy, $r(\tau)$, is obtained with $f^o_{22}(\tau)$ defined by (III.12).

With notation given in Fig. 8, the calculation of the orientation autocorrelation function of the transition moment $\vec{M}$,

$$P_2(\vec{M}(0) \cdot \vec{M}(t)) = \; <(3\cos^2 \Theta(t) - 1)/2 >$$

can be performed by using the angular functions $N(\Omega_o,0)$ and $P(\Omega,t|\Omega_o,0)$ introduced in III.2 and their expansions in a series of spherical harmonics. One obtains :

$$(IV.2) \quad P_2(\vec{M}(0) \cdot \vec{M}(t)) = (4\pi/K)f^o_{22}$$

which leads to :

(IV.3) $r = (2/5)\ P_2(\vec{M}(0)\ .\ \vec{M}(t))$

When absorption and emission moments are not parallel, which is the common case (see II), the limit anisotropy value (2/5), obtained from (IV.3) for motionless molecules, must be substituted by the "fundamental emission anisotropy" r_o, given by (III.5). Thus, the emission anisotropies are expressed by :

(IV.4) $r(t) = r_o < (3\ \cos^2\ \Theta(t) - 1)/2 >$

(IV.5) $r(\tau) = r_o\ .\ \tau^{-1} \int_o^\infty < (3\ \cos^2\Theta(t) - 1)/2 > e^{-t/\tau}\ dt$

It is clear that luminescence polarization measurements yield information on the molecular motion undergone during the luminescence lifetime, τ. In the case of fluorescence, τ values are of the order of few nanoseconds and solely high frequency motions contribute. On the contrary, with phosphorescence, the involved lifetimes range from 10^{-4} to several seconds, in such a way that lower frequency motions can be studied.

IV.2 - Motional models

Further investigation of molecular mobility requires to consider motional models in order to derive analytical expressions of the orientation autocorrelation function, $P_2(t)$.

1 - Isotropic motion

The simplest model corresponds to the motion of a sphere of volume V in a medium of viscosity η, and leads to :

$$P_2(t) = \exp(-\ t/\tau_R)$$

where $\tau_R = V\eta/kT$ is the characteristic correlation time.

Thus, the instantaneous emission anisotropy, corresponding to a short flash excitation, has an exponential decay :

(IV.6) $r(t) = r_o \exp(-\ t/\tau_R)$

and the mean anisotropy under continuous excitation is expressed by :

(IV.7) $1/r(\tau) = r_o^{-1}\ (1 + \tau/\tau_R)$

2 - Flexible macromolecules

Molecular motions performed by a bond inside a polymer chain cannot usually be described by an isotropic motion. One possibility consists to take an empirical distribution function of correlation times. Another approach has been proposed, based on stochastic jumps of bonds inside a polymer chain, each jump affecting a very small part of the chain. Using the jump model in the diamond lattice represented in Fig. 9, we have derived relations valid at short times (9, 10, 11). Within the diamond lattice, the

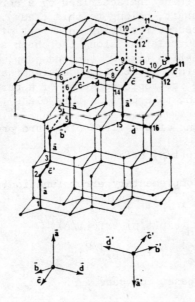

Fig. 9 - Representation of a chain on a tetrahedral lattice.
Examples of trans conformation (t) :
sequence 1 - 2 - 3 - 4; gauche + conformation (g_-^+) :
sequence $4 \backsim 5' - 6' \backsim 7$, gauche-conformation (g^-) :
sequence $4 \backsim 5 \backsim 6 \backsim 7$; three bond motion

$$4 - 5 - 6 - 7 \ (g^-) \longleftrightarrow 4 - 5' - 6' - 7 \ (g^+)$$
$$(b' \quad a \quad c') \qquad\qquad (c' \quad a \quad b')$$

four bond motion :

$$9 - 10 - 11 - 12 - 13(g^+g^-) \longleftrightarrow$$
$$(d \quad b' \quad c \quad d')$$
$$9 - 10' - 11' - 12' - 13(g^-g^+)$$
$$(a \quad b' \quad c \quad a')$$

smallest motions inside the chain correspond to 3 and 4 bond
jumps. As shown in Fig. 9 and described in the caption, a 3-bond
motion does not create any new orientation but it leads to diffuse
an orientation along the chain (b' and c' orientation are exchan-
ged in Fig. 9). In the case of a 4-bond jump, the orientations of
the two internal bonds remain unchanged, while a bond orientation
is lost (d and d' become a and a' in Fig. 9). In a real chain,
many other motions occur and it seems important to classify them
in two groups. The first class would correspond to any process
leading to a diffusion of a chain bond orientation along the
chain, each of them being characterized by a correlation time ρ_i.
(3-bond jump, crankshaft, belong to this class).The second class
would deal with processes resulting in a loss of chain bond orientation without any correlation along the chain sequence, each
process has a correlation time Θ_i. (4-bond jump, departure from the
diamond lattice, thermal fluctuations of internal rotation angles,
motions involving an isotropic rotation of a part of the chain
or of the whole chain, take part in this class).

When only motions of the first class are present, one obtain:

$$(IV.8) \qquad P_2(t) = \exp(t/\rho) \, \mathrm{erfc}(t/\rho)^{1/2}$$

where, erfc is the complementary error function, and ρ is the
harmonic average of $\rho_i'^{s}$. It results in :

$$r(t) = r_o \exp(t/\rho) \, \mathrm{erfc}(t/\rho)^{1/2}$$
$$(IV.9) \qquad 1/r(\tau) = r_o^{-1} (1 + (\tau/\rho)^{1/2})$$

In the general case, one gets :

$$(IV.10) \qquad P_2(t) = \exp(-t/\Theta) \exp(t/\rho) \, \mathrm{erfc}(t/\rho)^{1/2}$$

$$(IV.11) \qquad 1/r(\tau) = r_o^{-1}(1 + \tau/\Theta + (\tau/\rho + \tau^2/\rho\Theta)^{1/2})$$

$$\text{with} \quad \Theta^{-1} = \sum_i \Theta_i^{-1}$$

Expression (IV.10) has been successfully applied to dynamics
of polymers in solution (12, 13, 14). It has not yet been actually
tested in bulk polymers but there is no basic reason that it
shoud fail in this case.

IV.3 - Instrumentation for studying molecular mobility of isotro-
       pic samples.

Two types of equipments are convenient for mobility studies.

.1. Continuous excitation measurements.

These are the easiest to perform and a basic diagram convenient for studying polymer samples is shown in Fig. 10a. More
details on components can be found in (2). Straight-through as well

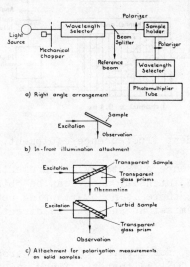

a) Right angle arrangement

b) In-front illumination attachment

c) Attachment for polarization measurements
on solid samples.

Fig. 10 - Arrangements for luminescence polarization measurements

as right-angle arrangements can be used. With the latter, in the
case of polymer films or thin sheets, a particular attachment
(Fig. 10b) has been designed in our laboratory in order to avoid
polarization changes at the air-polymer interface. Of course, in
the case of phosphorescence, additional mechanical shutters are
required.

This technique leads to the mean emission anisotropy, $r(\tau)$,
defined by (IV.5) and requires independent measurements of the
lifetime, τ, of the luminescent excited state and of the fundamental anisotropy r_0. No information can be obtained on the type of
motion undergone by the luminescent molecule (isotropic, anisotropic...). Furthermore, only one characteristic of molecular mobility is available. If isotropic motion or orientation diffusion
motional model are considered, only one correlation time (either
τ_R or ρ) is involved, and it can be derived from obtained measurements. For motional models implying several correlation times,
this method is less interesting. Indeed it leads only to an average
correlation time.

2. Time-dependent measurements.

In the case of fluorescence, this technique uses a flash
pulse light source, with a width at half-height of 2 ns, and a

single-photon counting system. A schematic diagram is shown in Fig. 11.

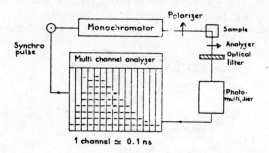

Fig. 11 - Time-dependent fluorescence equipment.

For phosphorescence anisotropy measurements, due to the long lifetime involved the intensity decay can be obtained on an oscilloscope. Furthermore, there is no overlapping between the flash and phosphorescence emissions and $r(t)$ is directly derived from intensity measurements.

On the contrary, for fluorescence anisotropy, the lifetime is so short (10^{-10} to 10^{-7} s) that the flash and fluorescence emissions overlap. It requires a deconvolution treatment for deriving $r(t)$ from the measured intensities $i_\parallel(t)$ and $i_\perp(t)$. This is the main difficulty in this technique for no deconvolution method is completely satisfactory.

It should be stressed that this technique represents the most powerful tool for studying molecular mobility. Indeed, the luminescence lifetime and the fundamental anisotropy, r_o, are obtained in each experiment. Furthermore the orientation autocorrelation function $P_2(\cos \Theta(t))$ is directly given by $r(t)$ (see IV.1) and yields information about the type of motion undergone by the luminescent molecule. Finally, when several relaxation times are involved they can be determined from a fitting of $r(t)$.

V - MOLECULAR ORIENTATION IN UNIAXIAL SYMMETRIC SYSTEMS AS STUDIED BY FLUORESCENCE POLARIZATION

Fluorescence emission is the most convenient for studying molecular orientation for it occurs in a large temperature range with a good quantum yield and in various physical conditions.

More details about this part are given in (7, 15).

Due to the uniaxial symmetry, simplification occurs in the general treatment presented in III.2. Thus, in the intensity expression (III.11), only six independent $f_{k\ell}^{mn}$ quantities remain :

$$f_{00}^{o}; \ f_{20}^{o}; \ f_{02}^{o}; \ f_{22}^{o}; \ f_{22}^{1}; \ f_{22}^{o}$$

All of them are proportional to the instrumental constant K and in addition, one gets :

$$f_{00}^{o} = K/4 \ \pi$$

It is convenient to introduced the following quantities :

$$G_{20}^{o} = (1/5)^{1/2} \ (f_{20}^{o}/f_{00}^{o}) = (1/2) < 3 \ \cos^2 \alpha_{o} - 1 >$$

$$G_{02}^{o} = (1/5)^{1/2} \ (f_{02}^{o}/f_{00}^{o}) = (1/2) < 3 \ \cos^2 \alpha - 1 >$$

(V.1) $\quad G_{22}^{o} = (1/5)(f_{22}^{o}/f_{00}^{o}) = (1/4) < (3 \ \cos^2 \alpha_{o} - 1)(3 \ \cos^2 \alpha - 1) >$

$$G_{22}^{1} = (3/40)(f_{22}^{1}/f_{00}^{o}) = (9/16) < \sin \alpha_{o}.\cos \alpha_{o}.\sin \alpha. \cos \alpha$$
$$. \cos(\beta-\beta_{o}) >$$

$$G_{22}^{2} = (3/40)(f_{22}^{2}/f_{00}^{o}) = (9/64) < \sin^2 \alpha_{o}.\sin^2 \alpha.\cos^2(\beta-\beta_{o}) >$$

the various involved angles are those defined in Fig.8 and the angular brackets denote an ensemble average. The quantities G_{20}^{o} and G_{02}^{o} represent the second moments of the distribution of vectors $\vec{M}$ and $\vec{M}$ respectively and they describe the molecular orientation indepently of molecular mobility. The functions G_{22}^{m} depends on both orientation and mobility.

Further investigation requires to consider if during the fluorescence lifetime motions can occur or not. We will hereafter refer to these situations as "mobile" or "frozen" systems, respectively.

V.1 - Uniaxial frozen systems

Assume that absorption and emission moments are parallel; the effect of electronic delocalization is treated in V.3. Thus, in expressions (V.1), $\alpha = \alpha_{o}$, $\beta = \beta_{o}$ and the quantities $G_{20}^{o}(= G_{02}^{o})$ and G_{22}^{m} can be rewritten with only two independent quantities : $\cos^2 \alpha_{o}$, $\cos^4 \alpha_{o}$. All the information on the fluorescence intensities may be displayed in a 3 x 3 tensor I, whose elements I_{ij} can be easily obtained by reporting G's expressions in the general expressions (III.14) :

$$(V.2) I=K \begin{vmatrix} (3/8)<\sin^4\alpha> & (1/8)<\sin^4\alpha> & (1/2)<\sin^2\alpha\cos^2\alpha> \\ (1/8)<\sin^4\alpha> & (3/8)<\sin^4\alpha> & (1/2)<\sin^2\alpha\cos^2\alpha> \\ (1/2)<\sin^2\alpha\cos^2\alpha> & (1/2)\sin^2\alpha\cos^2\alpha> & <\cos^4\alpha> \end{vmatrix}$$

This is identical to the result given in (16). The apparatus constant can be obtained from :

$$K = \sum_i \sum_j I_{ij} = (8/3)I_{XX} + 4\ I_{XZ} + I_{ZZ}$$

in such a way that the second and fourth moments of the orientation distribution can be derived from fluorescence intensity measurements by the relations :

$$(V.3) \quad \begin{aligned} <\cos^2\alpha> &= (I_{ZZ} + 2\ I_{XZ})/((8/3)I_{XX} + 4\ I_{XZ} + I_{ZZ}) \\ <\cos^4\alpha> &= I_{ZZ}/((8/3)I_{XX} + 4\ I_{XZ} + I_{ZZ}) \end{aligned}$$

The function $N(\alpha)$ describing the orientation distribution of vectors $\vec{M}$ (Fig. 8) can be expanded in terms of Legendre polynomials in $\cos\alpha$, as follows.

$$N(\alpha) = \sum_\ell b_\ell\ P_\ell(\cos\alpha)$$

with $b_\ell = (1/2\pi)((2\ell + 1)/2\ <P_\ell(\cos\alpha)>$

where $<P_\ell(\cos\alpha)>$ is the value of $P_\ell(\cos\alpha)$ averaged over the distribution :

$$(V.4) \quad \begin{aligned} <P_2(\cos\alpha)> &= (1/2)<3\cos^2\alpha - 1> \\ <P_4(\cos\alpha)> &= (1/8)<35\cos^4\alpha - 30\cos^2\alpha + 3> \end{aligned}$$

It has been recently shown (17) that for an uniaxial distribution, the determination of $<P_2>$ and $<P_4>$ is sufficient in most cases. It should be pointed out that for either amorphous polymers or the amorphous phase of crystalline polymers, in addition to fluorescence polarization, Raman polarization and N.M.R. yield to a determination of $<P_4>$.

V.2 - Uniaxial mobile systems

In this case both orientation and mobility contribute to the fluorescence polarization. Nevertheless such systems can be treated (7). Indeed, if we assume that during the fluorescence lifetime (10^{-10} to 10^{-7} s) the orientation distribution does not change, it results from expressions (V.1) :

$$G^o_{20} = G^o_{02} = (1/2) < 3 \cos^2 \alpha_o - 1 > = < P_2(\cos \alpha_o) >$$

On the other hand, for mobile systems, G^o_{22} depends on both orientation and mobility, in such a way that $< P_4(\cos \alpha_o) >$ cannot be obtained. Furthermore, intensity measurements corresponding to polarizer and analyzer directions along the X and Z axis are not sufficient for deriving G^o_{20}, as it can be shown from expressions (III.14). The set of quantities G^o_{20} and G^m_{22}, as defined in (V.1), must be obtained from the measurement of five intensities, i(P, A), corresponding to P and A orientations which are not contained in the same plane, excluding the use of only a straight-through optical arrangement.

   Concerning the mobility, from the available G^o_{20} and G^m_{22} quantities, one can derive :

1/ The mean mobility amplitude, M

$$M = < (3 \cos^2 \theta - 1)/2 > = G^o_{22} + (16/3)G^1_{22} + (16/3)G^2_{22}$$

describing the motion performed during the considered time interval. Thus, for a Dirac-pulse excitation :

$$M(t) = < (3 \cos^2 \theta (t) - 1)/2 >$$

whereas for a continuous excitation, one gets :

$$(V.5) \qquad M(\tau) = \int_o^\infty M(t) \exp(- t/\tau)dt$$

where τ is the fluorescence lifetime. The emission anisotropies are directly derived (see IV.1).

2/ Three angular correlation functions which contain information on the anisotropy of the motion (7).

V.3 - Effect of electronic delocalization

   The effect of non-parallel absorption and emission moments on orientation measurements have been considered by several authors (16, 18). For uniaxial systems, an easy correction has been derived (7). The delocalization results in the fact that the measured intensities i(P,A) do not lead to the G quantities defined in (V.1), but to γ coefficients which are related to them by the relations :

$$(V.6) \qquad \gamma^o_{20} = (5 \ r_o/2)^{\frac{1}{2}} \ G^o_{20}$$

$$\gamma^m_{22} = (5 \ r_o/2) \ G^m_{22} \qquad \text{with m = 0,1,2}$$

where r_o is the fundamental emission anisotropy, which must be determined on an unoriented sample from an additional measurement of emission anisotropy.

V.4 - Birefringence and light-scattering corrections

The birefringence effect occurs when the direction of either the polarizer or the analyzer does not coincide with a principal direction of the refractive index tensor. It has been shown (7) that only the quantity γ_{22}^1 of (V.5) is modified and becomes :

$$\gamma_{22}^1 = (5 \ r_o/2) \ b \ G_{22}^1$$

where b is a correction factor which can be experimentally determined (15).

When dealing with crystalline polymers, light-scattering occurs and modifies the state of polarization of both the excitation and emitted light. The principle of a method of correction has been given in (18) and applied to straight-through measurements (19).

V.5 - Instrumentation for measuring orientation in uniaxial systems.

For systems in which the orientation does not change in time, successive fluorescence intensity measurements can be performed for the required polarizer and analyzer directions, leading to a determination of $< P_2(\cos \alpha) >$ and $< P_4(\cos \alpha) >$.

Measurements are performed under continuous excitation. For frozen systems, a straight-through optical arrangement perpendicular to the sample is very convenient. In front illumination has been proposed (20) but the effect of refraction of the light inside the sample must be considered (15). Measurements under optical microscope have been performed either on stretched samples (18) or during stretching (19).

When dealing with mobile systems a more elaborated equipment is required and it has been developed in our laboratory (15). This apparatus permits simultaneous measurements of the intensities required for determining orientation and mobility even during stretching. The optical system is represented in Fig. 12.

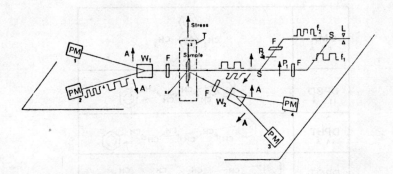

Fig. 12 - Optical equipment : L, mercury lamp (HBO 200 W);
          S, beam splitter; f_1, f_2, modulation frequencies of the
          mechanical choppers, F, optical filter; P_1, P_2 polari-
          zers; W_1, W_2, Wollaston prisms; A, analyzing direction;
          PM, photomultiplier; T, temperature chamber.

VI - MOLECULAR MOTIONS IN BULK POLYMERS

For many years, fluorescence polarization has been used
for studying molecular dynamics of polymers in solution and the
results have been recently reviewed (21, 22). On the contrary,
very few studies have been performed on bulk polymers.

VI.1 - Fluorescence polarization

Due to the very short fluorescence lifetime (10^{-10} to 10^{-7}s)
only high frequency molecular motions are observed with this
technique. It implies that studies must be performed in a temperature range fairly above the glass-rubber transition temperature.

The first quantitative study has been performed in our laboratory (23) on polyisoprene networks using either probes free in
the polymer matrix or labelled polymer chains. The chosen probes
are shown in Table 1 and the labelling is represented in Fig.13.
Notice that the fluorescent molecule concentration in the polymer
is as small as 1 p.p.m.. For the label and the DP probes, the
onset of mobility occurs at about - 10°C, that is 50°C above Tg
measured at 1 Hz. On the contrary, the DMA probe has a mobility
which does not vanish, even at - 60°C. Furthermore, all the probes
undergo an isotropic motion (see IV.2) which is characterized by
a correlation time τ_R given by (IV.7). The temperature dependence
of τ_R is shown in Fig. 14a for the various probes. The agreement
of the slopes of log τ_R for the three DP probes with that derived
from the free volume theory (24) means that these large probes
reflect molecular motions of the polymer which are involved in
the glass-rubber relaxation. In the case of the DMA probe, because

DMA 6.5 Å	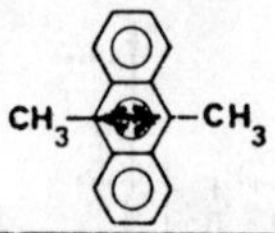
DPBD 11.5 Å	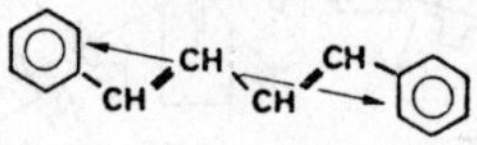
DPHT 14 Å	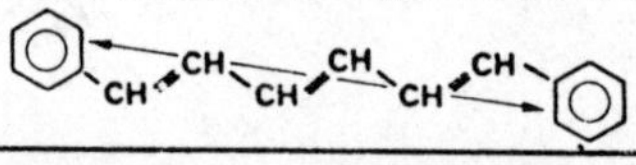
DPOT 16.5 Å	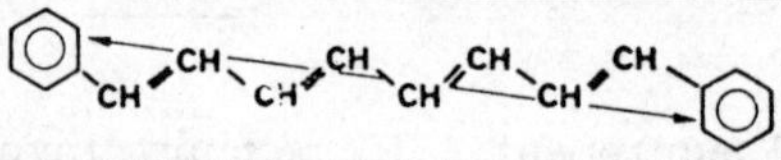

Table 1 - Formulas, approximative lengths and transition moments (double arrows) of the various probes.

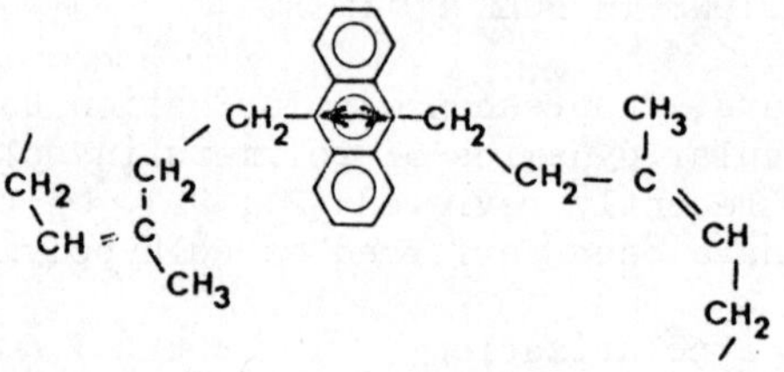

Fig. 13 - Position of DMA label in the middle of a chain skeleton; the double arrow indicates the transition moment.

of its smaller size, its behaviour would be related to the secondary transition of polyisoprene.

For labelled polyisoprene, shown Fig. 13, diluted in polyisoprene, the label motion is not isotropic but corresponds to the model of flexible chain (see IV.2) with an orientation diffusion along the chain backbone characterized by a correlation time ρ given by (IV.9). The temperature dependence of ρ is presented in Fig. 14b as well as the expectation from the free-volume theory. The rather good agreement in the slopes, obtained between + 20°C and + 80°C, indicates that in this temperature range the label reflects polymer motions involved in the glass-rubber relaxation. At lower temperature a secondary transition could contribute.

These probe and label results have been recently confirmed in our laboratory on a series of polybutadiene and butadiene-styrene random copolymers. From these results it is clear that the free-volume theory can be quantitatively applied to the high frequency

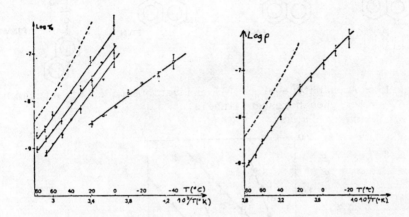

Fig. 14 - Log plot of correlation times vs 1/T.
          a/ probes, b/ DMA labels. The slope of the dotted
          line represents the expectation from free-volume
          theory

region (10^8 - 10^{10} Hz) with the same parameters as those determined from low-frequency mechanical measurements.

Further informations can be obtained by comparing label and probe mobilities. Thus, it seems that the smallest sequences involved in the glass-rubber relaxation of polyisoprene would correspond to about five monomers in average conformation.

VI.2 - Phosphorescence polarization

It is first interesting to point out that phosphorescence intensity, involving the triplet excited state, is very sensitive to changes in the surrounding medium as well as to small amounts of oxygen quencher. These effects have been used for evidencing secondary transitions of various polymers (25).

The phosphorescence polarization method has been proposed many years ago for studying molecular motions in a frequency range closer to mechanical experiments (10^{-1} to 10^3 Hz). Due to experimental difficulties, these measurements have been achieved only very recently on polymethyl acrylate (26) and methacrylate (27, 28). Probes and labels were studied, some of them are shown in Fig. 15. Assuming an isotropic motion for probe and labels, the correlation times, τ_R, can be derived by (IV.7) from phosphorescence polarization and intensity measurements. The temperature dependence of log τ_R in PMMA is shown in Fig. 16. and the transi-

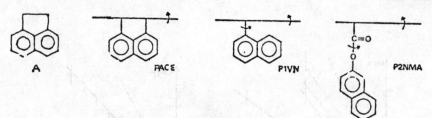

Fig. 15 - Acenaphtene probe (A) and various modes of attachment
of phosphorescent labels.

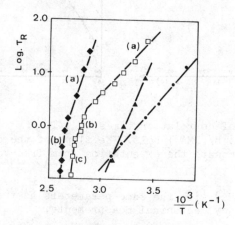

Fig. 16 - Log τ_R vs 1/T for phosphorescent probe and labels in
PMMA. (● A; ▲ P2NMA; ◆ PACE; ◻ P1VN) (from ref.28)

-tion temperatures as well as the activation energies are reported in Table 2. It is clear that the onset of mobility largely

Sample	T(°C)	E(kJ mole^{-1})
P/A	- 20	50
P1VN (a)	- 8	44
P1VN (b)	70	107
P1VN (c)	94	478
P2NMA	- 8	94
PACE (a)	70	115
PACE (b)	105	460

Table 2 - Transition temperatures and activation energies estimated from phosphorescence polarization data in PMMA
(from ref. 28)

depends on the probe or label, but it always occurs above the β
transition temperature of PMMA (≃ 250° K). For A probe and P1VN
in the initial part a, the activation energy is smaller than that
obtained by mechanical and dilectric relaxation for the β-process

(70 to 90 KJ.mole^{-1}), suggesting that the naphtyl label has a relatively unrestricted rotation about its bound of attachment to the chain backbone. On the contrary, the naphtyl species in P2NMA appears to be cooperative with that of the ester group, the rotation of which is usely associated with the β-process. In the case of PACE there is a strong coupling of label motion with chain backbone motion. In Fig. 16, two regions of thermal dependence are shown. The transition temperature at 105°C, and the activation energy at higher temperatures agree with the glass-rubber relaxation observed by low frequency techniques. But a question remains about the interpretation of mobility at lower temperatures. Indeed, the activation energy and the onset of mobility (70°C) are too high for a β-process, it could correspond to the α' transition which has been reported. Due to the "out of plane" polarization . of the $T_1 \rightarrow S_0$ transition, one could tentatively assigned the motion observed between 70°C and 105°C to rotation about the local chain axis whereas the α transition would reflect rotation motions of this local chain axis. The appearence of such a motion at a temperature higher than that of the β-process would suggest that the ester group motion could occur independently of the chain backbone. The motion observed from 70°C to 105°C could be related to the earliest stage of the largest chain motions which occur at the glass-rubber transition.

VII - EXPERIMENTAL STUDIES ON ORIENTATION OF POLYMERS

As above mentioned, only fluorescence polarization technique (F.P.) is interesting for studying molecular orientation. Either fluorescent probes or labels can be used and their concentration is always very small (1 p.p.m.). In the case of crystalline materials, the fluorescent molecules remain in the amorphous phases. This statement is supported first by the temperature dependence of their mobility, which agree with the high frequency glass-rubber relaxation of the amorphous phase, secondly by the fact that their orientation behaviour is different of this of crystalline regions as deduced from X ray measurements.

In V it has been shown that different information and measurements are obtained depending on the fact that the fluorescent molecule undergoes a motion during the fluorescence lifetime or not.

VII.1 - Orientation of stretched samples

Nishijima and his coworkers were the first to use this technique in the polymer field, their results have been reviewed in (20, 29). However, most of them have been of a qualitative or semi-quantitative nature and will be not considered in this paper; they have been discussed in (30).

The most extensive study, in a quantitative way, has been performed on polyethylene terephtalate (18, 31, 32), using as a probe the following molecule :

The orientation measurements were done on samples either as-drawn or shrunk and crystallized in various temperature and time conditions. Optical birefringence, U.V., I.R. and Raman dichroisms, X-ray diffraction were used in addition to fluorescence polarization. All measurements have been performed at room temperature below the glass-rubber transition temperature (around 70°C).

The results show that U.V. dichroism and F.P. lead to the same values of $< P_2(\cos \alpha_A) >$, defined by (V.4) where α_A is the angle between the absorption transition moment and the draw direction. Furthermore, it has been proved from comparison with I.R. and Raman dichroisms that the probe molecule is preferentially oriented as the trans conformation sequences of the polymer chain. Shrinkage experiments (31) give a good example of the interest of F.P., evidencing that at temperatures below 100°C shrinkage is mainly due to the disorientation of the amorphous regions whereas crystallization occurs at a later stage.

VII.2 - Orientation studies performed during stretching

Recent experiments have been performed in our laboratory during stretching, first on polypropylene under optical microscope (19) looking at orientation during the necking process, secondly on polystyrene stretched above the glass transition temperature (33, 34). The results obtained on the latter case are summarized hereafter.

The optical equipment shown Fig. 12 and described in (15) has been adapted to a tensile machine, designed in our laboratory, which performs stretching at constant strain rate $\dot{\varepsilon}$ (from 2.10^{-3} to 2.10^{-1} s^{-1}) with a cross-head displacement of 40 cm and a temperature homogeneity along the stretching axis equals to 1/30 th of a degree in a temperature range from ambient to 150°C.

Labelled polystyrene shown in Fig. 17 with $\overline{M}_n$ = 290.000 has

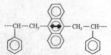

Fig. 17 - Position of DMA label in the middle of a polystyrene chain, the double arrow indicates the transition moment

been used in a narrow dispersed polystyrene ($\overline{M}_n$ = 190.000, $\overline{M}_w$ = 210.000). At studied temperature, the label does not occur any motion during the fluorescence lifetime. Typical examples of stress and orientation curves versus extension ratio λ, are given in Fig. 18. Two parts are clearly shown in stress curves.

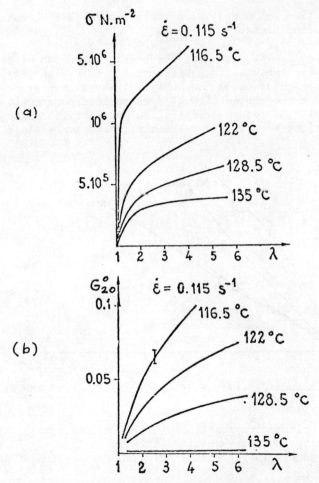

Fig. 18 - Strain and temperature dependences a/ of the stress σ, b/ of the orientation G_{20}^o. λ is the extension ratio.

The first one, at small extension ratios, is attributed to the glassy deformation, whereas the second one at higher extension ratios, reflects a rubber-like deformation with a possible contribution of flow. From comparison of stress and orientation behaviours, it appears that orientation does not originate from the glassy deformation but is related to the rubber-like deformation.

This statement is supported by the strain-rate dependences observed at 116.5°C and 122°C. In addition shrinkage experiments performed at 120°C for 24 hours on stretched samples show that the
flow contribution is very small even at an extension ratio as
high as 6.

VII.3 - Orientation and mobility in mobile, uniaxially stretched
        polyisoprene networks

    As described in V.2, fluorescence polarization can be applied
to uniaxial mobile systems. Such measurements have been performed
in our laboratory (35), using the equipment shown in Fig. 12.
Labelled polyisoprene chains (Fig. 13) where incorporated in polyisoprene Shell IR 307 before crosslinking by dicumylperoxide.
As an illustration, Fig. 19 shows the strain and temperature depen-

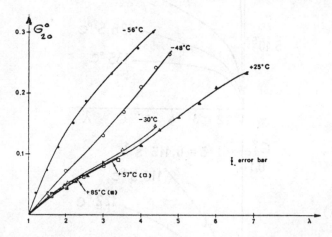

Fig. 19 - Strain and temperature dependences of G_{20}^{O} vs the exten-
          sion ratio λ.

-dences of the orientation fonction G_{20}^{O} ($\equiv$ < P_2(cos α) >), defined
in V.2. Concerning the mean mobility amplitude defined by (V.5),
its strain dependence is reported in Fig. 20 for various crosslinking densities.

VIII - CONCLUSION

    This paper shows that luminescence polarization techniques
are interesting tools for studying both molecular mobility and
orientation. The extension of the theory to uniaxial mobile systems and the development of a convenient equipment leads to a
large number of possible applications in the polymer field.

    In the case of molecular mobility studies, fluorescence and
phosphorescence polarization measurements, using either probes or

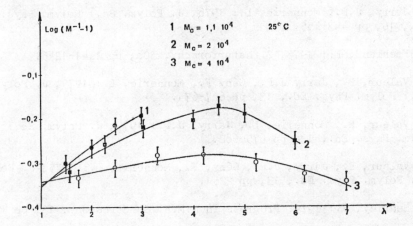

Fig. 20 - Mean mobility amplitude at 25°C for DMA labelled polyisoprene networks vs the extension ratio λ. M_c denotes
the average molecular weight of chain strands between
adjacent cross-links.

labels, permit new insight in this area and a quantitative test
of the free-volume expectation. Studying labelled chains of
various low molecular weights will lead to an estimation of the
length of mobile chain sequences involved in the glass-rubber
relaxation.

The chain orientation determination by fluorescence polarization in crystalline polymers is a unique means of looking at the
molecular behaviour in the amorphous phase. When this technique
is applied during stretching, it can provide a way of studying
polymer viscoelasticity at a molecular scale (34).

REFERENCES

1. Perrin, F.: 1929, Ann. Phys., 12, pp 169-275

2. Parker, C.A.: 1968, "Photoluminescence of solutions", Elsevier.

3. Becker, R.S.: 1969, "Theory and interpretation of fluorescence
and phosphorescence", Wiley.

4. Schulman, S.G.: 1977, "Fluorescence and phosphorescence spectroscopy", Pergamon.

5. Guilbault,G.G.: 1973, "Practical fluorescence Theory, Methods
and Techniques", Marcel Dekker.

6. Pesce, A.J., Rosen, C.G., Pasby, T.L.: 1971, "Fluorescence
Spectroscopy", Marcel Dekker.

7. Jarry, J.P., Monnerie, L.: 1978, J. Polym. Sc.: Polym. Phys. Ed., 16, pp 443-455.

8. Frehland, E.: 1975, Z. Naturforsch., 30a, pp 1241-1246.

9. Valeur, B., Jarry J.P., Gény F., Monnerie, L.: 1975, J. Polym. Sc.: Polym. Phys. Ed., 13, pp 667-674.

10. Valeur, B., Monnerie, L., Jarry, J.P.: 1975, J. Polym. Sc.: Polym. Phys. Ed., 13, pp 675-682.

11. Valeur, B., Jarry, J.P., Gény, F., Monnerie, L.: 1975, J. Polym. Sc.: Polym. Phys. Ed., 13, pp 2251.

12. Heatley, F.: 1979, Progress in N.M.R. Spectroscopy, 13, pp 47-85.

13. Gény, F., Monnerie, L.: 1977, J. Polym. Sc.: Polym. Phys. Ed., 15, pp 1-9.

14. Valeur, B., Monnerie, L.: 1976, J. Polym. Sc.: Polym. Phys. Ed., 14, pp 11-27.

15. Jarry, J.P., Sergot, P., Pambrun, C., Monnerie, L.: 1978, J. Phys. E: Sci. Instrum., pp 702-706.

16. Desper, C.R., Kimura, I.: 1967, J. Appl. Phys., 38, pp 4225-4233.

17. Bower, D.I.: 1981, J. Polym. Sc.: Polym. Phys. Ed., 19, pp 93-107.

18. Nobbs, J.H., Bower, D.I., Ward, I.M., Patterson, D.: 1974, Polymer, 15, pp 287-300.

19. Pinaud, F., Jarry, J.P., Sergot, P., Monnerie, L.: Polymer, in press.

20. Nishijima, Y., Onogi, Y., Asai, T.: 1966, J. Polym. Sc., Part C, 15, pp 237-250.

21. Chapoy, L.L., Du Pré, D.B.: 1980, Methods Exper. Phys., 16 A, pp 404-441.

22. Anufrieva, E.V., Gotlib, Yu. Ya.: Adv. Polym. Sc., 40, pp 1-68.

23. Jarry, J.P., Monnerie, L.: 1979, Macromolecules, 12, pp 927-932.

24. Ferry, J.D.: 1970, "Viscoelastic Properties of Polymers", 2 nd ed., Wiley.

25. Somersall, A.C., Dan, E., Guillet, J.E.: 1974, Macromolecules, 7, pp 233-244.

26. Rutherford, H., Soutar, I.: 1977, J. Polym. Sc.: Polym. Phys. Ed., 15, pp 2213-2225.

27. Rutherford, H., Soutar, I.: 1978, J. Polym. Sc.: Polym. Let. Ed., 16, pp 131-136.

28. Rutherford, H., Soutar, I.: 1980, J. Polym. Sc.: Polym. Phys. Ed., 18, pp 1021-1034.

29. Nishijima, Y.: 1970, J. Polym. Sc., Part C, 31, pp 353-373.

30. Bower, D.I.: 1975, in "Structure and Properties of Oriented Polymers" ed. by I.M. Ward, Applied Science, pp 187-218.

31. Nobbs, J.H., Bower, D.I., Ward, I.M.: 1976, Polymer, 17, pp 25-36.

32. Nobbs, J.H., Bower, D.I., Ward, I.M.: 1979, J. Polym. Sc.: Polym. Phys. Ed., 17, pp 259-272.

33. Fajolle, R., 1980, Thèse Docteur-Ingénieur, Université Pierre et Marie Curie, Paris.

34. Monnerie, L.: 1981, Polymer Preprints 22, pp 96-97

35. Jarry, J.P., Monnerie, L.: 1980, J. Polym. Sc.: Polym. Phys. Ed., 18, pp 1879-1890.

INFRARED AND RAMAN STUDIES OF THE POLYMERIC SOLID STATE

Paul C. Painter and Michael M. Coleman

The Pennsylvania State University. Polymer Science
Section, University Park, PA 16802.

ABSTRACT

Developments in the discipline of vibrational spectroscopy
are occurring at a rapid rate due to the impact of lasers, new
detectors, interferometers and computers. For molecules as complicated as polymers these advances are welcomed with enthusiasm,
since polymer spectroscopists need all the help they can get. In
order to fully benefit from these advances in instrumentation,
however, it is necessary to also obtain a knowledge of the theory
of vibrational spectroscopy as applied to polymers. Consequently,
in this review we will first consider the origin of the vibrational spectrum and the conditions necessary for infrared absorption and Raman scattering. We will then extend this discussion to chain molecules and consider how translational symmetry leads to a simplification of the predicted spectrum. The
basis of normal coordinate analysis of polymers will be reviewed
and a specific example, a vibrational analysis of isotactic poly-
stryene, will be considered.

Having laid this foundation we will turn our attention to
the advances made possible by recent improvements in instrumentation. Because of the constraints of time and space we will concentrate most of our discussion on the application of Fourier
transform infrared (FT-IR) instruments to the study of polymers,
since this allows us to consider the optical advantages of interferometers and the new information that can be extracted from the
spectrum through the use of computer methods. We will consider
the application of this technique to a number of problems in
polymer science, including studies of defects in polymer crystals,
chain folding, the characterization of the gel form of crystal-

R. A. Pethrick and R. W. Richards (eds.), Static and Dynamic Properties of the Polymeric Solid State, 415–455.
Copyright © 1982 by D. Reidel Publishing Company.

lizable polymers, and studies of ionomers, an example of a complex multicomponent polymer system.

1.0   INTRODUCTION.

Spectroscopy has often been defined as the interaction of electromagnetic radiation with matter or, in the words of Gerhard Herzberg, "the science of discovery." The nature of the interaction depends upon the wavelength or frequency of the radiation, so that regions of the electromagnetic spectrum have become associated with various types of spectroscopy. Vibrational spectroscopy is somewhat unusual in that it involves two different effects, the absorption of radiation in the infrared region of the spectrum and the inelastic scattering of light, usually in the visible. As the name suggests, both involve the vibrational energy levels of a molecule.

Infrared and Raman spectroscopy have been widely applied to the characterization of polymeric materials. In the early 1970's, however, the use of vibrational spectroscopy seemed to be in general decline as other techniques that were more sensitive to chain structure (eg NMR) came to the fore. This situation prompted H. A. Laitinen, in an editorial in Analytical Chemistry (1), to draw an analogy between Shakespeare's seven ages of man and the corresponding seven ages of an analytical technique, using infrared spectroscopy as a prime example. Senescence and an early demise seemed assured. Fortunately, for analytical techniques there is always the possibility of reincarnation and at the same time as Laitinen's depressing prognosis appeared commercial Fourier transform infrared (FT-IR) instruments were being delivered to a number of laboratories. The subsequent resurgence of infrared spectroscopy has been remarkable and has opened up new possibilities for studies of polymeric materials (2,3).

Advances in vibrational spectroscopy have not been confined to infrared instrumentation. A number of developments in Raman instrumentation (eg. improved lasers) and sampling techniques (eg the surface-enhanced Raman effect) are extremely promising. Theoretical developments have included the application of electro-optical parameters to the calculation of the infrared and Raman intensities of polyethylene (4,5). However, space does not permit a consideration of all these advances and, in addition, there is not yet a substantial body of work in the literature concerning the application of these latter developments to the specific study of polymeric materials. In contrast, during the last five years or so a considerable amount of research with regard to the application of computer assisted vibrational spectroscopy, particularly FT-IR, has been reported by a number

of groups. Consequently, in this review, we will discuss FT-IR
studies at some length, but in order to lay the foundation for
this discussion we will first consider the theory of vibrational
spectroscopy with specific reference to macromolecules.

2.0 BASIC THEORY; THE ORIGIN OF THE VIBRATIONAL SPECTRUM.

A change in the total energy of a molecule occurs upon
interaction with electromagnetic radiation. This change is reflected in the observed spectrum of the material. In order to
describe this interaction and formulate a useful mathematical
model certain assumptions are usually made. The total energy of
a molecule consists of contributions from the rotational,
vibrational, electronic and electromagnetic spin energies. This
total energy can be approximated to a sum of the individual
components and any interaction treated as a perturbation. The
separation of the electronic and nuclear motions, known as the
Born-Oppenheimer approximation, depends upon the large difference
in mass between the electrons and nuclei. Since the former are
much lighter they have relatively greater velocities and their
motion can be treated by assuming fixed positions of the nuclei.
Conversely, the small nuclear oscillations (compared to interatomic distance) occur in an essentially averaged electron
distribution. The change in energy of these nuclear vibrations
upon interaction with radiation of suitable frequency is the
origin of the vibrational spectrum. This energy change is, of
course, quantized. In addition, an absorption band or Raman
line nearly always corresponds to discrete vibrational transitions in the ground electronic state. Absorption of higher
energy visible or UV light is required to produce changes in
electronic energy.

Any vibration of an atom can be resolved into components
parallel to the x,y and z axis of a cartesian system. The atom
is described as having three degrees of freedom. A system of N
nuclei therefore has 3N degrees of freedom. The number of fundamental vibrational frequencies or normal modes of vibration
of a molecule is equal to the number of vibrational degrees
of freedom. However, for a non-linear molecule six of these
degrees of freedom correspond to translations and rotations of
the molecule as a whole and have zero frequency, leaving 3N - 6
vibrations. For strictly linear molecules, such as carbon
dioxide, rotation about the molecular axis does not change the
position of the atoms and only two degrees of freedom are required to describe any rotation. Consequently, linear molecules
have 3N - 5 normal vibrations. For a theoretically infinite
polymer chain only the three translations and one rotation have
zero frequency and there are 3N - 4 degrees of vibrational
freedom. Each normal mode consists of vibrations (although not

necessarily significant displacements) of <u>all</u> the atoms in the
system.

Since molecular vibrations cannot be observed directly a
model of the system is required in order to describe these
normal modes. The nuclei are considered to be point masses and
the forces acting between them springs that obey Hooke's Law.
The motion of each atom is assumed to be simple harmonic. Even
with these assumptions it is intuitively obvious that a system
of N atoms is capable of innumerable different complex vibrations, each involving a range of displacements of the various
nuclei, rather than a number equal to the vibrational degrees of
freedom. However, in the harmonic approximation any motion of
the system can be resolved into a sum of so-called fundamental
normal modes of vibration, just as displacements can be represented
by components parallel to a set of cartesian coordinates. In a
normal vibration each particle carries out a simple harmonic
motion of the same frequency and in general these oscillations
are in phase; however, the amplitude may be different from
particle to particle. It is these normal modes of vibration
that are excited upon infrared absorption or Raman scattering.
Naturally, different types of vibrations will have different
energies and so absorb or inelastically scatter radiation at
different frequencies.

It is possible to briefly demonstrate the major factors
that influence the frequency of a normal mode using a simple
model system consisting of the harmonic oscillations of a diatomic molecule. It is convenient to consider the classical
solution and then account for the conditions imposed by quantum
mechanics. A simple system is illustrated in Figure 1.

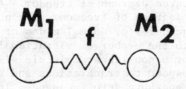

Figure 1. The coordinates of a model diatomic molecule.

The frequencies of the normal modes of vibration can be simply calculated in the harmonic approximation from expressions for the kinetic and potential energies. A simple form of the equations of motion will be considered later when we discuss the vibrations of polymer chains. Here we will simply state the result. If the bond joining masses M_1, M_2 in Figure 1 behaves as a Hookean spring with stiffness, f, then the frequency of vibration, ν, of this simple one dimensional system is given by

$$\nu = \frac{1}{2\pi}\sqrt{\frac{f}{M_r}} \tag{1}$$

where M_r is the reduced mass;

$$M_r = \frac{M_1 M_2}{(M_1 + M_2)} \tag{2}$$

This is the vibrational equation for the harmonic oscillator. Although the model considered is extremely simple, it does demonstrate that the vibrational frequency depends inversely on mass and directly on the force constant, f.

In the classical mechanical solution a continum of vibrational energy is allowed. However, the quantum mechanical solution requires quantized energy levels. Fortunately, it can be shown that in the limit of the harmonic approximation the classical vibrational frequency corresponds to that determined by quantum mechanics, and with this assumption the vibrational frequencies of a molecule can be determined by classical methods. Normally, the force constants are defined in terms of internal coordinates, i.e. those forces resisting the stretching, bending and torsions around bonds. This description is not only intuitively satisfying from a chemical viewpoint, but has also allowed the calculation of the frequencies and the form of the normal modes of complex systems using the Wilson GF matrix method.(6)

In the Wilson GF method, the potential energy of a molecule is defined in terms of the force constants by a matrix F, while the kinetic energy, which is defined in terms of the geometry of the molecule, is defined by a matrix G. Using the methods of classical mechanics the following equation can be derived;

$$[GF - \lambda I]L = 0 \tag{3}$$

where the eigenvalues λ and the eigenvectors L are matrices of the vibrational frequencies and displacements, respectively. Even though the size of the GF matrix can be very large for even simple molecules, solutions are now readily obtained by computer matrix diagonalization methods. However, the force constants of a molecule are not known 'a priori.' It is the vibrational frequencies that are directly obtained from experimental observation of the Raman and infrared spectra. Hence, vibrational or normal coordinate analysis consists of the calculation of force constants from a given data set. This allows the form of the vibration or normal mode (the columns of the L matrix) to be determined for each vibrational frequency. In essence, it is assumed that the force constants describing the motions of similar types of bonds, (e.g., the C-H stretch of simple paraffins) have the same value, so that a force constant set derived from a group of simple molecules can be transferred to more complicated molecules (e.g. polyethylene). This assumption has allowed the normal modes of a number of polymers to be determined. We will discuss a specific example, isotactic polystyrene, below.

3.0   CONDITIONS FOR INFRARED ABSORPTION AND RAMAN SCATTERING.

We have briefly discussed one basic and inherent condition for infrared absorption and Raman scattering, namely that the frequency of the absorbed radiation or the frequency shift of the scattered light must correspond to the frequency of a normal mode of vibration and hence a transition between vibrational energy levels. However, this condition is not by itself sufficient. There are additional so-called selection rules that determine activity. Naturally, an understanding of the selection rules can only be attained through the methods or theories capable of successfully describing the interaction of radiation with matter. It is easier to obtain a mental picture of these interactions by first considering the classical interpretation. The quantum mechanical description will then be simply stated and compared without deriving the appropriate equations.

Infrared absorption is simply described by classical electromagnetic theory; an oscillating dipole is an emitter or absorber of radiation. Consequently, the periodic variation of the dipole moment of a vibrating molecule results in the absorption or emission of radiation of the same frequency as that of the oscillation of the dipole moment. The classical approach breaks down in describing details of the spectrum. For instance, there is a rotational fine structure superimposed upon the vibrational bands of gaseous molecules. Classically, rotational energy is a continuous function of angular velocity, so that these bands are not directly explained. Many problems in the classical approach such as the one described above are simply accounted for in quantum mechanical terms; i.e. rotational energies are also quantized.

The requirement of a change in dipole moment with molecular vibration is fundamental, and it is simple to illustrate that

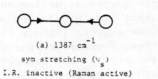
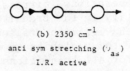

(a) 1387 cm^{-1} (b) 2350 cm^{-1}

sym stretching (ν_s)   anti sym stretching (ν_{as})

I.R. inactive (Raman active)   I.R. active

Figure 2.   The in-plane stretching modes of CO_2 (a) symmetric stretching 1387cm^{-1}(b) anti-symmetric stretching 2350 cm^{-1}.

certain normal modes do not result in such a change. Consider the two in-plane stretching vibrations of CO_2 shown in Figure 2. The net dipole moment of the symmetrical unperturbed molecule is zero. In the totally symmetric vibration 2a, the two oxygen atoms move in phase successively away and towards the carbon atom. The symmetry of the molecule is maintained in this vibration and there is no net change in dipole moment, hence no interaction with infrared radiation. Conversely, in vibration 2b the symmetry of the molecule is perturbed and there is a change in dipole moment with consequent infrared absorption.

Infrared selection rules are expressed quantum mechanically by the following condition

$$\int \psi_i \mu \psi_f dt \neq 0 \ . \ . \ . \ . \tag{4}$$

This integral (which is over the whole configuration space of the molecule) is called the transition moment and its square is a measure of the probability of the transition occurring. μ is the dipole moment vector and ψ_i and ψ_f are wavefunctions describing the initial and final states of a molecule. The dipole moment, μ, can be expressed as the sum of three components μ_x, μ_y and μ_z in a cartesian system, so that the condition for absorption is that any one of the three corresponding transition moment integrals does not vanish. Note that the size of the integral need not be calculated in order to determine activity, we only need to know whether or not its value is zero.

The selection rules for the symmetric and antisymmetric stretching modes of CO_2 were determined above by inspection. For this simple molecule it is easy to see that the symmetric stretch does not result in a change in the dipole moment. It is intuitively clear that this behavior is due to the symmetry of the CO_2 molecule and it turns out that the number and activity of the normal modes can be predicted from symmetry considerations alone. All molecules can be classified into groups according to the symmetry operations (mirrors, rotation axes etc.) they possess. Each quantity in the transition moment integral has a clearly defined behavior with respect to these symmetry operations. Consequently, the vanishing or non-vanishing of the integrals are the same for all transitions between states of two particular symmetry classes. The activity of each class of vibrations of a given symmetry group has already been determined, so that the infrared activity of the normal modes of a molecule can be determined solely from a knowledge of its symmetry.

Raman activity can also be determined from symmetry arguments in a corresponding manner. In contrast to infrared absorption, the Raman effect is not concerned with the <u>intrinsic</u> dipole moment of the molecule. For Raman scattering to occur the electric field of the light must <u>induce</u> a dipole moment by a change in what is termed the polarizability of the molecule. Under conditions that are usually fulfilled by most molecules the quantum mechanical analogue of the polarizability can be used:

$$\alpha_{nm} = \int \psi_n \alpha \psi_m \, d\tau \tag{5}$$

This treatment was described by Placsek and in its developed form simplifies spectral analysis by allowing the experimental determination of the symmetry species of Raman active modes. The selection rules imposed by the harmonic approximation (only transitions between adjacent energy levels are allowed) are again imposed, in addition to those determined by symmetry.

Since both the induced dipole moment μ_I and the electric field of the light E are vector quantities, α_{ik} is a tensor defined by an array of nine components as:

$$\mu_{Ix} = \alpha_{xx} E_x + \alpha_{xy} E_y + \alpha_{xz} E_z \tag{6}$$

$$\mu_{Iy} = \alpha_{yx} E_x + \alpha_{yy} E_y + \alpha_{yz} E_z \tag{7}$$

$$\mu_{Iz} = \alpha_{zx} E_x + \alpha_{zy} E_y + \alpha_{zz} E_z \tag{8}$$

The tensor is symmetric, i.e.

$$\alpha_{xy} = \alpha_{yx}; \; \alpha_{xz} = \alpha_{zx}; \; \alpha_{yz} = \alpha_{zy} \tag{9}$$

so that the selection rule for Raman scattering is that at least one of the six integrals

$$\int \psi_n \alpha_{ik} \psi_m \, d\tau$$

is totally symmetric. In an analagous fashion to the infrared selection rules, a fundamental is allowed in Raman scattering only if it belongs to the same symmetry species as a component of the polarizability tensor.

4.0 ELEMENTARY TREATMENT OF THE LATTICE VIBRATIONS OF CHAIN MOLECULES.

At first glance, the enormous number of atoms present in a polymer chain would seem to make the calculation of its vibrational frequencies an impossible task. However, it can be shown that for a theoretically infinite polymer chain in a regular, well-defined conformation, the problem can be reduced by symmetry to the determination of the normal modes of the translational repeat unit. In effect, an ordered polymer chain can be considered to be a one-dimensional crystal or linear lattice. Physically, this is because the intermolecular forces between chains packed in a three-dimensional lattice are an order of magnitude less than the intramolecular interactions between repeat units in the same chain. Consequently, their effect on the vibrational spectrum can be neglected or treated as a perturbation. Before proceeding to a consideration of polymer chains, it is useful to first discuss the vibrations of simple model systems. This will provide a basis for defining concepts and terms frequently used to describe the normal modes of polymers and additionally allows us to gain some insight into the vibrations of a chain of identical units. We will start with the simplest model; infinite, one dimensional chain of point masses. Consider an infinite linear array of atoms, mass m, separated by a distance d as shown in Figure 3. Let the longitudinal displacement of the nth particle from the equilibrium position be x_n and the potential energy of the displacements of the atoms be determined by the force, f, acting between adjacent particles. In this treatment we will assume that only nearest neighbor forces are significant. Applying Newton's second law of motion, the force f acting on a given atom n, is given by:

$$F_n = m\ddot{x}_n = f(x_{n+1} - x_n) - f(x_n - x_{n-1}) \tag{10}$$

$$m\ddot{x}_n = fx_{n+1} - 2fx_n + fx_{n-1} \tag{11}$$

These equations can be more formally derived by expanding the potential energy, which is a function of the distance between adjacent particles, as a Taylor series in the displacements and

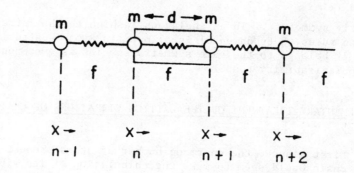

Figure 3. Schematic representation of a monatomic chain.

imposing the harmonic approximation. The displacements are defined relative to the equilibrium position of the atoms. Since the atoms are spaced uniformly at a distance d from each other, the coordinate of the nth atom is related to the origin by a factor nd. The displacements are therefore propagated as a wave along the chain if the physical problem admits a solution of the type

$$x_n = Ae^{-2\pi i (\nu t - knd)} = Ae^{-i(\omega t - \phi n)} \tag{12}$$

Where ν is the frequency; t represents time; ω is the circular frequency = $2\pi\nu$; k is the wavevector; ϕ is the phase angle = $2\pi kd$ and A is the amplitude. The quantity ϕ defines the difference in phase between the displacements of particle n and those of particle n + 1. The magnitude of the wave vector k is defined here as equal to the reciprocal of the wavelength, $1/\lambda$. (It should be noted that in many texts the definition $k = 2\pi/\lambda$ is used.) It is a vector quantity and its direction is the direction of propagation of the wave along the lattice.

Using equation 12, equation 11 becomes (after division by x_n)

$$-4\pi^2\nu^2 m = f [e^{2\pi ikd} - 2 + e^{-2\pi ikd}] = -2f(1 - \cos 2\pi kd) \tag{13}$$

and the positive root is given by

$$\nu = \frac{1}{\pi} \left(\frac{f}{2m}\right)^{1/2} (1 - \cos 2\pi kd)^{1/2} \tag{14}$$

which can be written:

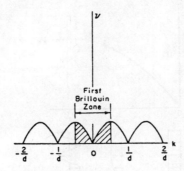

Figure 4.   Plot of dispersion curve of the monatomic chain showing the first Brillouin zone.

$$\nu = \frac{1}{\pi} \left(\frac{f}{m} \right)^{1/2} |\sin \pi k d| = \frac{1}{\pi} \left(\frac{f}{m} \right)^{1/2} |\sin \phi/2| \qquad (15)$$

Figure 4 shows a plot of ν against k demonstrating that ν is periodic in k. The period is equal to 1/d and a series of maxima occur at a frequency of

$$\nu_c = \frac{1}{\pi} \sqrt{\frac{f}{m}} \qquad (16)$$

where the first maxima being at a point k = 1/2d from the origin.

On account of the periodic properties of the chain, it is sufficient to discuss the values of ν inside one period of k or ϕ. The most convenient choice is

$$-\pi \le \phi \le \pi \qquad (17)$$

$$-1/2d \le k \le 1/2d \qquad (18)$$

This range of values is referred to as the first Brillouin zone of the linear chain. Note that the symmetry of the zones implies that k can take positive and negative values of equal magnitude, so that waves associated with any zone can travel in the positive or negative x-direction along the chain. The plot of ν against k or ϕ is the dispersion curve and it is usual to show only the positive half, as in Figure 5. Also shown in

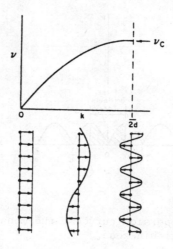

Figure 5. Plot of half first Brillouin zone showing the form of
the normal modes for certain values of k.

Figure 5 is the form of the vibrations for various values of k.
Naturally, the displacements are longitudinal i.e. in the
lattice direction, since we are only considering the one dimensional case. However, in order to demonstrate the sinusoidal
nature of the lattice vibrations the displacements are displayed
perpendicular to the axis of the chain. If k = 0, the displacements of the atoms have the same amplitude and direction
and are in-phase. This vibration represents a translation of
the lattice and has zero frequency. If k = 1/2d, it can be
shown that the displacements represent a standing wave in which
all the particles have the same amplitude but alternate ones are
out of phase by π radians. Travelling wave motion is exhibited
at intermediate values of k and a typical example is also shown
in Figure 5.

     A lattice consisting of two or more types of atoms displays
new features in the dispersion relationship that throw additional light onto the vibrations of polymer chains. Consider a chain
consisting of two different atoms arranged alternately, as shown
in Figure 6.

     In the general case the atoms are linked by bonds of
different type, characterized by force constants f_1 and f_2.
If we again assume only nearest neighbor forces we can write down
the equations of motion in the same manner as for the single

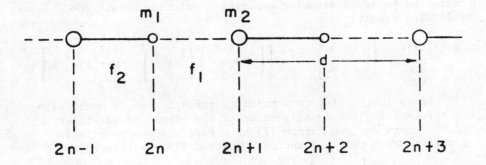

Figure 6. Schematic representation of a diatomic chain.

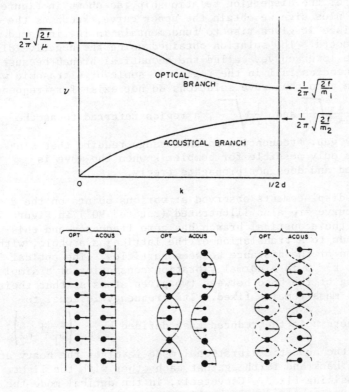

Figure 7. Dispersion curve of diatomic chain showing the form of the normal modes for certain values of k.

chain and by again assuming a periodic solution obtain the following result;

$$\nu^2 = \frac{f_1 + f_2}{8\pi^2}\left[\frac{1}{m_1} + \frac{1}{m_2}\right] + \frac{1}{4\pi^2}\left[\left(\frac{f_1 + f_2}{2}\right)^2\left(\frac{1}{m_1} + \frac{1}{m_2}\right)^2 - \frac{4f_1 f_2}{m_1 m_2}\sin^2\pi kd\right]^{1/2} \quad (19)$$

It is instructive to consider a special case, a model with $f_1 = f_2 = f$ implies equal spacing of the atoms and was first used by Born and von Karman (7) as a model of a simple ionic crystal. With this assumption, Equation (19) is reduced to:

$$\nu^2 = f/4\pi^2\left\{\frac{1}{m_1} + \frac{1}{m_2}\right\} \pm \left[\left\{\frac{1}{m_1} + \frac{1}{m_2}\right\}^2 - \frac{4}{m_1 m_2}\sin^2\pi kd\right]^{1/2} \quad (20)$$

The two solutions of this equation result in two frequency branches of the dispersion relationship, as shown in Figure 7. With the plus sign we obtain the upper curve, known as the optical branch since it gives rise to fundamentals in the infrared and Raman spectra. The solution obtained using the negative sign gives the lower curve, called the acoustical branch because its frequencies fall in the region of sonic or ultrasonic waves. It can be seen that wave solutions do not exist for frequencies between $\frac{1}{2\pi}\sqrt{\frac{2f}{m_1}}$ and $\frac{1}{2\pi}\sqrt{\frac{2f}{m_2}}$, a region referred to as the frequency gap. Frequencies in this range require that $\sin\pi kd > 1$, which are only possible for complex k when the wave is attenuated and does not propagate freely.

The displacements observed at various points on the dispersion curve are also illustrated (rotated 90°) in Figure 7. At $k = 0$ the acoustical branch has zero frequency and this mode corresponds to a translation of the lattice as a whole, without alteration of the distance between particles. The optical branch at $k = 0$ is a normal vibration consisting of a simple stretching of the bond between two given atoms so that their center of mass remains fixed. Its frequency is equal to $\frac{1}{2\pi}\sqrt{\frac{2f}{\mu}}$ where μ is the reduced mass defined by $\frac{1}{\mu} = \left(\frac{1}{m_1} + \frac{1}{m_2}\right)$. At $k = 1/2d$ the acoustic vibrational mode involves the heavy atoms vibrating back and forth against each other with the light atoms remaining fixed. Conversely, in the optical mode the heavy atoms are fixed and the light atoms vibrate. Clearly, in the special case of $m_1 = m_2$ there is no frequency gap and these two points on the frequency branches coincide. Typical waves observed for values of k intermediate between zero and $1/2\ d$ are also shown in Figure 7.

The significance of the dispersion curve derived for a diatomic lattice is that it indicates the existence of frequency banding for the normal modes of periodic structures. Only those frequencies located in the optical and acoustical branches can propagate along the chain. If we had a more complex chain containing N atoms per unit cell there would be N bands (for vibrations in one dimension). A vibrational frequency located in a forbidden region or frequency gap corresponding, for example, to a normal mode of a defect (a unit with a different mass or structure to the rest of the chain), would normally dampen very quickly with increasing distance from the defect site and would not propagate along the chain.

5.0 THE NORMAL MODES OF VIBRATION OF ORDERED POLYMER CHAINS.

We have seen that for a simple chain the normal modes of vibration can be considered in terms of the motions of the translational repeat unit (e.g. the atoms n and n+1 in the diatomic chain) and a phase angle ϕ or wave vector k relating motions in adjacent units. Further simplifications can be obtained by applying symmetry arguments. (8,9) Essentially, if the polymer chain is ordered and if we assume it is infinitely long, then only those vibrations which are totally in phase (i.e. k = 0, ϕ = 0) are allowed by the selection rules to be Raman or infrared active. Consequently, the problem is reduced to a consideration of the motions of the translational repeat unit. In fact, a further simplification to a consideration of the vibrations of the chemical repeat unit is usual in polymer normal coordinate analysis.

Consider a general helical polymer where the rotation angle (around the helix axis) relating adjacent chemical units is Ψ.

Let Θ be the phase angle between vibrations in adjacent chemical units. The phase angle Θ is separate and distinct and should not be confused with the angle ψ defined above as a symmetry operator determined by the geometry of the chain. To satisfy the k = 0 condition, vibrations in the first chemical unit of a translational repeat unit must be in phase with the corresponding modes in the first chemical unit of the adjacent translational unit. For example, isotactic polystyrene has 3 chemical units in one turn of a helix. For the above phase condition to hold we could obviously have Θ = 0° i.e. the vibrations in adjacent chemical repeat units are all in phase. However, when Θ = 120° the condition also holds, since the phase angle between the first chemical unit in each translational repeat unit would then be 3 x 120° = 2π, i.e. they would also be in phase.

Generalizing this condition, $n\Theta/m$ must be 2π times a whole number;

$$n \cdot \Theta/m = 2\pi \cdot r \tag{21}$$

or,

$$\Theta = (2\pi m/n)r = \psi \cdot r \tag{22}$$

with $r = 0, 1, 2, - - n-1$ (for $n > r-1$ we get nothing new since $r = n$ is equivalent to $r = 0$ and so on). This condition is more usually defined with $r = 0, \pm 1, \pm 2, --- \pm (n-1)/2$ for n odd and $r = 0, \pm 1, \pm 2 - - -$ for n even, so that Θ now takes values $-\le \Theta \le \pi$.

Consequently, the frequency of a normal mode ν_i in an isolated chemical repeat unit gives rise to a band of frequencies $\nu_i(\Theta)$ in a helix, where (Θ) is the phase difference between the motions in adjacent units. This is the phonon dispersion curve similar to that derived above for a simple lattice. However, Θ is not continuous but has discrete values given by equation (22)

The selection rules for a polymer chain have been derived by Higgs (8) and it can be shown that infrared active modes belong to symmetry species A and E_1 while Raman active modes belong to species A, E_1 and E_2. For symmetry species A, E_1 and E_2 the vibrations of motions in adjacent chemical repeat units are phase shifted 0, ψ and 2ψ respectively.

The periodicity of an ordered polymer chain will naturally determine the form of the secular equation, whether we express it in terms of cartesian displacement coordinates, as in the simple monatomic and diatomic chains, or in terms of internal coordinates using the more useful Wilson GF matrix method. Higgs (8) originally derived the phase relations by a reduction of the infinite F and G matrices using internal symmetry coordinates. An alternative derivation, used by Piseri and Zerbi (10) is to assume a perodic solution, an approach discussed above for vibrations of a simple lattice. As might be intuitively expected from our discussion of a simple lattice, we obtain an expression for the secular equation in terms of the vibrations of the chemical repeat unit and a phase angle relating vibrations in adjacent units.

$$|G(\Theta) F(\Theta) - \lambda(\Theta) I| = 0 \tag{23}$$

where

$$G(\Theta) = \sum_{n=-\infty}^{\infty} G_n e^{in\Theta} \tag{24}$$

$$F(\Theta) = \sum_{n=-\infty}^{\infty} F_n e^{in\Theta} \tag{25}$$

and n designates the repeat unit.

As in the case of the simple monatomic and diatomic chains we have obtained a dispersion relationship, the difference being only in the level of complexity. For a polymer molecule having 3N atoms in every chemical repeat unit, we calculate 3N branches. The dispersion relation has a period of 2π, i.e, $\nu(\Theta) = \nu(\theta+2\pi)$, and the first Brillouin zone is defined by $-\pi < \Theta < \pi$. In addition it can be shown that $\nu(\Theta) = \nu(-\Theta)$ so that it is possible to limit calculations to half of the brillouin zone, $0 < \Theta < \pi$. An examination of the selection rules for an infinite chain demonstrates that the infrared and Raman active fundamentals of a helical polymer can be determined from three equations of the infinite set, namely, those corresponding to $\Theta = 0, \psi$, and 2ψ. However, it is often of value to plot the dispersion curve by obtaining, for example, solutions every 5°. Unfortunately, there can be difficulties or ambiguities in determining the dispersion curves because they can cross or repel and often a knowledge of the eigenvectors is required in order to trace the curves accurately.

5.1 Aspects of the Normal Coordinate Analysis of Isotactic Polystyrene.

The infrared spectrum of polystyrene is perhaps the most familiar and easily recognized of any polymeric material. The isotactic polymer has been the subject of numerous vibrational spectroscopic studies, but a complete knowledge of band assignments requires normal coordinate calculations (11,12). Initial calculations (11) were limited by a paucity of data in the low frequency region of the spectrum and by discrepancies in the force field. Furthermore, dispersion curves were not calculated. Dispersion curves allow a greater understanding of chain dynamics and are useful in the interpretation of the results of structural studies. For example, on the basis of model compound work Jasse et al.(13,16) have postulated a correlation between the frequencies of certain bands and local conformations such as TT (trans, trans) and TG (trans, gauche). The transferability of such correlations to the polymer depend upon the vibrational modes in question maintaining a localized character. The dispersion curves provide a knowledge of the degree of coupling along the chain together with an understanding of the dependence of the frequency of a given mode upon the sequence length of ordered conformations.

In addition to these spectroscopic reasons for recalculating the normal modes of isotactic polystyrene, there is a more

fundamental structural reason. A 'gel' form of this polymer
has been reported (17-19). This gel has an ordered component
that has a different melting temperature and x-ray diffraction
pattern compared to the usual 3_1 helical crystalline form. The
infrared spectrum of the gel is also significantly different (20),
as shown in Figure 8. This figure compares the difference
spectrum obtained by subtracting the amorphous contribution from
the spectra of the gel and crystalline forms. The application of
FT-IR instruments to this type of study will be discussed more
completely below, but it is pertinent to point out here that
the use of spectral subtraction procedures indicates that
approximately 35% of the chains participate in the ordered gel
component, (20), effectively ruling out structural defects as
the origin of this phase, as had been previously proposed (17,18).
The difference in the spectra (the spectrum of atactic polystyrene is also shown for purposes of comparison) suggest that
the chain is in a different conformation and is consistent with
the nearly extended structure suggested by Atkins et al.(19).

It is unfortunate that in all but a few cases specific
conformations cannot yet be determined from vibrational spectroscopic results and normal coordinate calculations with any degree
of confidence. This is because any change in conformation will
not only affect the calculated frequencies through corresponding
changes in the so-called G matrix, but will also affect the force
field (F matrix). For hydrocarbon chains interaction force
constants are only defined for conformations in which hydrogen
atoms are trans or gauche to one another, thus precluding a
definitive characterization of other possible structures, such as
the 12_1 helix proposed by Atkins et al (19). However, it is
possible to calculate the normal modes of an extended all-trans
structure with some confidence. To a first approximation we can
then determine the effect of small distortions in bond and torsional rotational angles to ascertain whether frequency shifts
relative to the 3_1 helical form are consistent with those
observed.

With these factors in mind, a new valence force field for
monosubstituted alkyl benzenes has been calculated and applied
to isotactic polystyrene (21). This normal coordinate analysis
resulted in some revised assignments for certain modes. For
example, on the basis of model compound studies the two most
prominent and conformationally sensitive, near 550 cm^{-1} in the
infrared spectrum and 225 cm^{-1} in the Raman spectrum, have often
been assigned to C-X out-of-plane and C-X in plane bending modes,
respectively, where C-X represents a bond between the aromatic
ring and the alkyl group in model compounds. However, Jasse
et al.(13,16) have made a detailed study of polystyrene oligomers and on the basis of the observed conformational sensitivity of these modes suggested that there must be a degree of

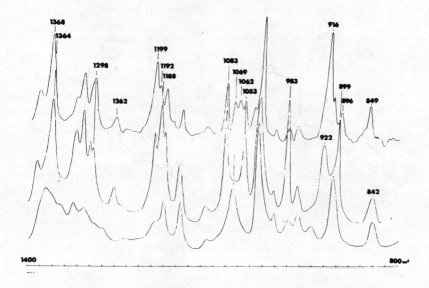

Figure 8. Comparisons of the infrared spectrum of the ordered
gel component of IPS, the 3_1 helical crystal, and atactic
polystyrene. Top spectrum, gel; middle spectrum, crystal form;
bottom spectrum, atactic polystyrene.

coupling between ring and backbone vibrations. This conclusion
is supported by the normal mode calculations, as illustrated by
the cartesian displacements coordinates of these modes shown in
Figures 9 and 10, respectively. The dispersion curves for these
modes also demonstrate that they are coupled along the chain, as
illustrated in Figure 11.

We have mentioned above that normal coordinate calculations
have to be used with great care in structural studies, mainly
because of limitations on the available force fields. In
addition, for the specific case of isotactic polystyrene we have
only observed frequency shifts for many modes of the order of
10 cm^{-1} or so between the spectra of the ordered gel component
and the 3_1 helical form. For many of these modes the agreement
between observed and calculated frequencies in the 3_1 helical form
is of this order of magnitude, so that we could scarcely have
much confidence in any predictions of conformation that are based
on apparently 'good' agreement between the observed and calculated

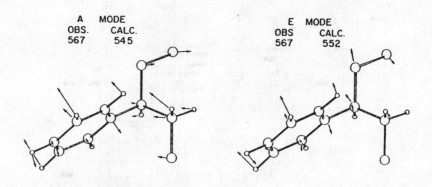

Figure 9.   Cartesian displacement coordinates of IPS for 567 cm^{-1} mode.

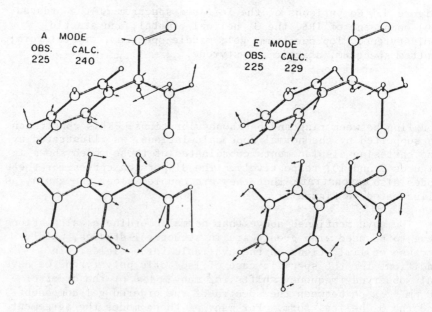

Figure 10.   Cartesian displacement coordinates of IPS for 225 cm^{-1} mode.

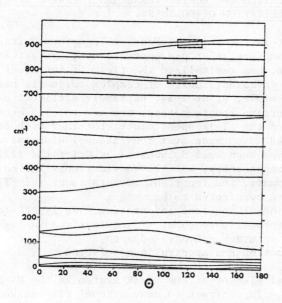

Figure 11.   Dispersion curves for single IPS chain in the 3_1 helical conformation.

frequencies of some other helical form. However, there is one region of the infrared spectrum where there are large differences between the observed spectra of the gel and the crystal, namely the group of bands near 1050 cm^{-1}. In the 3_1 helical form there is a characteristic doublet at 1083 and 1052 cm^{-1}, discussed above. In the spectrum of the ordered gel component this doublet is replaced by four bands at 1083, 1069, 1062 and 1053 cm^{-1} (20). In addition, the 12_1 helical chain conformation proposed by Atkins et al (19) is fairly close to all-trans. In spite of the reservations mentioned above, we therefore considered it useful to perform certain calculations to determine the trends in this region of the spectrum on going from the 3_1 helical to nearly extended conformations. In these calculations the doublet near 1083 and 1052 cm^{-1} in the spectrum of the crystal form was determined to be split into four components in the spectra of extended structures. Although the results have to be treated with circumspection, we believe that they indicate that the observed changes are at least consistent with a conformation close to extended, such as the 12_1 helix.

6.0 THE APPLICATION OF FOURIER TRANSFORM INFRARED SPECTROSCOPY TO THE CHARACTERIZATION OF POLYMERS.

The recent introduction of Fourier Transform Infrared (FT-IR) spectrometers, that are operable over the entire infrared frequency range, has revitalized the field of infrared spectroscopy. Infrared studies of complex polymeric materials that were impossible, or at least extremely difficult using conventional dispersive spectrometers, are now readily accomplished. A description of the FT-IR spectrometer together with the underlying theory and advantages over conventional dispersive spectrometers has been well documented by Griffiths (22) and Koenig (2). Consequently, we will consider only the major points. In summary, the important advantages of the FTIR spectrometer are considered to be:
1. Multiplexing of spectral information - Fellgett's advantage
2. Enhanced optical throughput - Jacquinot's advantage
3. Frequency accuracy - Connes advantage
4. Computerization and spectral manipulation

The central element of the FT-IR system is the Michaelson interferometer. In contrast to conventional dispersive spectrometers, where the infrared radiation is divided into frequency elements by the use of a monochromator and slit system, the total spectral information is contained in an interferogram from a single scan of the movable mirror in the interferometer. The interferometer contains no slits and the amount of infrared energy falling on the detector is greatly enhanced compared to that in a dispersive spectrometer. In fact, for the mid infrared region of the spectrum at a resolution of 2 cm^{-1}, the amount of infrared radiation reaching the detector at any time in a grating spectrometer is < 0.1% (22). Theoretically, Jacquinot's advantage, depending upon the resolution, can be as high as 80 to 200 times greater than that of a comparable dispersive instrument. However, this is rarely accomplished in commercial FTIR instruments due to limitations of sample and source geometries. Nevertheless, in energy starved situations such as those encountered when examining optically dense materials (e.g. carbon black filled; highly colored, degraded or oxidized polymers), any enhancement of energy throughput is highly desirable. The multiplexing or Fellgett's advantage arises from the fact that the data from all spectral frequencies are measured simultaneously in one scan of the interferometer. Consequently, a considerable increase of signal to noise ratio is gained in comparison to that observed from a dispersive instrument at identical measurement times. To increase the signal to noise ratio of the resultant spectrum signal averaging techniques are employed and it is essential that the frequency accuracy of each individual spectrum be kept

between strictly defined limits (typically 0.02 cm^{-1}). This is
readily accomplished by the use of a small monitoring laser
incorporated into the FT-IR system (Connes advantage).

        At first glance, the necessity of a computer in the FTIR
system to convert the interferogram (using the Cooley-Tukey Fast
Fourier Transform algorithm), to the normal frequency/intensity
spectrum could be considered a disadvantage, if only from the
economic point of view. However, the very fact that it was
essential to employ a computer resulted in the development of
signal averaging techniques and sophisticated software that enables the storage of digitized spectral data and mathematical
manipulation of the spectra.

        Viewed critically, the major advantages of FT-IR to the
polymer scientist that are apparent from the recently published
literature are a direct result of the data collection, handling
and manipulation techniques, which are not strictly limited to
the FT-IR system (3). The vast majority of the published FT-IR
studies on polymeric materials in the mid infrared region of the
spectrum could have equally been obtained on a high quality,
stable, conventional grating spectrometer equipped with a
computer and programmed to permit signal averaging and analogous
data handling and manipulation. The routine use of data processing methods, however, has gone hand in hand with the deployment
of FT-IR instruments, so that in many reviews the use of
sophisticated programs is considered one of the advantages of
FT-IR. Among the techniques that are now being routinely applied
to polymer analysis are difference methods, derivative spectroscopy, curve resolving, ratio methods and factor analysis. Space
does not permit a review of all these procedures. Instead we
will consider the application of FTIR to the study of specific
problems in polymer science and consider the use of these
routines as they arise.

6.1 The Application of FT-IR to the Characterization of the Microstructure of Polychloroprenes.

        Polychloroprene is a polymer with a relatively large monomer
repeating unit that consequently exhibits little, if any, vibrational coupling between the chemical units along the chain (23).
It is an outstanding example of a material where the digital absorbance subtraction method may be successfully employed to obtain
significant information that was previously unattainable. Thus,
it is feasible to separate the infrared bands due to the
crystalline phase of the material from those of the amorphous (24)
and, in addition, accentuate and identify infrared bands due to
the presence of structural irregularities present in the otherwise predominantly _trans_ -1,4 configuration (25).

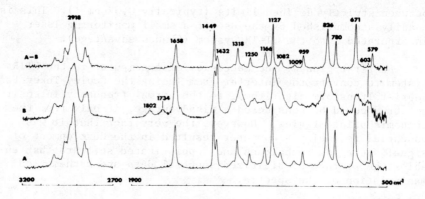

Figure 12.   FTIR spectra in the range 500–3200 cm^{-1} of chloroprene polymerized at –150°C; (A) spectrum recorded at room
temperature; (B) spectrum recorded at 80°C; and (A–B) difference
spectrum obtained by subtracting (B) from (A).

Initially, the subtraction method was used to obtain a
"purified" spectrum of the crystalline phase of the material (24).
This technique considerably simplifies the assignment of the
individual infrared bands arising from the crystalline and
amorphous components of the spectrum and is important for subsequent theoretical normal coordinate calculations (23).   The procedure is illustrated in Figure 12 where the amorphous contribution of a chloroprene polymer polymerized at – 150°C was subtracted from that of the semi-crystalline material.   This polymer
consists of essentially pure <u>trans</u> -1,4 units with only
approximately 2% as head to head placements.   The amorphous
spectrum was recorded at 80°C in a heated vacuum cell and the
amorphous band at 602 cm^{-1} was employed to determine the correct
subtraction parameter.   A slight and unavoidable oxidation of the
heated polymer is evident from the bands at 1734 and 1802 cm^{-1}
but causes an insignificant effect to the resulting difference
spectrum because of the relatively small concentration of the
amorphous component of this polymer.

Previous studies on chloroprene polymers polymerized at
temperatures in the range –20 to + 40°C revealed that specific
crystalline bands (obtained by the subtraction technique described
above) were sensitive to the number of structural irregularities
present in the polymer (26).   It is well known that as the polymerization temperature is increased, the concentration of head to
head <u>trans</u>-1,4; <u>cis</u>-1,4; 1,2 and 3,4 units also increases (27).
From these studies, and those of a copolymer of chloroprene and

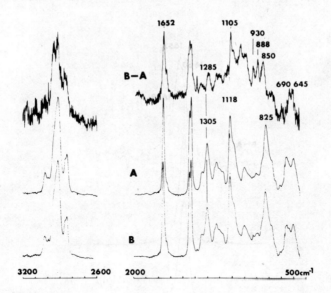

Figure 13. FTIR spectra at 70°C in vacuum in the range 500-3200 cm^{-1}; (A) polychloroprene polymerized at -20°C; (B) polychloroprene polymerized at +40°C; (B-A) difference spectrum obtained by subtracting (B) from (A). (From [9].)

2,3-dichlorobutadiene, it was concluded that specific structural irregularities are incorporated into the crystalline lattice causing a perturbation of the vibrational force field which results in specific frequency shifts. The most sensitive bands are those at 959, 1009, 1250 and 1318 cm^{-1} (-150°C polymer). These are all highly coupled normal modes (23) involving the C-C stretch and various CH$_2$ bending vibrations. It was originally concluded that it was probable that the cis-1,4 unit was incorporated into the crystalline phase causing the observed frequency shifts. However, recent ^{13}C NMR results (28) suggest that the concentration of cis -1,4 units in the chloroprene polymers has been considerably over-estimated (e.g. 5.2% by ^{13}C NMR compared to 13% by infrared for the +40°C polymer). It is now considered more likely that it is the inverted (head to head) trans-1,4 units and not the cis-1,4 units that are incorporated into the crystalline lattice. This conclusion is supported by more recent FT-IR studies of chloroprene copolymers (29).

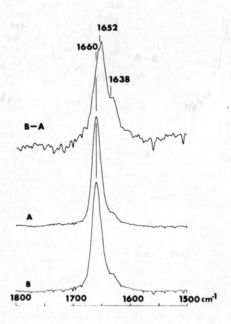

Figure 14.   Expanded FTIR spectra at 70°C in vacuum in the range
1500–1800 cm^{-1}; (A) polychloroprene polymerized at −20°C; (B)
polychloroprene polymerized at +40°C; and (B–A) difference spectrum obtained by subtraction (B) from (A) (From [10].)

        Additional information may be obtained by studying the
amorphous spectra of the various polychloroprenes prepared at
different polymerization temperatures. By digitally subtracting
the amorphous spectrum (acquired at 70°C in vacuo) of a more
structurally regular polychloroprene from that of a polymer containing a higher concentration of structural irregularities, it
is possible to accentuate the infrared bands due to these
structural irregularities (25). Figure 13 illustrates this
procedure.   Spectra A and B are those obtained at 70°C from the
polychloroprene samples polymerized at −20°C and 40°C respectively.
The scale expanded difference spectrum (B−A) was obtained using

the bands attributable to trans-1,4 units at 1660, 1305, 1118
and 825 cm^{-1} to determine the correct subtraction parameter. In
terms of experimental technique, it should be emphasized that to
obtain high signal to noise difference spectra it is essential
that the films employed should be of comparable thickness
(0<.5 mil) and be within the absorbance range where the Beer-
Lambert law is obeyed. The infrared bands observed in the
difference spectrum are attributable to the cis-1,4;1,2 and 3,4
structural units and represent a difference of approximately 4,1
and 1% respectively (28). By examining the infrared spectra of
the model compounds cis and trans-4-chlorooctene, 3-chlorobutene-1,
1-chlorobutene-2 and 2-chlorobutene-1 the major bands occurring in
the difference spectrum can be tentatively assigned. Of particular interest is the fact that it was possible to resolve the
C=C stretching vibration of the cis-1,4 unit at 1652 cm^{-1} from
that of the very strongly absorbing 1660 cm^{-1} C=C stretch of the
predominant trans-1,4 unit, as shown in Figure 14. The major
bands due to the cis-1,4 unit, (i.e. 1652, 1440, 1430, 1285, 1225,
1205, 1105, 1090, 1030, 850, 690 and 654 cm^{-1}), observed in the
difference spectrum are consistant with the infrared spectrum reported previously for the predominantly cis-1,4-polychloroprene.
The 1,2 unit is readily identified by the well isolated bands at
985 and 930 cm^{-1} while the 3,4 unit has a useful band at 888 cm^{-1}.
There are no isolated bands that can be unambiguously assigned
to the isomerized-1,2 unit although a weak band occurring at
785 cm^{-1} is consistant with a medium intensity band occurring in
the model compound.

It is feasible to use the above method to obtain seim-
quantitative data on the concentration of structural irregularities in polychloroprene. If an amorphous spectrum of the structurally regular -150°C polymer is employed as a standard and subtracted from that of the polychloroprene under investigation,
the concentration of structural irregularities could be calculated
from a knowledge of the subtraction parameter. This would, however, necessitate accurately weighing the sample contained in the
infrared beam.

Finally, it has proved possible to use the results of
FT-IR studies of polychloroprene microstructure to demonstrate
the involvement of 1,2-and 3,4-structural irregularities in the
initial stages of the thermal oxidation of polychloroprenes at
60°C(30). This observation suggests that polychloroprenes could
possibly be stabilized against oxidative degradation by the
elimination of the 1,2 and 3,4 structures by chemical modification.

6.2 FTIR Studies of Amorphous and Fold Structures in Poly-
ethylene Single Crystals.

Polyethylene has a special place in polymer science. As the simplest polymer in terms of chemical structure it has been the preferred subject of many studies aimed at elucidating the general physical laws that govern the behavior of all polymers. In spite of these years of effort, however, there is still a great deal of controversy concerning aspects of the structure of this macromolecule. In particular, the arguments concerning the nature and concept of chain folding have recently taken on an interesting degree of rancor (31).

A number of different structures have been proposed, ranging from models consisting of tight folds with adjacent reentry to those in which there is a loosely looped random reentry of chains. Krimm and co-workers (32-34) obtained evidence for chain folding with adjacent reentry through the infrared analysis of mixed crystals of polyethylene and deuteropolyethylene. But, the definition of the precise conformation of the fold surface remains elusive, even to the extent of determining if the fold is a loose, largely amorphous structure; consists of a tight regular conformation, or is perhaps somewhere between these two extremes.

Many infrared studies have concentrated on the methylene wagging mode observed between 1300 cm^{-1} and 1400 cm^{-1}. Bands in this region of the spectrum have been assigned to conformations other than the all-trans sequences found in the crystal phase. Two major approaches have been used to assign such bands to specific conformations. Snyder (35) analyzed the spectrum of polyethylene on the basis of detailed normal coordinate calculations of several rotamers of normal hydrocarbons, assuming that the vibrations in defect structures or folds are predominantly localized and remain unaffected by inclusion in a long chain. In contrast, Zerbi et. al. (36) argued that all band modes are activated and it is therefore necessary to analyze the dynamics of the whole chain and determine the density of states of the regular crystal. Calculations were performed on a model system of 200 CH_2 units of the type:

$$-(T)_m-X-(T)_n-$$

where $(T)_m$ refers to all-trans sequences consisting of m CH_2 units, while X corresponds to conformations such as G (gauche), GG, GTG, etc. It was noted that the observed spectrum is probably due to contributions from both localized modes and the activated density of states, but the relative contributions cannot be determined without a quantitative knowledge of intensity factors. However, despite disagreement in the assignment of certain bands, both authors assigned modes near 1350 cm^{-1} to GG

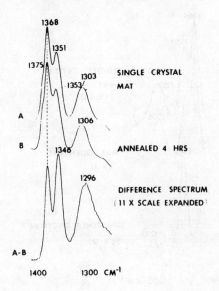

Figure 15.   (A) Single crystals; (B) annealed 4 hr; (A-B) 1:1.
(Italicized ratios denote the subtraction parameters.)

structures.

Experimental investigations aimed at the detection of bands
that could be assigned to regular fold structures have usually
foundered on the difficulty of separating the possible contributions of fold and amorphous conformations. The data processing
methods of FTIR can make a major contribution to solving problems
of this type through the use of spectral subtraction or curve
resolving. For example, Figure 15 compares the methylene wagging
region of the infrared spectrum of a single crystal mat to that
of the same mat after annealing 4 hours at 125°C(37). Both
spectra display the characteristic bands at 1368 cm^{-1}, near
1350 cm^{-1} and near 1303 cm^{-1}. The difference spectrum presented
in the same figure, shows a new band at 1346 cm^{-1} together with
the band at 1368 cm^{-1}. Since both bands are positive in the
difference spectrum (original mat minus annealed) they must
represent conformations that are reduced by annealing. Both
amorphous and fold bands would be expected to display this type
of behavior. However, the relative intensities of the difference
bands changed significantly with time of annealing (37), suggesting that they originate in conformations found in different
structures. It was found that the 1368 cm^{-1} band maintained its
sensitivity to heat treatment longer than the 1346 cm^{-1} band.
This is precisely the relationship expected between bands associated with conformations found in an amorphous phase and those

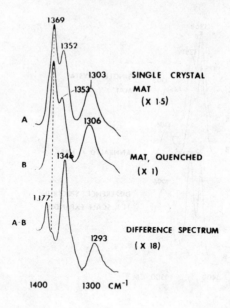

Figure 16.   (A) Single crystals; (B) quenched; (A-B) 1:0.65.

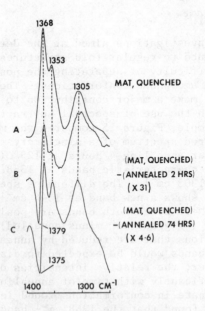

Figure 17.   Single crystals – (A) quenched, 1:0.65; (B) quenched, annealed 2 hr, 1:0,95; (C) quenched, annealed 74 hr, 1:0.98.

associated with a unique regular fold structure. As the fold
period increases it requires significantly more energy to reduce
the number of folds. However, since the amorphous component of
the crystal is related to excluded chain ends and possibly an
absorbed surface layer, the energy limitations imposed on the
fold period increase do not affect changes in amorphous bands.
Furthermore, the conformation responsible for the 1346 cm^{-1} band
is apparently removed by heating the single crystals above the
melt temperature followed by quenching. Figure 16 shows that in
a difference spectrum obtained from subtracting the spectrum of
the quenched mat from that of the original single crystals the
1348 cm^{-1} band can be eliminated leaving a strong band at 1346 cm^{-1}.
However, the difference spectra shown in Figure 17 obtained by
subtracting the spectrum of the quenched mat from that of the
same sample after annealing does not reveal a 1346 cm^{-1} band at
all, even when the spectra are oversubtracted. Since the annealing of melt quenched material also produced an increase in fold
period it was concluded that the conformation of the folds in a
solution grown crystal are different to those in bulk-crystallized
material. The conformations present in the latter structure are
apparently of the same type as those found in the amorphous
regions, so that their absorption in the methylene wagging region
of the spectrum is distributed amongst the 1368 cm^{-1}, 1353 cm^{-1}
and 1305 cm^{-1} bands.

It has been proposed that the unique band at 1346 cm^{-1}
assigned to a regular fold conformation is possibly due to a distorted GG structure. It was mentioned above that both Snyder
(35) and Zerbi et. al., (36), assigned bands near 1350 cm^{-1} to
a regular GG conformation. The cyclic paraffin $C_{34}H_{68}$ has tight
folds of the type GGTGG where the GG structures, as determined
by x-ray diffraction, are distorted to the extent that the methylene wagging band appears near 1342 cm^{-1}(38). It therefore seems
reasonable to assign bands near 1346 cm^{-1} to conformations with
an intermediate degree of distortion. If this assignment is
correct it follows that the single crystal fold is tight, since
there is no apparent reason why a GG conformation in a loose fold
should differ from the same sequence in an amorphous surface
layer or collapsed cilia.

6.3 FTIR Studies of Ionomers.

Copolymers of ethylene and methacrylic acid are an interesting macromolecular system, capable of existing in a number of
structural forms. If the proportion of methacrylic acid is small
(on the order of 5%) and the ethylene sequences are predominantly
linear, then the neutralized polymer or ionomer is a complex
multiphase system consisting of crystalline and amorphous poly-
ehtylene phases and ionic domains. The structure of the ionic
domain is sensitive to the degree of neutralization and thermal

history of the polymer and a number of models have been proposed
in order to account for various experimental observations (39-48).

In principle, vibrational spectroscopy should be an excellent
technique with which to study these complex multiphase systems,
although Neppel, et. al. (49) have stated that this method has
rarely contributed to elucidation of supermolecular structure.
Nevertheless, these authors reported changes in the intensities
of certain low frequency Raman lines in the spectra of various
ionomers as a function of ion concentration and proposed an empirical correlation with so-called multiplet and cluster formation. There has been a number of other spectroscopic studies of
ionomers (50-58) and this interpretation has been challenged by
Tsujita et.al. (50), who pointed out that in ethylene-methacrylic
acid copolymers the observed behavior of the low frequency Raman
lines could also be explained in terms of defect structures and
the density of states of the polyethylene component. Far-infrared
studies were also performed by Tsatsas, et. al. (51). These
authors report the cation and temperature dependence of several
low frequency infrared bands and concluded that ionic domains
were present in their samples. They did not, however, elaborate
on possible structures of these domains.

The assignment of bands in the 1800 to 1500 cm^{-1} region of
the infrared spectra of ionomers are, broadly speaking, more
certain. Low molecular weight model compounds such as the salts
of organic acids, have been extensively studied by infrared
spectroscopy (59). It has been shown that the frequency of the
asymmetric stretching mode of the COO^- group (near 1550 cm^{-1}) is
sensitive to the nature of the ionic bond and to the type of
metallic counter-ion present. Consequently, it would seem
reasonable to assume that this region of the infrared spectrum of
ionomers is sensitive to the structure of the ionic domains.

Ernest and MacKnight (57) have studied the 1800 to 1500 cm^{-1}
region of the infrared spectrum of the ethylene-methacrylic acid
system, but were principally concerned with the temperature
dependence of hydrogen bond formation between carboxylic acid
groups. However, these authors also reported spectra obtained
from the partially neutralized copolymer. They observed that the
COO^- assymetric stretching absorption consists of at least two
overlapping peaks but cast doubts on the interpretation of similar
infrared studies by Andreeva et.al. (58), who had assigned infrared bands at 1550 and 1565 cm^{-1} to ionic multiplet and clusters
respectviely.

In part, the problems encountered in using infrared
spectroscopy to study this type of ionomer are due to the overlap
and superposition of the various components of the asymmetric
COO^- stretching mode. In this respect, the data handling

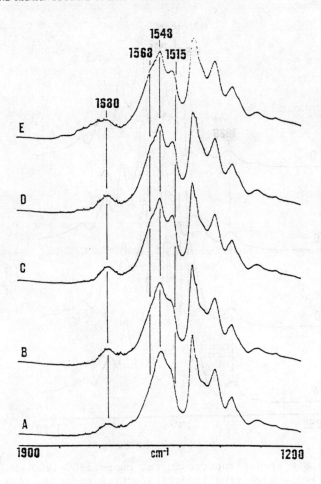

Figure 18.   FTIR spectra in the range 1900–1200 cm^{-1}.   Completely
ionized calcium ionomer at (A) room temperature after quenching
in liquid nitrogen from 190°C; (B) 40°C after 30 minutes; (C) 70°C
after 30 minutes; (D) 130°C after 15 minutes (after 15 minutes at
90°C) (E) 150°C after 5 minutes.

capabilities of FT-IR instruments offer distinct advantages.   We
have made a study of the calcium and sodium ionomers of an
ethylene-methacrylic acid copolymer (60).   In the first series of
experiments we prepared a sample of the calcium ionomer that was
quenched from the molten state and then subjected to a temperature
study.   Figure 18 shows the infrared spectra (from 1900–1200 cm^{-1})
of the quenched sample recorded at room temperature (A) and the
same film recorded at 40, 70, 130 and 150°C (B to E respectively).
As the temperature is raised above 40°C two prominent bands appear

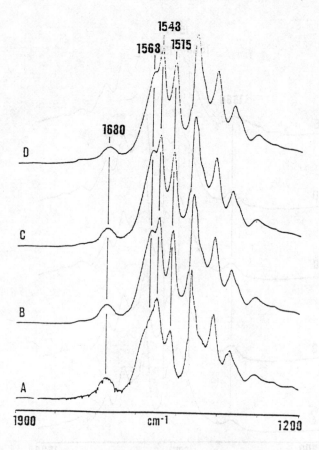

Figure 19. Difference spectra in the range 1900–1200 cm^{-1} of the completely ionized calcium ionomer spectrum taken at room temperature subtracted from spectrum taken at (A) 40°C; (B) 70°C; (C) 90°C; (D) 130°C.

at approximately 1515 and 1548 cm^{-1}. These bands can be revealed more clearly by difference spectroscopy as shown in Figure 19. Although these difference spectra are obtained from spectra recorded at different temperatures, there is no evidence of a significant distortion of the band shapes. This procedure accentuates the relatively sharp doublet at 1515/1548 cm^{-1} in the spectra of the elevated temperature samples. This doublet is characteristic of an interaction or vibrational splitting that could well occur between the modes of COO^{-} groups that exist in well defined pairs.

It should be apparent that if this interpretation is correct
then we would anticipate observing only one analogous band in
the spectrum of the sodium ionomer. This is indeed the case as
we will demonstrate below. A prominent band at 1568 must also
be accounted for, but we will defer an assignment until after we
have described the next set of experiments.

Following the temperature study described above, the calcium
ionomer sample at 150°C was allowed to cool to room temperature
over a period of about 5 minutes and was maintained at this
latter temperature for periods up to 87 days. Figure 20 shows
the room temperature spectra obtained after 2,7, 14 and 87 days
(A thru D respectively). The spectrum recorded after 2 days at
room temperature closely resembles that of the parent 150°C
spectrum (Figurd 18E). In contrast, upon extended periods of
standing at room temperature the doublet at 1515/1548 progressively becomes less well defined and finally a broad band centered
at 1553 cm^{-1} is observed. Once again, difference spectroscopy is
informative. Figure 21 shows the difference spectrum obtained by
subtracting the spectrum obtained after 2 days from that obtained
after 7 days. Although this difference spectrum is not
particularly 'clean' it does demonstrate that the broad band at
approximately 1552 cm^{-1} is being formed at the expense of the
1515/1548 cm^{-1} doublet.

The morphology of ethylene/methacrylic acid ionomers is
complex and depends on at least five major variables; time,
temperature, thermal history, ion characteristics and degree of
ionization. In addition to both crystalline and amorphous phases
associated with the ethylene units, the distribution and type of
ionic domains vary. Although we can confidently assign the
1515/1548 cm^{-1} doublet to the calcium carboxylate dimer   ,
the assignments of the remaining infrared bands that we have observed can only be considered speculative at this stage. It has
been suggested that a number of possible structures associated
with carboxylic acid groups and their salts occur in these
ionomers and we have attempted to interpret our results in terms
of contemporary theories of ionomer structure. In this respect
the band observed at 1568 cm^{-1} is important as it is observed to
vary in a systematic fashion during the heating and 'annealing'
studies. Although it is masked by other absorbances occurring
in the region it does appear to be relatively narrow which infers
a specific structure. Furthermore, we can be confident from the
frequency of this band that it is associated with a COO$^-$ type
structure. We postulate that a structure containing a specific
number of COO$^-$ groups, determined by the coordination number of
the cation (six in the case of alkali metal ions) is consistent
with the relatively narrow singlet observed at 1568 cm^{-1}. This
structure, which could also be described as a multiplet,
represents an intermediate state between the isolated carboxylic

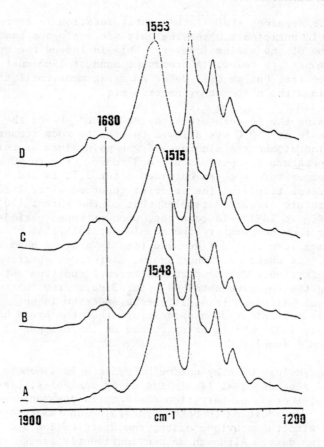

Figure 20.   FTIR spectra in the range 1900–1200 cm^{-1}.   Completely
ionized calcium ionomer (after slow cool from heating study) at
room temperature for (A) 2 days; (B) 7 days; (C) 14 days; (D) 87
days.

acid dimers and the relatively large ionic domains referred to as
clusters by Eisenburg (39).   Finally, in the spectrum of a
material consisting of relatively large ionic clusters one would
anticipate a relatively broad band such as the one observed near
1553 cm^{-1} in the spectrum of the polymer 'annealed' at room
temperature for 87 days (Figure 20D).

    If we assume that the above assignments are reasonable, we
must now reconsider the results of the heating and 'annealing'
studies.   Returning to Figure 18 it can be seen that upon quenching
the calcium ionomer from the melt(190°C) into a liquid nitrogen
bath, a broad featureless band centered at ∿ 1548 cm^{-1} is observed.

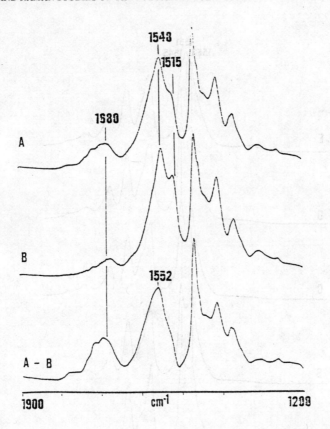

Figure 21. FTIR spectra in the range 1900-1200 cm^{-1}. (A) Completely ionized calcium ionomer after 7 days at room temperature after heating study; (B) Completely ionized calcium ionomer after 2 days at room temperature after heating study; (A-B) Difference spectrum obtained by subtracting B from A.

Presumably, in the molten state, sufficient chain mobility is present to ensure a random distribution of ionic groups. This distribution is apparently preserved by the quenching technique. Upon heating the quenched ionomer from 40°C to 150°C (Figures 18 and 19) the development of calcium carboxylate dimers and multiplets are favored, as indicated by the doublet at 1515/1548 cm^{-1} and the singlet at 1568 cm^{-1}. Upon relatively slow cooling of the ionomer from 150°C to room temperature this structure is maintained. After 'annealing' at room temperature for periods of up to 87 days however, the calcium carboxylate dimer and the multiplets are eliminated in favor of cluster formation, as indicated by the broad absorbance centered near 1553 cm^{-1}.

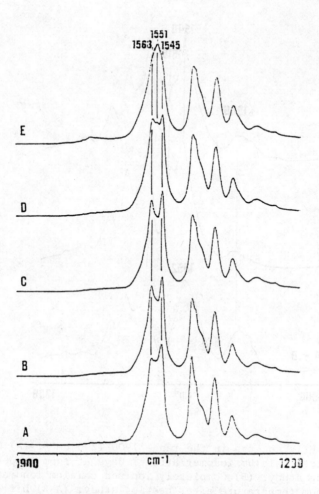

Figure 22.   FTIR spectra in the range 1900-1200 cm^{-1}.   Completely
ionized sodium ionomer at (A) room temperature after quenching
in liquid nitrogen from 190°C; (B) 70°C after 30 minutes (after
30 minutes at 40°C); (C) 90°C after 15 minutes; (D) 130°C after
15 minutes; (E) 150°C after 5 minutes.

    Similar studies to those previously described for the calcium
ionomer were performed on the fully ionized sodium ionomer.
Figure 22 shows the spectrum of the sodium ionomer which was
quenched from the melt at 190°C into liquid nitrogen.   Two major
infrared bands at 1568 and 1541 cm^{-1} are observed in the 1500-
1600 cm^{-1} region.   In this case we do not believe we are detecting
a characteristic doublet analogous to the 1515/1548 cm^{-1} bands

seen in the spectra of the calcium ionomer. The relative intensities of the 1515/1548 cm^{-1} doublet in the calcium ionomers appears constant regardless of sample preparation. However, the relative intensities of the 1568 and 1541 cm^{-1} bands in the sodium ionomers change with sample preparation and history (60). The 1568 cm^{-1} band occurring in the sodium ionomer appears analogous to the band observed at the same frequency in the calcium ionomer which we have tentatively assigned to multiplets. On the other hand, the 1541 cm^{-1} band appears analogous to the 1515/1548 cm^{-1} doublet. This is a reasonable interpretation as carboxylate dimers would not be formed with monovalent sodium cations and we would anticipate a singlet from a Na^+ ^-OOC structure.

The results presented in this preliminary study demonstrate that FT-IR is potentially a powerful tool for investigating the structure of ionomers (60). The changes observed in the spectra of calcium and sodium ionomers are readily interpreted in terms of contemporary theories of ionomer structure. Calcium ionomers differ from their sodium counterparts in that a characteristic doublet associated with ion pairs can be observed under certain conditions. In the spectra of both the sodium and calcium salts, whose behavior is consistent with the formation of multiplets and clusters can be identified.

ACKNOWLEDGEMENTS

The authors wish to acknowledge the financial support of the National Science Foundation, Grant DMR 7910841, Polymers Program.

REFERENCES

1. Laitinen, H.A: 1973, Analytical Chemistry, 45 (14), p. 2305.
2. Koenig, J.L: 1975, Applied Spectroscopy, 29, pp 293-308.
3. Coleman, M.M. and Painter, P.C: 1978, J. Macromol. Sci., Revs. Macromol. Chem., C16(2), pp. 197-313.
4. Gussoni, M., Abbate, S. and Zerbi.G: 1979, J. Chem. Phys., 71(8), pp 3428-3439.
5. Abbate, S., Gussoni, M., Masetti,G. and Zerbi, G: 1977, J. Chem. Phys., 67, pp 1519-1531.
6. Wilson, E.B., Decius, J.C. and Cross, P.C: 1955, "Molecular Vibrations", McGraw-Hill, New York.
7. Born, M. and von Karman, Th: 1912, Physik Zeitshr, 13, pp 297-309.
8. Higgs, P.W: 1954, Proc. Roy. Soc. (London). A220, pp 472-485.
9. Turrell, G: 1972, "Infrared and Raman Spectra of Crystals," Academic Press, New York.
10. Piseri, L. and Zerbi, G: 1968, J. Chem. Phys., 48, pp 3561-3572.

11. Painter, P.C. and Koenig, J.L: 1977, J. Polym. Sci., Polym. Phys. Ed., 15, pp 1885-1903.
12. Snyder, R.W. and Painter, P.C: 1981, Polymer, accepted for publication.
13. Jasse, B., Lety, A. and Monnerie, L: 1975, Spectrochim. Acta, 31A, pp 391-398.
14. Jasse, B. and Monnerie, L: 1975, J. Phys. D. Appl. Phys., 8, pp 863-871.
15. Jasse, B., Lety, A. and Monnerie, L: 1973, J. Mol. Struct., 18, pp 413-420.
16. Jasse, B., and Monnerie, L: 1977, J. Mol. Struct., 39, pp 165-173.
17. Girolamo, M., Keller, A., Miyasaka, K. and Okerbergh, N: 1976, J. Polym. Sci., Polym. Phys. Ed., 14, pp 39-61.
18. Atkins, E.D.T., Issac, D.H., Keller, A., Miyasaka, K: 1977, J. Polym. Sci., Polym. Phys. Ed., 15, pp 211-226.
19. Atkins, E.D.T., Issac, D.H. and Keller, A: 1980, J. Polym. Sci., Polym. Phys. Ed., 18, pp 71-82.
20. Painter, P.C., Kessler, R.E. and Snyder, R.W: 1980, J. Polym. Sci., Polym. Phys. Ed., 18, pp 723-729.
21. Snyder, R.W. and Painter, P.C: 1981, Polymer, accepted for publication.
22. Griffiths, P.R: 1975, Chemical Infrared Fourier Transform Spectroscopy, John Wiley and Sons, New York.
23. Petcavich, R.J. and Coleman, M.M: 1980, J. Macromol. Sci., Phys. B 18 (1), pp 47-71.
24. Coleman, M.M., Painter, P.C., Tabb, D.L. and Koenig, J.L: 1974, J. Pol. Sci., Polymer Letters Ed., 12, pp 577-581.
25. Coleman, M.M., Petcavich, R.J. and Painter, P.C: 1978, Polymer, 19, pp 1243-1248.
26. Tabb, D.L., Koenig, J.L. and Coleman, M.M: 1975, J. Polym. Sci., Polym. Phys., 13, pp 1145-1158.
27. Maynard, J.T. and Mochel, W.E.: 1954, J. Polym. Sci., 13 pp 235-250; ibid 13, pp 251-262.
28. Coleman, M.M., Tabb, D.L. and Brame, E.G: 1977, Rubber Chem. Technol., 50(1), pp 49-62.
29. Coleman, M.M., Petcavich, R.J. and Painter, P.C: 1978, Polymer, 19, pp 1253-1258.
30. Petcavich, R.J., Painter, P.C. and Coleman, M.M: 1978, Polymer, 19, pp 1249-1252.
31. Organization of Macromolecules in the condensed phase: 1979, Faraday Discussions of the Royal Society of Chemistry, No.68.
32. Bank, M.I. and Krimm, S: 1969, J. Pol. Sci., A2, 7, pp 1785-1809.
33. Bank, M.I. and Krimm, S: 1969, J. Appl. Physl. 40, pp 4248-4253.
34. Tasumi, M. and Krimm, S: 1968, J. Pol. Sci., A2, 6 pp 995-1010.
35. Snyder, R.G: 1967, J. Chem. Phys. 47, 1316-1360.

36. Zerbi, G., Piseri, L., and Cabassi, F: 1971, Mol. Phys. 22, 241-256.
37. Painter, P.C., Havens, J., Hart, W.W. and Koenig, J.L: 1977, J. Pol. Sci. 15, pp 1223-1235.
38. Schonhorn, H. and Luongo, J.P: 1969, Macromols., 2, pp 366-369.
39. Eisenburg, A: 1970, Macromolecules, 3(a), pp 147-154.
40. Eisenburg, A: 1974, J. Polym. Sci., Polym. Symp., 45, pp 99-111.
41. Longworth, R: 1975, Chapter 2 in Ionic Polymers, L. Holliday Ed., John Wiley and Sons, New York.
42. Cooper, W: 1958, J. Polym. Sci., 28, pp 195-206.
43. Otocka, E. and Kwei, T: 1968, Macromolecules, 1, pp 401-405.
44. MacKnight, W.J., Taggart, W.P. and Stein, R.S: 1974, J. Polym. Sci., Polym. Symp., 45, pp 113-128.
45. Binsberger, F.L. and Kroon, G.F: 1973, Macromolecules, 6, p 145.
46. Marx, C.L., Confield, D.F. and Cooper, S.L: 1973, Macromolecules, 6, pp 344-353.
47. Longworth, R. and Vaughan, D.J: 1968, Am. Chem. Soc., Polymer Preprints, 9, pp 525-526.
48. Bonotto, S. and Bonner, E: 1968, Macromolecules, 1, pp 510-515.
49. Neppel, A., Butler, I.S. and Eisenberg, A: 1979, Macromolecules, 12, pp 948-952.
50. Tsujita, Y., Hsu, S.L. and MacKnight, WL: 1981, private communication.
51. Tsatsas, A., Reed, J. and Risen, W.M., Jr: 1971, J. Chem. Phys., 55, pp 3260-3269.
52. Neppel, A., Butler, I.S. and Eisenburg, A: 1979, J. Polym. Sci., Polym. Phys. Ed., 17, pp 2145-2150.
53. Rouse, G.B., Risen, W.M., Jr., Tsatsas, A.T. and Eisenburg, A: 1979, J. Polym. Sci., Polym. Phys. Ed., 17, pp 81-85.
54. Read, B.E. and Stein, R.S: 1968, Macromolecules, 1(2), pp 116-126.
55. Uemura, Y., Stein, R.S. and MacKnight, W.J: 1971, Macromolecules, 4(4), pp 490-494.
56. MacKnight, W.J., McKenna, L.W., Read, B.E. and Stein, R.S: 1968, J. Phys. Chem., 72(4), pp 1122-1126.
57. Earnest, T.R., Jr. and MacKnight, W.J: 1980, Macromolecules, 13, pp 844-849.
58. Andreeva, E.D., Nitkitin, V.N. and Boyartchuk, Y.M: 1976, Macromolecules, 9, pp 238-243.
59. Bellamy, L.J: 1975, The Infrared Spectra of Complex Molecules, Third Edition, John Wiley and Sons, New York.
60. Painter, P.C., Brozoski, B.A. and Coleman, M.M: 1981, J. Polym. Sci., Polym. Phys. Ed., submitted for publication.

SEMINARS

AN ORDER-DISORDER THEORY OF STRESS STRAIN BEHAVIOUR OF GLASSY POLYMERS

Jeffrey Skolnick
Louisiana State University
Department of Chemistry
Baton Rouge
Louisiana 70803

ABSTRACT

The stress strain behaviour of glassy polymers deformed in uniaxial tension is viewed as an order disorder transition in which flow results from increased orientation of polymer segments responding to an external strain field. In our treatment, we envisage the polymer chain to be composed of rods of length, P, linked together end-to-end by universal joints. P is of the order of a persistence length, the length over which local main chain motions damp out. (In the theoretical development, the actual value of P never enters). Let the order parameter that characterizes the lining up of the ith segment with the external strain field and b

$$M_i = \cos\theta_i \qquad\qquad (1)$$

with θ_i the angle between the ith segment and the applied field. Now M_i is assumed to be a monotonic function of the applied strain, i.e. increasing the strain results in increasing order in the glass.

As the temperature, T, is raised to Tg, the glass transition temperature, experimentally it requires a smaller stress, σ, to achieve a given strain, and by assumption, a smaller σ to achieve a given amount of order in the material. Thus we identify, Tg with a critical temperature in the limiting sense that as

459

R. A. Pethrick and R. W. Richards (eds.), Static and Dynamic Properties of the Polymeric Solid State, 459–460.

T→Tg from below, M_i becomes non zero even in the
absence of a strain field. Of course the limit point
of the model, T=Tg is nonphysical: the model has
physical reality only for T < Tg. Consequently, the
theory can be cast in the form of a corresponding
states approach and requires two universal reduced
parameters: the ratio of the temperature to the glass
transition temperature and a coupling constant, μ,
with the external strain field.

Using a mean field theory, an expression can be
derived for the mean orientation of a polymer segment,
$\bar{M}$, as a function of the strain. Furthermore the stress
is related to the strain by

$$T = \frac{K}{x_O} (1-\bar{M}^2)$$ (2)

where K is a constant characteristic of the polymer
and x_O is the intrinsic susceptibility to flow,
$\left(\frac{dM}{d\epsilon}\right)_{\epsilon=0}$. Eq (2) can be expressed in terms of reduced
variables as a universal curve applicable to
all polymers. The derived stress strain curves
incorporate the fact that polymers tend to line up in
the direction of applied stress and contain all the
qualitative features required by experiment. That is

(1)   For sufficiently low strain rates, the true
       stress strain curves have two linear Hookian
       regions connected by a curved region that
       includes the yield point.

(2)   The initial slopes of stress-strain curves
       increase as a function of strain velocity.

(3)   The yield stress, σ_y decreases with increasing
       temperature and effectively vanishes at Tg.

Defining the yield point by the Consideré construction, universal curves for the yield stress and
yield strain have been calculated. The former quantity
is in agreement with extensive experimental results on
poly(methyl methacrylate).

ON THE ADSORPTION OF A POLYMER CHAIN

Marios K. Kosmas
University of Ioannina
Department of Chemistry
Ioannina
Greece

ABSTRACT

Two chain conformations have been used to describe the behaviour of a polymer molecule in the
vicinity of a liquid-solid interface. The one comes
from the study of the chain in the presence of an
absorbing wall where the probability for the polymeric units to reach the wall is zero while the second
one appears if a reflecting wall is present where the
probability for the polymeric units to reach the wall
is large. In the real situation a lot of other conformations can be realised depending on the interaction
energy between the polymeric units and the interface.
For high repulsions the probability to find the monomers
at the interface gets reduced and a picture reminding
the absorbing wall conformation emerges. Reducing the
repulsion more and more units are adsorbed. There is
a critical energy below of which the majority of
polymeric units go to the adsorbed state. At this
critical point reflecting wall statistics are suitable
to describe the conformation of the chain. We use the
reflecting wall probability as a zeroth order solution
in order to describe the general problem in the
presence of polymer-interface interactions. We work
in the neighbourhood of the adsorption-desorption phase
transition point and we give expressions for the
average amount of polymeric units adsorbed at the
interface and the extension of the adsorbed chain into
the solution.

461

R. A. Pethrick and R. W. Richards (eds.), Static and Dynamic Properties of the Polymeric Solid State, 461.
Copyright © 1982 by D. Reidel Publishing Company.

Department of Chemistry

ABSTRACT

SMALL ANGLE NEUTRON SCATTERING FROM STYRENE-ISOPRENE BLOCK COPOLYMERS

R.W. Richards & J.L. Thomason
University of Strathclyde
Department of Pure & Applied Chemistry
Glasgow G1 1XL
U.K.

ABSTRACT

A variety of styrene-isoprene block copolymers, covering a wide range in composition, have been investigated using small angle neutron scattering (SANS). Investigations of hydrogenous copolymers enabled the domain separation and packing to be obtained, it was found that spherical domains form a face centred cubic structure and cylindrical domains a hexagonally close packed arrangement. The relation between interdomain spacing and copolymer molecular weight is,

$$d_{spheres} \propto M^{0.48}$$

$$d_{cylinders} \propto M^{0.7}$$

$$d_{lamellar} \propto M^{0.52}$$

Domain sizes have been obtained from the SANS from copolymers with fully deuterated styrene domains and using the knowledge that single particle form factors are governed by Bessel functions. For these measurements the relations between domain size D (radius for spheres and cylinders, thickness for lamellae) and styrene block molecular weight are:-

$$D \propto M^{0.65}$$

R. A. Pethrick and R. W. Richards (eds.), Static and Dynamic Properties of the Polymeric Solid State, 463–464.

This last relationship exactly matches theoretical predictions[1] but at present cannot be reconciled with the exponents found for the interdomain distances.

Deviations from Porod's law[2] at high scattering vector were used to determine the thickness of the diffuse boundary at the domain surface. Such experiments are a stringent test of current experimental equipment and theory since background scattering is not negligible here. Nonetheless, interfacial thicknesses of 25 ± 10 A were obtainable, this value being independent of domain morphology and molecular weight. On the other hand the volume fraction of the interface decreases with molecular weight in a manner consistent with Meier's predictions.[3]

(1) Helfand, E. and Wasserman, Z.R. in "Developments in Block Copolymers" editor L. Goodman. To be published by Applied Science Publishers.

(2) Ruland, W., J.App.Cryst. (1971), 4, 70.

(3) Meier, D.J., Polym.Prepr. (1974), 15, 171.

SMALL ANGLE NEUTRON SCATTERING STUDIES OF POLYMER DIMENSIONS IN BINARY BLENDS

J.S. Higgins
Imperial College
Department of Chemical Engineering
London   SW7 2BY

ABSTRACT

Most polymer pairs are incompatible, or at best partially compatible over limited ranges of temperature and concentration. A number of systems showing lower or upper critical solution temperatures have been studied in our laboratory by a variety of techniques and the phase boundaries determined. Labelling by partial deuteration of one component allows us to investigate chain dimensions and interaction parameters by small angle neutron scattering from the single phase mixture close to the phase boundary and thus to extend our understanding of the limited compatibility to a molecular level. It has been found that both chain expansion and chain contraction may occur in different systems as the phase boundary is approached from the single phase side.

R. A. Pethrick and R. W. Richards (eds.), Static and Dynamic Properties of the Polymeric Solid State, 465.

INELASTIC & QUASIELASTIC NEUTRON SCATTERING STUDIES OF POLYMETHYL METHACRYLATE

K. Ma
Imperial College
Department of Chemical Engineering
London SW7 2BY

ABSTRACT

    Neutron scattering experiments have been useful
in studying the vibrational and rotational motion of
the two methyl sidegroups in poly(methyl methacrylate).
It is shown that detailed information about torsional
frequencies and barrier heights to rotation for these
groups may be obtained from inelastic scattering
measurements.

    Quasielastic scattering measurements allow the
calculation of a function known as the Elastic
Incoherent Structure Factor (EISF), which is a direct
measure of the spatial distribution of the moving
protons, and as such, characterises completely the
actual geometry of rotation. Experimental determination of the EISF for both the α-methyl and ester
methyl group is discussed in some detail. In
addition, measurements relating rotational frequencies
(obtained directly from the rotational quasielastic
scattering laws) to temperature in order to calculate
activation energies are presented, and compared to
values of the barrier height obtained from inelastic
measurements. A series of supplementary experiments
concerned with the measurement of coherent scattering
from poly(methyl methacrylate) is mentioned briefly,
and it is indicated how these results may be interpreted to give details about local structure in the
amorphous material.

R. A. Pethrick and R. W. Richards (eds.), Static and Dynamic Properties of the Polymeric Solid State, 467.
Copyright © 1982 by D. Reidel Publishing Company.

INDEX

469